Agri-Nanotechnology: Innovations for Sustainable Agriculture and Environmental Restoration

Sukhvinder Singh Purewal
Vijay Gahlaut • Sandeep Kumar Dash
Editors

Agri-Nanotechnology: Innovations for Sustainable Agriculture and Environmental Restoration

Springer

Editors
Sukhvinder Singh Purewal
University Centre for Research and Development
Chandigarh University
Mohali, Punjab, India

Vijay Gahlaut
University Centre for Research and Development
Chandigarh University
Mohali, Punjab, India

Sandeep Kumar Dash
Department of Physiology
University of Gour Banga
Malda, West Bengal, India

ISBN 978-981-96-9755-7 ISBN 978-981-96-9756-4 (eBook)
https://doi.org/10.1007/978-981-96-9756-4

This Springer imprint is published by the registered company Springer Nature Singapore Pte Ltd.
The registered company address is: 152 Beach Road, #21-01/04 Gateway East, Singapore 189721, Singapore

Preface

Agriculture today stands at a crossroads, balancing the urgent need for increased food production with the imperative to protect and restore the environment. This book, *Agri-Nanotechnology: Innovations for Sustainable Agriculture and Environmental Restoration*, brings together meaningful scientific information and practical applications from diverse domains within agricultural science and nanotechnology. Each chapter provides a distinct perspective on how nanomaterials and nanoscale techniques are revolutionizing different aspects of agriculture and environmental stewardship. Chapter 1, "Recent Trends in Nano-Fertilizers for Sustainable Agriculture," introduces the fundamentals of nano-fertilizers, discussing their synthesis, controlled release mechanisms, uptake efficiency, and potential to reduce nutrient losses. It presents recent research and field applications that demonstrate improved crop yields with minimal environmental impact. How Nanotechnology is helpful in the Restoration of Degraded and Polluted Soil is discussed in detail in Chap. 2. Chapter 3 explains how nanoscale innovations contribute to real-time decision-making and resource efficiency. Chapters 4 and 5 collectively provide a dual focus on pest control strategies. Chapter 6, "Nanotechnology and Agribusiness: Part I," explores the commercial and economic implications of nanotechnology in agriculture. It evaluates market trends, product development, and commercialization pathways, laying the groundwork for understanding the agribusiness ecosystem. Chapter 7, "Nanotechnology in Crop Protection: Innovations for Sustainable Agriculture," examines how nanomaterials enhance the efficacy of fungicides, herbicides, and protective agents. Chapter 8, "Engineered Nano-Biochar from Lignocellulosic Residues: Innovative Approaches for Sustainable Agriculture and Environmental Remediation," focuses on the synthesis and application of nano-biochar derived from agricultural waste. It discusses its dual role in improving soil fertility and sequestering pollutants, showcasing a circular economy approach. Chapter 9, "Nanotechnology and Agribusiness: Part II," continues the discussion on commercial prospects by addressing regulatory frameworks, consumer perceptions, risk assessments, and future opportunities. This book is designed for researchers, students, industry professionals, and policymakers seeking to understand the transformative role of nanotechnology in sustainable agriculture and environmental restoration.

Mohali, Punjab, India — Sukhvinder Singh Purewal
Mohali, Punjab, India — Vijay Gahlaut
Malda, West Bengal, India — Sandeep Kumar Dash

Contents

About the Editors

Sukhvinder Singh Purewal, Ph.D., is currently serving as an Assistant Professor at the University Centre for Research and Development (UCRD), Chandigarh University, Mohali, Punjab, India. He was awarded a Major Research Project (2018–2022) by the SEED Division, Department of Science and Technology (DST), New Delhi. His research interests include solid-state fermentation, extraction of bioactive compounds from natural sources, antioxidant research, and fruit processing. He has authored over 50 research papers in peer-reviewed international journals and has received multiple best poster awards at both national and international conferences. He is an active member of several professional organizations such as the Association of Microbiologists of India (AMI), Mycological Society of India (MSI), Association of Food Scientists and Technologists (India) (AFSTI), and the Indian Science Congress Association (ISCA). In 2023, he was recognized with the prestigious "Young Scientist Award" by AMI for his notable contributions to Food Microbiology. He has authored the book *Millets: Properties, Processing, and Health Benefits* and edited six other titles.

Vijay Gahlaut, Ph.D., is currently working as an Associate Professor at the University Centre for Research and Development, Chandigarh University, Mohali, Punjab, India. He previously served as a DST INSPIRE Faculty at the CSIR-Institute of Himalayan Bioresource Technology, India. With over 12 years of experience in research and teaching, he specializes in crop biotechnology, bioinformatics, and genomics. His work has focused on identifying key genomic regions, epialleles, and candidate genes on wheat chromosomes associated with tolerance to water and heat stress, using diverse bioinformatic and genomic approaches. He has authored more than 45 research articles in reputed international journals and contributed to several book chapters. In recognition of his scientific contributions, he was recently elected as a Member of the National Academy of Sciences, India (NASI).

Sandeep Kumar Dash has been serving as an Assistant Professor in the Department of Physiology at the University of Gour Banga, Malda, since 2016. He earned his Ph.D. in 2016 from the Department of Human Physiology with Community Health at Vidyasagar University, specializing in immunology and microbiology. He has published 74 research articles in internationally reputed journals, 3 articles in national journals, and 3 book chapters with prominent publishers. His work spans multiple disciplines including nanotechnology, cancer biology, immunology, microbiology, pharmacology, and toxicology. He has developed several nanoformulations for targeted therapy against various cancer cells and for the treatment of clinically isolated multidrug-resistant bacteria. His current research focuses on the modification of plant-derived triterpenoids and short peptide-based nanocomposites, enhancing their selectivity for cancer cells, cancer stem cells, and drug-resistant pathogens. He is also engaged in the green synthesis of metal and organic nanoparticles and evaluating their efficacy against drug-resistant superbugs. Beyond laboratory research, he conducts community-based surveys on socio-demographic, nutritional, and anthropometric aspects of underserved populations. He is an active life member of the Indian Science Congress Association and Chirantan Rasayan Sanstha and serves as a regular reviewer for numerous esteemed journals.

Contributors

Muhammad Adeel BNU-HKUST Laboratory of Green Innovation, Advanced Institute of Natural Sciences, Beijing Normal University at Zhuhai, Guangdong, China

Zakaria Abdelhalim Attia Department of Pests and Plant Protection, National Research Centre, Cairo, Egypt

Rogelio Carrillo-González Programa de Edafología, Colegio de Postgraduados, Montecillo, México

Cintia Maria Sousa Ferreira Postgraduate of the Bioprocess Engineering and Biotechnology Program (PPGEnBio), Federal University of São João del-Rei (UFSJ), São João del Rei, Brazil

Ma. del Carmen A. González-Chávez Programa de Edafología, Colegio de Postgraduados, Montecillo, México

Norma L. Gutiérrez Ortega Departamento de Ingeniería, Civil y Ambiental, División de Ingenierías, Campus Guanajuato, Universidad de Guanajuato, Guanajuato, Gto, México

Eshani L. Lokuge Department of Materials and Mechanical Technology, Faculty of Technology, University of Sri Jayewardenepura, Homagama, Sri Lanka

Disha Mallick Department of Entomology, Gazipur Agricultural University, Gazipur, Bangladesh

Gabriela Medina-Perez ICAP-Institute of Agricultural Sciences, Autonomous University of the State of Hidalgo, Tulancingo de Bravo, Hidalgo, Mexico

Jorge L. Mejía-Méndez Programa de Edafología, Colegio de Postgraduados, Montecillo, México

Xu Ming BNU-HKUST Laboratory of Green Innovation, Advanced Institute of Natural Sciences, Beijing Normal University at Zhuhai, Guangdong, China

Ana Luíza Schuenck Alberice Monteiro Department of Chemical Engineering (DEQUI), Federal University of São João del-Rei (UFSJ), São João del-Rei, Brazil

Alma Hortensia Serafin Muñoz Departamento de Ingeniería, Civil y Ambiental, División de Ingenierías, Campus Guanajuato, Universidad de Guanajuato, Guanajuato, Gto, México

Mughees Mustafa Department of Agronomy, Pir Mehr Ali Shah Arid Agriculture University, Rawalpindi, Pakistan

Kalaivani K. Nadarajah School of Biological Sciences and Biotechnology, Faculty Science and Technology, Universiti Kebangsaan Malaysia, Bangi, Selangor, Malaysia

Elton Barreto Department of Chemical Engineering (DEQUI), Federal University of São João del-Rei (UFSJ), São João del-Rei, Brazil

Yazmín Stefani Perea-Vélez Programa de Edafología, Colegio de Postgraduados, Montecillo, México

Hermes Pérez-Hernández Instituto Nacional de Investigaciones Forestales, Agrícolas y Pecuarias (INIFAP). Campo Experimental Edzná, Campeche, C.P, México

Gayan Priyadarshana Department of Materials and Mechanical Technology, Faculty of Technology, University of Sri Jayewardenepura, Homagama, Sri Lanka

Md. Mamunur Rahman Department of Entomology, Gazipur Agricultural University, Gazipur, Bangladesh

Patrícia Avelar Ribeiro Postgraduate of the Physics and Chemistry of Materials Program (FQMAT), Federal University of São João del-Rei (UFSJ), São João del-Rei, Brazil

Al-kazafy Hassan Sabry Depaartment of Pests and Plant Protection, National Research Centre, Cairo, Egypt

Igor J. B. Santos Department of Chemical Engineering (DEQUI), Federal University of São João del-Rei (UFSJ), São João del-Rei, Brazil

Andersen E. Schlogl Postgraduate of the Physics and Chemistry of Materials Program (FQMAT), Federal University of São João del-Rei (UFSJ), São João del-Rei, Brazil

Noman Shakoor College of life science and oceanography, Shenzhen University, Shenzhen, China

Marcos Sousa Silva Postgraduate of the Chemical Engineering Program (PPGEQ), Federal University of São João del-Rei (UFSJ), São João del-Rei, Brazil

Millena Alves Máximo Vaz Department of Chemical Engineering (DEQUI), Federal University of São João del-Rei (UFSJ), São João del-Rei, Brazil

Edgar Vázquez-Núñez Grupo de Investigación sobre Aplicaciones Nano y Bio Tecnológicas para la Sostenibilidad (NanoBioTS), Departamento de Ingenierías Química, Electrónica y Biomédica, División de Ciencias e Ingenierías, Universidad de Guanajuato, Campus León, Loma del Bosque, León, Guanajuato, Mexico

Ileana Vera-Reyes Centro de Investigación en QuímicaAplicada (CIQA), Departamento de Biociencias y Agrotecnología, Saltillo, Coahuila, México

Abbreviations

Ag NP	Silver Nanoparticles
AI	Artificial Intelligence
Au NRs	Gold Nanorods
BPH	Brown Plant Hopper
CFA	Conventional Fertilizer Application
CMCS	Carboxymethyl Chitosan
CRF	Control Release Formulation
CRFs	Controlled Release Fertilizers
CRN	Controlled Release Nitrogen
DBM	Diamond Back Moth
EDAX	Energy Dispersive X-ray Spectroscopy
e-Nose	Electronic Nose
e-Tongue	Electronic Tongue
FAO	Food and Agriculture Agency
GHG	Green House Gas
IONPs	Iron Oxide Nanoparticles
IPR	Iprodione
K	Potassium
LC50	Lethal concentration of 50% of insect population
LCA	Life Cycle Assessment
MIC	Minimum Inhibitory Concentration
mNPs	Magnetic Nanoparticles
N	Nitrogen
NMs	Nanomaterials
NNMs	Nematocidal Nanomaterial
NPs	Nanoparticles
NUE	Nutrient Use Efficiency
OFA	Optimized Fertilizer Application
P	Phosphorous
PPNs	Plant Parasitic Nematodes
SCRFs	Controlled Release Fertilizers
SEM	Scanning Electron microscopy
SFTs	Smart Fertilizer Technologies
SiO2	Silicon dioxide powder

SNPs	Silica Nanoparticles
TiO2	Titanium Dioxide
VRT	Variable Rate Technology
WSNs	Wireless sensors networks
WUE	Water Use Efficiency
ZnO NPs	Zinc Oxide Nanoparticles

Recent Trends in Nano-fertilizers for Sustainable Agriculture

1

Jorge L. Mejía-Méndez, Rogelio Carrillo-González, Yazmín Stefani Perea-Vélez, and Ma. del Carmen A. González-Chávez

Abstract

This chapter explores recent trends in developing and applying nano-fertilizers, emphasizing their potential to enhance nutrient use efficiency, reduce environmental impact, and promote crop yield and soil health. The nano-fertilizers addressed included polymeric nanomaterials for nutrient-encapsulation and efficient delivery, and green-synthesized nanostructures to improve plant growth and physiological/biochemical plant functions. Furthermore, we compiled novel evidence highlighting the mechanisms of action of the presented nanomaterials, focusing on their capacity to orchestrate molecular events such as enzymatic and non-enzymatic reactions. This chapter also considered the challenges associated with the regulatory hurdles, toxicology, and public perception of nano-fertilizers. The role of nanotechnology in precision agriculture was critically analyzed, and how nano-fertilizers can be integrated into smart farming practices to optimize nutrient delivery and minimize waste. By exploring recent research findings and industry advancements, this chapter provided a comprehensive overview of the nano-fertilizers state of the art and their potential to transform sustainable agriculture. Given the experience of our research group working with nanotechnology and soil systems, future directions regarding research and rational development of nano-fertilizers are included.

J. L. Mejía-Méndez · R. Carrillo-González · Y. S. Perea-Vélez
Ma. del Carmen A. González-Chávez (✉)
Programa de Edafología, Colegio de Postgraduados, Montecillo, México
e-mail: mejia.jorge@colpos.mx; crogelio@colpos.mx; yazmin.perea@colpos.mx; carmeng@colpos.mx

S. S. Purewal et al. (eds.), *Agri-Nanotechnology: Innovations for Sustainable Agriculture and Environmental Restoration*,
https://doi.org/10.1007/978-981-96-9756-4_1

Keywords
Nano-enabled agriculture · Nanomaterials · Nutrient-encapsulation · Plant nutrition · Polymeric nanomaterials · Precision agriculture

1.1 Introduction

Fertilizers are substances applied to soil or plants to provide nutrients that promote growth and enhance crop yield. Given their composition, fertilizers are cataloged as chemical and organic. Chemical fertilizers are synthetically manufactured products that contain macro- and micronutrients such as nitrogen, phosphorus, potassium, or their aminated or halogenated forms (de la Nuez Pantoja et al. 2024). On the contrary, organic fertilizers are derived from natural sources, including compost, manure, seaweed, or wood ash (Yılmaz et al. 2024). Chemical fertilizers have been advantageous in improving nutrient concentration and soil health over the last decades; however, their prolonged utilization can decline organic matter and soil biodiversity, while inducing nutrient runoff and potential toxicity risks (Wang et al. 2024a). In contrast, the application of organic fertilizers has been shown to improve soil structure, organic matter content, and microbial activity. Still, their use can be associated with several challenges since they pose risks to plant health or food safety when improperly processed. In addition, their effectiveness can vary due to their nutrient content variability, which depends on the source utilized for their synthesis and influences their availability (Wu et al. 2024). Nano-biotechnology involves synthesizing and applying nanomaterials (NMs) to provide innovative and efficient approaches to resolve agriculture, medicine, and environmental hurdles. NMs are objects with nanometric architecture, and contrary to bulk materials, they exhibit unique physical, chemical, and biological performance. The development of NMs is achieved through top-down and bottom-up methods. The selection of synthesis routes relies on the type of material, desired properties, cost and resource availability, and environmental and safety considerations. NMs are classified as organic and inorganic.

Nano-fertilizers are the nanometric counterparts of fertilizers that can be synthesized from organic and inorganic precursors under controlled experimental conditions, yielding structures with tailoring capability and functionality. Contrarily to chemical and organic fertilizers, nano-fertilizers have been exploited because of their small size, high surface area to volume ratio, and low toxicity, favoring their capability to improve beneficial soil microorganisms' activity, nutrient mobilization, stress tolerance, together with enhanced crop yield and quality. Over the last decades, the design and experimental evaluation of NMs for agricultural applications have constituted a highly activity research field; however, few studies have focused on their possible toxic effects on human health and the environment. In addition, the lack of regulatory policies regarding their synthesis, evaluation, and commercialization has occurred in their indiscriminate use and documentation. Given the need to compile the existent evidence about nano-fertilizers, this chapter contains various strategies used to encapsulate fertilizers into two significant groups of NMs: polymeric and green synthesized nanostructures. In addition, it reviews how the molecular interactions of NMs with plants can improve their growth and functionality when transformed by oxidation, reduction, complexation, and dissolution phenomena. The capability of nano-fertilizers to interact with biological or environmental components is included by presenting information about the formation of bio- and eco-coronas and their environmental fate, transport, and toxicity. Nano-enabled agriculture practices are also reviewed as an innovative approach to reduce economic losses while providing benefits for farmers. The potential toxicity of nano-fertilizers is discussed based on the impact of other NMs on human health and ecosystems. International policies to regulate the development of NMs are presented with a particular focus on current Mexican agencies and their role in minimizing the negative impact of devices. Public perception of nano-fertilizers is included, and examples of current agencies devoted to providing guidance are mentioned.

1.2 Polymeric Nanomaterials and Fertilizer Encapsulation Technologies

Currently, NMs that can be used as nano-fertilizers encompass carbon nanotubes, liposomes, quantum dots, polymeric nanoparticles (NPs), hydrogels, and metal-based NMs. Even though they can be obtained from distinct synthesis routes and differ in their physical, chemical, and biological features, they are advantageous as nano-fertilizers as they can enhance chlorophyll content, promote hormone synthesis, and induce the synthesis of stress-related proteins. In addition, they are convenient as they can promote enzyme activity (e.g., catalase, superoxide dismutase, or glutathione peroxidase) while occurring in nutrient mobilization and soil microdiversity (Fig. 1.1).

Polymers are large molecules composed of repeating structural units known as monomers, such as amino acids, nucleotides, or saccharides. In the nanobiotechnological pipeline, polymers are highly utilized since their tailorable features, self-assembly, biocompatibility, mechanical properties, and thermal resistance are convenient for producing NMs (e.g., NPs, nanofibers, hydrogels, or dendrimers) for the encapsulation and delivery of nutrients. Common natural polymers utilized to encapsulate fertilizers include alginate, chitosan, starch, cellulose, and gelatin (Ma et al. 2023; Channab et al. 2023). However, synthetic polymers have also been considered; some of them encompass polylactic acid (Das and Satapathy 2024), polyvinyl acetate (Ahmad et al. 2024), and polyurethane (Sitthisuwannakul et al. 2023). The encapsulation technologies of nutrients are associated with experimental processes required to entrap desired substances into organic or inorganic matrices through defined temperature, stirring, or pH conditions. The techniques implemented for encapsulating fertilizers into polymeric nanostructures comprehend the coacervation, spray drying, sol-gel, and supercritical fluid technologies.

The coacervation technique is convenient for encapsulating fertilizers since it yields NMs with high encapsulation and controlled release efficiency. In contrast,

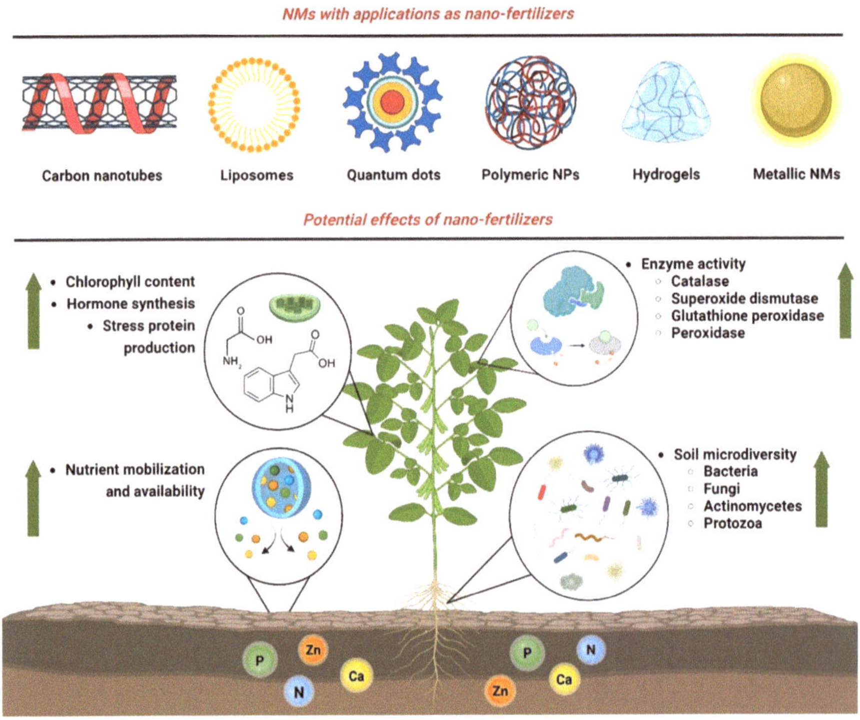

Fig. 1.1 Nanomaterials with application as nano-fertilizers and their potential effects

the spray drying approach is suitable for obtaining rapidly cost-effective polymeric NMs with controlled particle size, enhanced stability, and improved release profiles (Lodi et al. 2024). The sol-gel technique is utilized to synthesize biocompatible, stable, and versatile thin films or coatings to encapsulate fertilizers (Nam and Do 2024). The supercritical fluid method involves the use of substances that exhibit properties of both liquids and gases that can improve the solubilization of fertilizers and, hence, promote their effective incorporation into polymeric matrixes (Abd-Elsalam et al. 2024). Depending on the selected raw materials, polymeric NMs can release their cargo through various mechanisms such as diffusion, erosion, and swelling.

The diffusion mechanism is based on the capacity of the entrapped molecules to move from the inside of the polymer matrix to the surrounding medium. The release of cargo through diffusion follows Fick's first law, which associates the rate of diffusion to the concentration gradient (Lu et al. 2025). However, the erosion release mechanism occurs when the surface of the polymer matrix is degraded by water penetration or as a response toward physical (e.g., stretching, moisture, or radiation) and chemical (e.g., pH, oxidative agents, or enzymes) agents (Mateen et al. 2024). When exposed to physical or chemical stimuli, the surface of the polymeric nanostructures can be degraded controllably to release the entrapped substance (Viswanadha et al. 2024).

Since the erosion of the surface can occur abruptly after exposure to environmental conditions, the release of cargo frequently follows a zero-order kinetics, which is of great importance for the application of fertilizers since it ensures its efficacy by maintaining optimal nutrient levels in the soil while minimizing the risk of nutrient leaching (Elsabagh et al. 2024). Given the capability of polymeric NMs to absorb water during erosion events, they are also prone to release their cargo by the swelling release mechanism. In contrast to the diffusion and erosion mechanisms, the swelling capacity of polymer matrixes is related to their affinity toward solvents and degree of cross-linking, where highly cross-linked polymers swell less, while poorly cross-linked structures present enhanced swelling levels. The absorption of solvents by the polymeric nanostructures results in the formation of pores, which enables the diffusion of the entrapped substances out of the matrix.

1.3 Green-Synthesized Nanostructures

Green synthesis is the term designated to the implementation of methods for producing NMs by utilizing renewable resources and minimizing hazardous chemicals. In nanobiotechnology, green-synthesized NMs are applied for the development of delivery systems, pollution control, solar cell manufacturing, and fertilizing purposes. Commonly, the green synthesis of NMs is limited to the obtention of metal-based nanostructures such as NPs, nanoflowers, or nanorods; however, it also encompasses the manufacturing of polymer-, silica-, and carbon-based NMs through cost-effective, simple, and energy-efficient conditions (Saini et al. 2024).

The routes to green synthesize NMs are categorized into top-down (e.g., laser ablation, etching, or lithography) and bottom-up (e.g., co-precipitation, biogenesis, or chemical vapor deposition). For the synthesis of NMs with applications as fertilizers, the use of top-down routes is advantageous as they yield nanostructures with controlled sizes and morphologies (Padhan et al. 2024). However, their utilization can be energy-intensive and generate significant waste. Conversely, bottom-up techniques are characterized by versatility to occur in NMs with tailored features, enhanced biocompatibility, and upregulated performance. Still, they are also limited by scalability challenges, longer synthesis rates, and NMs with heterogeneous structural arrangements, sizes, and surface charges (Atinafu et al. 2021).

Besides the entrapment of fertilizers into polymeric structures, NMs that can be green synthesized for fertilizing comprehend metal oxide NPs, nano-biochar, and nano-clay. Metal oxides are compounds composed of metal elements combined with oxygen; representative examples include zinc oxide (ZnO), copper oxide (CuO), and iron oxide (Fe_2O_3). The use of ZnO, CuO, and Fe_2O_3 for the development of nanofertilizers is prefered since they can be manipulated through biological, hydrothermal, co-precipitation, and sol-gel routes to fabricate nanostructures to promote plant growth, photosynthesis, enzyme activity modulation, and chlorophyll synthesis. When applied in synergy with plant-promoting microorganisms such as

Pseudomonas spp., ZnO-based NMs can also alleviate drought and heat stress while enhancing the levels of antioxidant enzymes and osmoprotectants together with the biomass, nutrients, and pigments content (Azmat et al. 2022).

Biochar is a carbon-abundant material retrieved from the pyrolysis of organic biomass, such as agricultural residues (e.g., corn stover, rice husk, or wheat straw), forestry waste (e.g., tree bark, wood shavings, or sawdust) or plant materials (e.g., coconut shells, switchgrass, or bamboo) under conditions of limited oxygen. As fertilizer, biochar contributes to nutrient retention, carbon sequestration, beneficial soil microorganisms' diversity, and soil acidity reduction. To produce nano-fertilizers, biochar is initially processed through dry or wet milling to obtain its nanoparticulated form (Atinafu et al. 2021; Song et al. 2022). The fabrication of nano-fertilizers from biochar is predominantly focused on synthesizing carbon nanotubes, as it can act as a precursor to ensure their complete formation. However, only a few studies have focused on evaluating their fertilizing activity in their nanoparticulated form. Still, it has been suggested that due to its increased surface area and reactivity, it can ameliorate the leaching of nutrients to crops under stress caused by drought (Abd El-Moaty et al. 2024), and salinity (Tourajzadeh et al. 2024).

Clays are natural soil materials composed of phyllosilicate minerals with varying amounts of other minerals (e.g., Al or Mg) and water. Common examples encompass kaolinite, illite, bentonite, and chlorite. Given their cation-exchange capacity, thermal stability, and structure, clays can improve soil structure by enhancing water retention, porosity, pH, aggregation, pollutant removal, and fertility. In green synthesis routes, clays are transformed into their nanometric counterparts by energy-efficient physical approaches and, hence, integrated with additional materials (e.g., starch hydrogel), metal oxides, or reducing agents to obtain NMs with suitable features to entrap or deliver fertilizers such as urea and nitrogen (Saini et al. 2024). Some properties such as surface area can be improved after modification (Ali et al. 2024).

1.4 Molecular Interactions of NMs with Plants to Improve Their Growth and Functionality

The interactions between NMs and plants are complex (Fig. 1.2) and vary due to the number of factors that influence them, including plant species, growth stage, physiological status of plant, composition, and properties of NMs (Kandhol et al. 2024). Nevertheless, to optimize and guarantee the safe utilization of NMs in agriculture, it is essential to find the principal factors and the mechanisms of interaction between NMs and plants that influence their yield and functionality.

Among plant-NMs interplays, the transformation of NMs exerts a pivotal influence on the plant response because, upon entry into the plant cell, they can interact with cytoplasmic constituents, cellular organelles, or biomolecules, leading to changes in plant metabolism and cell signaling (Kandhol et al. 2024; Wei et al. 2024). NMs transformation encompasses alterations in their surface morphology causing corona formation, aggregation, or structural defects. In addition, it is also

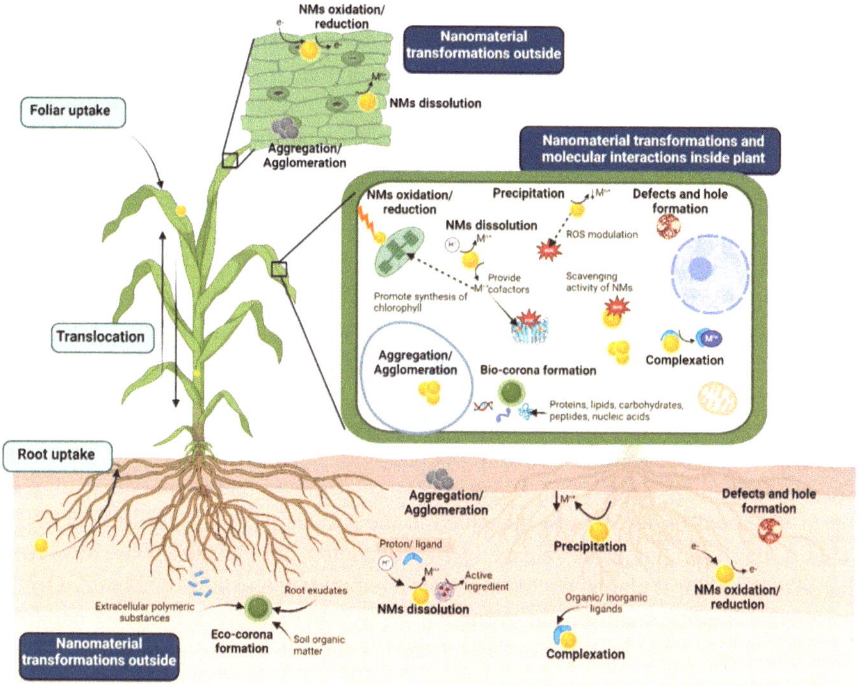

Fig. 1.2 Schematic illustrations of interactions between nanomaterials and plants (green and blue boxes). The blue boxes highlight the molecular interactions that occur outside and inside the plant. Nanomaterials can be transformed (bold letters) before entering the plant. Once inside the cells, nanomaterials can be transformed, modify the dynamics of reactive oxygen species, induce changes in the plant metabolism, and be a source of cofactors and nutrients

related to alterations in their dissolution, transformation to organic forms, complexation, oxidation, or reduction (Wei et al. 2024). These events are illustrated in Fig. 1.2.

The transformation of NMs can occur both inside and outside the plant. The corona is a natural interface between nanostructures and living matter, and it is a robust shell (Monopoli et al. 2013). When inside the plant, NMs can bind with proteins, lipids, carbohydrates, or nucleic acids, forming bio-coronas, layers of biomolecules formed around NMs that can influence their interaction with plant cells and tissues and determine their toxicity. When transformed outside the plant, NMs can bind with biomolecules secreted from plants (proteins, lipids, and sugars) (Monopoli et al. 2013) and microorganisms and form an eco-corona, layers of organic molecules formed around NMs that influence their interaction with environmental components and affect their environmental fate and transport (Wei et al. 2024). NMs' transformation outside the plant affects their uptake, translocation, biodistribution, or toxicity (Mahmoudi et al. 2023; Kang et al. 2024; Wei et al. 2024).

Regarding surface morphology alterations, the corona formation seems to affect protein structure and function, as well as cellular processes like energy synthesis, root exudation, and enzymatic activity (Mahmoudi et al. 2023; Yu et al. 2023; Prakash et al. 2023). However, the analysis of corona inside the plants is still under investigation, and the molecular mechanisms of their interaction are scarce (Liu et al. 2013). Available information shows that the stimulation of root exudation and the formation of eco-corona may be a plant mechanism to protect against the toxic effects of NMs. In this regard, pristine graphene oxide (13–50 mg L^{-1} dose) has shown a genotoxic effect on the root tips of *Allium cepa*, reducing the cell viability by 73% compared to unexposed root cells to graphene oxide, inducing oxidative stress. In contrast, coronated graphene oxide (with extracellular-polymeric substances from rhizospheric soil) exhibits low toxic effects. The cell viability augmented from 7% to 32% compared to those from pristine material (Christudoss et al. 2023). Kang et al. (2024) investigated the effect of ZnO-NPs on the metabolic plant response of *Vigna radiata* sprouts and the formation of eco-corona on ZnO-NPs. The size (20 and 100 nm) and hydrophobicity (hydrophilic and hydrophobic NPs) of ZnO-NPs were the NMs properties evaluated. The results of this study showed that small-size NPs at a dose of 10 mg L^{-1} inhibited the root elongation by up to 37%, and this effect was attributed to the surface properties of the ZnO-NPs. This further stimulated the release of root exudates and elevated it by 74% compared to the control plants. On the other hand, small size and hydrophobic ZnO-NPs had higher binding stability and stronger affinity with biomolecules in root exudates than larger-sized and hydrophilic NPs, thus driving eco-corona formation.

The eco-corona formation also affects the dispersion and agglomeration/aggregation of NMs and, therefore, their bioavailability and plant uptake (Yu et al. 2023). In the case of metal and metal oxide-based NMs, it has been observed and suggested that low molecular weight of extracellular-polymeric substances from rhizosphere soil or root exudates inhibits the aggregation of NMs by enhancing the electrostatic repulsion between NPs (Yu et al. 2023; Kang et al. 2024). For polymeric NPs and organic NMs used as vehicles, the corona formation may affect the targeting efficiency of NMs, as has been observed in nano-drugs (Yu et al. 2023). Nevertheless, their plant transformation has not been systematically studied (Stolte Bezerra Lisboa Oliveira and Ristroph 2024).

The dissolution, oxidation, and reduction of NMs are other transformations that explain how the NMs can improve plant nutrition and performance. Dissolution of NMs can occur both inside and outside the plant, and two processes are responsible for that. The first process involves a proton-promoted mechanism, which is that organic acids inside the plant tissues or in the rhizosphere microzone provide H^+ (Wei et al. 2024). For instance, copper oxide (CuO) NPs are a very stable and low soluble compound due to the packed crystal structure and high lattice energy ($U_{CuO} = 4050$ kJ mol^{-1}). However, their dissolution has been observed and attributed to the H^+ provided for choline ascorbate, choline salicylate, malic acid, oxalic acid, succinic acid, and citric acid (Martins et al. 2024). Similarly, cobalt ferrite ($CoFe_2O_4$) NPs were observed, and their dissolution was promoted by malic acid and oxalic acid (Perea-Velez et al. 2022). The second process is ligand-assisted dissolution, meaning NMs react with surrounding ligands to form metal ion-ligand complexes and detach ions into the solution (Wei et al. 2024). In cobalt oxide (CoO) NPs, it has been observed that the adsorption of phosphate and large biomolecules (like mucin and lysozyme) on the NPs surface promotes their dissolution by 35–55% after 1 h (Mei et al. 2019). On the other hand, the oxidation of MoS_2 NPs has been described as the mechanism of molybdenum release (MoO_4^{2-}) (Li et al. 2024a). In this regard, ions released from metal- or metal oxide-based NPs of Mg, Mn, Zn, Cu, Fe, and Ni boost plant nutrition by increasing their bioavailability. Moreover, it has been observed that ions released from NMs can promote chlorophyll synthesis, act as cofactors, enhance the activity of enzymes, and modulate reactive oxygen species (ROS) production (Thangavelu et al. 2024; Wei et al. 2024).

Besides the NMs transformation, NMs can also modify the dynamics of ROS and nitric oxide (•NO), thereby improving plant performance by reducing oxidative damage to plant cells under stress conditions (Li et al. 2024b; Bao et al. 2024). A detailed review of this topic is provided by Kandhol et al. (2024) and Bao et al. (2024). However, we can summarize that NPs can eliminate ROS due to their scavenging capabilities, induce differential expression of ROS-related genes, induce increased antioxidant activity, induce low ROS, and •NO leading modifications in the plants' metabolism and influencing pathways related to nutrient acquisition and the synthesis of secondary metabolites. For instance, after the exposition of *Vigna radiata* sprouts to ZnO-NPs, augmentation by 0.8–1.0 folds of stress markers (malondialdehyde and hydrogen peroxide) was observed, as well as superoxide dismutase activity by 60% compared to control sprouts. The metabolomic analysis showed that ZnO (20 nm size and 10 mg L^{-1}) stimulated the flavonoid biosynthesis of *V. radiata*, which plays a role in nitrogen fixation and plant defense systems. While ZnO-NPs of 100 nm promoted the carbon/nitrogen metabolisms and increased the citrate levels in the tricarboxylic acid cycle, providing energy for stress resistance (Kang et al. 2024). Finally, it should be noted that a holistic understanding of the molecular interactions between plants and NMs opens the opportunity to optimize the use of NMs in agriculture while minimizing potential risks.

1.5 Nano-enabled Agriculture: Smart Farming Modeland Industry Innovations

Agriculture production faces challenges from exponential population growth, food production demand, climate change, and environmental issues (Chugh et al. 2021). Hence, new technologies and processes to improve agricultural production are vital. Innovative technologies like remote sensing, smart irrigation, and nano-fertilizersare proposed to boost agriculture efficiency, development, and sustainability. Smart irrigation and remote sensing are successfully implemented in solving many of the issues in conventional agriculture regarding several economic and environmental benefits of agricultural production, which may also transform this industry (Garg et al. 2024). On the other hand, nanotechnology has gained recognition in different commercial applications; however, its implementation in agriculture is in its infancy. Even though nano-enabled agriculture, an emerging hot issue, is gaining attention to the design of several materials for their use in several agricultural needs.

Development and implementation of innovative strategies to increase plant productivity and food security are crucial. Contrary to conventional agriculture, nano-enabled agriculture may enhance crop production while reducing the environmental impacts of agrochemicals (Xin et al. 2020; Chugh et al. 2021). Nano-fertilizers and nanosensors, an obligate step to modern agricultural technology, are envisioned as a powerful tool in precision agriculture based on farm efficiency and production. Precision farming seeks to maximize profit through the nutrient stewardship: right source, rate, time, and place. This ensures that both, plants and soil, receive exactly what they need for optimal productivity and health, reducing losses, improving crop performance, keeping environmental quality, and providing benefits for farmers (Garg et al. 2024). Nano-fertilizers revolutionize modern agriculture and are strongly linked to precision agriculture by improving nutrient delivery while minimizing waste. The nano-size of several formulations and designs gives unique advantages over conventional fertilizers, for example, more efficient nutrient delivery, controlled release, and improved nutrient uptake by plants.

Nano-coated fertilizers are designed from conventional fertilizer particles but are coated with additional layersbased on polymers, metal oxides, or clays. The coating slows nutrient release, protects it from leaching and adsorption, and enhances fertilizer efficiency. In the case of nano-chelated fertilizers, a chemical bond forms between NPs and a chelated nutrient, ensuring nutrient transport and preventing its interaction with soil components. Comparably, nano-emulsions have high nutrient uptake efficiency because nutrients are suspended in a nano-sized matrix of water and oil droplets with a high surface area that allows strong interaction with plant tissues. Hybrid formulations combine different NPs and may function synergistically to improve nutrient delivery and plant response. The direct use of NPs uploaded with nutrients is the most common formulation. Industrial innovations related to nano-fertilizers constitute a promising approach for transforming agricultural practices by enhancing nutrient efficiency, reducing environmental impact, and promoting sustainable farming. Besides the nano-formulations already mentioned, industrial innovations use slow or controlled-release nano-fertilizers, and biopolymeric hydrogels. Recent reports explained that slow-release nano-fertilizers (SRNF) and controlled-release nano-fertilizers (CRNF) are a novel class of nano-fertilizers, wherein organic materials (e.g., phosphogypsum, bentonite, and sulfur) or organic polymers (e.g., polyethylene, polyurethane, or alkyd resin) are frequently used for their development (Haydar et al. 2024). SRNF are nanocomposites with high molecular weight and complex matrix structures, embedding nutrients that will be released by water diffusion and microbial action from nanocomposite decay compounds. This nutrient release is slower in SRNF than in normal NPs since it is not controlled and depends on soil reactions and environmental factors such as pH, temperature, and soil moisture. In contrast, CRNF, also named coated, delayed, or metered fertilizers, are coated materials containing water-soluble nutrients that can exhibit different nutrient release patterns due to the type and coating material utilized during their synthesis.

Biopolymeric hydrogels are three-dimensional networks that use natural polymers such as alginate, Arabic gum, chitosan, or starch and can be loaded with nutrients. Plants can uptake this nutrient after gradual release. These novel fertilizers also enhance water retention, improve soil structure, and stimulate microbial activity (Jumpapaeng et al. 2023; Shanmugavel et al. 2024). The industrial innovations also refer that nanotechnology may be involved in the reduction of gases with the greenhouse effect, enhancing soil carbon sequestration and reducing nitrogen oxide emission, a new adoption called climate-smart agriculture. In this context, nano-technological alternatives protect soil, increase food security, and mitigate climate change effects (Chugh et al. 2021).

Nano-abled agriculture is a promising approach to participating in transgenic events, plant stress tolerance, sensing stress and early detection, and seed nano-priming (Wu and Li 2022; Raza et al. 2023). One of the recent biotechnological innovations is nanoparticle-mediated genome editing via clustered regularly interspaced short palindromic repeats (CRISPR/Cas) system, which aims to develop stress-smart crops representing a vital approach to ensure future global food security and modern agriculture's resilience, efficiency, and progress. This is a speed breeding method for crop production improvement, which transfers exogenous DNA for genome editing by the CRISPR/Cas system using novel molecular nano-carries, for example, polymeric nanostructures; however, the understanding of mechanisms involved and the potential risk of NPs used for genome editing needs more profound research as referred by Raza et al. (2023).

Nanotechnology can be used in the smart farming model for better adaptation/mitigation to climate change, reducing environmental impact, optimizing water use, facilitating water reuse, promoting land resilience, and improving crop yields. Smart farming, also known as smart agriculture, increases the quantity and quality of products and optimizes the human labor required (Mandal et al. 2024). In this context, nanotechnology and precision agriculture relate much more. This is possible by involving information and communication technology such as sensors, global positioning, geographic information systems, weather monitoring, drones, satellite imagery, data management, automation, and agricultural robots. Advanced technologies based on data-driven, artificial intelligence, and machine learning may also optimize and improve sustainability in agricultural production (Xin et al. 2020; Mandal et al. 2024).

1.6 Safe Concerns and Public Perceptions

The use of NMs to increase agricultural productivity has been a highly active research field over the last decade. However, their fate in the environment and potential risk to human health have been partially taken into account. The safety concerns of NMs are established considering their type of toxicity, which varies from very low to very dangerous, and must be assessed before their fabrication, storage, and, if applicable, commercialization. Despite the differences in their application, the safety concerns concerning using nano-fertilizers are shared with other nanostructures due to similarities regarding their physical, chemical, optical, and mechanical characteristics.

Organic and inorganic NMs pose a challenge to environmental safety since they are prone to accumulate in soil and water due to their large surface area, which enables their adsorption by particles, organic matter, or sediments. The fate of NMs in soil and water is driven by their small size, which allows their transport over long distances through atmospheric deposition, drainage, and runoff, ensuring their persistence in environmental compartments (Wang et al. 2024b). Their reactivity and stability also influence the potential toxicity of NMs in soil and water systems. The former suggestion is based on the chemical composition, surface area to volume ratio, and capability of NMs to undergo oxidation and reduction reactions. The latter refers to the capacity of NMs to break down under specific conditions such as pH and temperature. The evidence about the environmental fate of nano-fertilizers remains unexplored and it has been documented for NMs that are not nano-fertilizers. For example, it has been noted that the high reactivity of NMs such as cadmium selenide zinc sulfide quantum dots (Le et al. 2023), polyaniline nanocomposites, and metal-organic frameworks is problematic since they can form by-products with up- or down-regulated toxicity. In contrast, highly stable NMs such as carbon nanotubes, fullerenes, and silica NPs pose a significant challenge as their tendency to aggregate reduces their effective surface area but enhances their long-term accumulation within the environment.

Given their high reactivity and tendency to bioaccumulate, NMs can cause significant health concerns in humans since they can be deposited within organs or react with proteins, lipids, and genetic material. The interaction between NMs and biomolecules can lead to molecular and cellular aberrations, consequently causing the initiation and progression of cardiovascular, respiratory, and neurological disorders (Hermosillo-Abundis et al. 2024). Bioaccumulation is the process by which organisms accumulate compounds at a rate faster than they can eliminate them. Among the organs, the bioaccumulation of NMs has been identified in the lungs, liver, kidneys, spleen, bone marrow, adipose tissue, and skin (Jakic et al. 2024). The human safety concerns regarding the bioaccumulation of nano-fertilizers are challenging to discuss due to the lack of scientific evidence; however, it has been documented that chronic exposure to NMs based on metal oxides can compromise the functionality of the bronchi by causing inflammation (Di Ianni et al. 2024). Against the liver and kidneys, exposure to NMs can impair the detoxification and reabsorption process of nutrients, respectively (Jarrar et al. 2024).

The public perception of nanotechnology alludes to the awareness and understanding of the applications, benefits, and ethical implications of nanotechnology. Over the last decades, numerous nanoproducts have been extensively manufactured and commercialized as sunscreens, stain-resistant fabrics, and screens with enhanced brightness and energy efficiency (Vaudagna et al. 2023; Parlak et al. 2024). However, consumer knowledge regarding their existence is limited. Currently, various organizations intend to expand the knowledge about the agricultural, electronic, and medical implications of NMs to the audience, including potential regulatory policies and innovative research fields. Representative organizations that address these features encompass Gov4Nano, RiskGONE PROJECT, Graphene Flagship, and NanoHarmony, respectively.

1.7 Toxicology and Ecotoxicology of Nano-fertilizers

Toxicology includes the analysis of the hazards of chemical (e.g., pesticides, heavy metals, or phthalates; Hao et al. 2024), biological (e.g., aflatoxins, influenza virus, or *Staphylococcus aureus*; Chu and Zi 2024), and physical (e.g., radiation, heat, or noise) agents to humans, ecosystems, and wildlife (Ben Attia et al. 2024). The toxicity of chemical, biological, or physical agents is assessed through extensive investigation of their mechanisms of action, dose-response relationships, and exposure routes. In human health, fertilizers (nitrite) have been indirectly reported to cause toxicity in infants, adolescents, and the elderly by inducing methemoglobinemia (Kazmi et al. 2024), growth and development abnormalities, and neurodegenerative disorders (e.g, Alzheimer's disease; Aloizou et al. 2020), respectively. In water bodies, using excessive fertilizers can cause eutrophication and algal blooms, leading to aquatic organisms' death by depleting oxygen levels or exacerbating toxins production (Pepe-Vargas et al. 2024; Gobler et al. 2024). In soils, the over-application of fertilizers can acidify them, promote the accumulation of heavy metals, disrupt microbial communities, decrease compaction and aeration levels, and induce salt accumulation (Zhang et al. 2024). Given their negative effects on the environment, researchers have expected to minimize it using less fertilizers or through less reactive, safe, and functional approaches such as nano-fertilizers.

Nanoecotoxicology provides theoretical and practical references to the environmental consequences of the use of nanotechnology. It is a highly interdisciplinary field that correlates the influence of the size, shape, or solubility of nanostructured materials with their toxicity. The toxicity of nano-fertilizers has been evaluated in a low number of experimental models such as the soil invertebrates *Enchytraeuscrypticus* and *Folsomia candida* (Malheiro et al. 2024). However, there are no studies that explore their detrimental effects on major terrestrial or aquatic organisms. The use of metallic NPs, at a low rate, has positive effects on plant absorption, but at high concentrations may have toxic implications (Santás-Miguel et al. 2023). However, due to the low cost and short observation period, the effect of NPs on germination is one of the most common tests, and short-time experiments have been carried out.

1.8 Regulatory Hurdles of Nano-fertilizers

The regulation of the production and utilization of NMs for fertilizing applications is scarce since it considers the application of policies and standardization processes originally destinated for harmonizing the use of NMs (e.g., liposomes, micelles, and multiwalled carbon nanotubes) for drug delivery, clinical diagnostics, cytotoxicity, or in vivo toxicity purposes (Rodríguez-Gómez et al. 2024).

In the USA, the development, marketing, and potential toxicity evaluation of NMs are regulated by the Food and Drug Administration (FDA), a federal agency devoted to evaluating and approving the safety of food products, medical devices, cosmetics, vaccines, and medications (Anand 2024). On the other hand, the sustainability, potential environmental and health risks, and exposure pathways to NMs are covered by the Environmental Protection Agency (EPA), an independent agency of the US government that also aims to conduct policy decisions regarding air, water, and land contamination (Kim 2024). Every country or region has its own laws. In Mexico, research involving nanotechnology is indirectly regulated by several agencies such as the environmental secretariat (Secretariat of Environment and Natural Resources (SEMARNAT), Health Secretariat (Federal Commission for the Protection against Sanitary Risks; COFEPRIS), and National Service for Agro-Food Health, Safety, and Quality (SENASICA), which rules substances and their environmental impact, pharmaceuticals and food additives, and agricultural products, respectively. The National System for the Toxicological Evaluation of Nanomaterials (SINANOTOX) was also created in México to enlist the risks of NMs through the design and implementation of assays that allow the understanding of their toxicity from a molecular and cellular perspective; however, the implementation, effectiveness, and potential of SINANOTOX are inconclusive (Antonio Chávez-Hernández et al. 2024). The regulation should be equivalent to the rules of USA and Canada, the three partners in the North American Free Trade Agreement to avoid handicap conditions for México.

For academic institutions and industries, the awareness and understanding of the synthesis, characterization, and application of NMs are also regulated following the International Organization for Standardization (ISO) guidelines, which provide the necessary frameworks to comprehend standard definitions related to nanotechnology, safety concerns or risk management, and guidelines for the validation of NMs (Amutha et al. 2024). The technical specifications associated with standardized terminology for nanomaterials, characterization techniques, classification based on their properties of uses, and environmental management are the ISO/TS 27687:2008, ISO/TC 229, ISO/TR 13014:2012, and ISO 14001 specifications, respectively (Caputo et al. 2024). The data retrieved following international and national policies can aid academic and governmental institutions in establishing preventive protocols, public health guidelines, and training programs to use NMs and, if applicable, respond to their toxicity effects.

1.9 Conclusions

The development of nano-fertilizers constitutes a transformative approach in agriculture by maximizing nutrient efficiency while reducing detrimental effects on the soil, environment, and human health. The design of nano-fertilizers is based on polymeric materials and green synthesis methods. However, it is essential to consider that their manufacturing also constitutes challenges regarding their complexity in synthesis, yield, and scalability. Therefore, the integration of additional approaches and multidisciplinary collaborations are required to ensure their obtention, functionality, and effectiveness.

The understanding of the molecular interplays between NMs and plants is still in development due to the intricate nature of these interactions and some methodological challenges associated with the study of corona formation inside the plant. The transformation of NMs should be considered as part of designing a new generation of nano-fertilizers to optimize and guarantee the safe utilization of NMs in agriculture. In this regard, developing and harmonizing methodologies for studying nanostructured object transformations is critical. Knowledge about the molecular interactions between plants and NMs and the key factors in maximizing nano-fertilizers' positive effects are essential for fine-tuning nano-fertilizer design. Nanotechnology may help to enhance crop production in the context of food security and global climate change by reducing the excessive use of agrochemicals. Nano-enabled agriculture is a revolutionary step toward modern agricultural technology involving precision agriculture with features of producing more with less without dismissing productivity, health, waste, environment, or farmer benefits. Industrial innovations with nano-fertilizers using the innovative farm model will adopt climate-smart agriculture, genome editing, plant stress issues, and other relevant aspects to face challenges from the exponential population growth and food production demand. The regulation of nano-fertilizers is based on international and national policies designed to control the fabrication and quality of devices, especially the ones intended for biomedical applications. For this reason, political and academic frameworks to govern the fabrication of NMs for agricultural are still being determined. In the scientific community, they are based on review articles rather than original research. Based on the consulted evidence, it was also observed that few studies have focused on the knowledge, reception, and engagement of people with concepts related to nanotechnology, especially nano-fertilizers. The lack of interactive events and citizen science initiatives covering nanotechnology can result in communities with misconceptions and skepticism about the benefits and awareness of NMs. Despite the limited evidence about the toxicity of NMs, it is noteworthy to consider that their detrimental effects can be analyzed through various in vitro, in vivo, in silico, or *in chemico* methodologies that can represent novel approaches to consolidate innovative research fields. The toxicity of nano-fertilizers can be studied by determining their impact on seed germination rates, root elongation, or enzyme activity associated with nutrient cycling. Some nano-fertilizers can also cause significant toxicity effects among individuals, given their accumulation in plants, entry into the food chain, and occupational or environmental exposure. The toxicity of nano-fertilizers can be explored by analyzing their cytotoxicity, genotoxicity, and capability to cause inflammation or disrupt the anatomical integrity of in vivo models.

Acknowledgments J.L.MM. acknowledges the support provided by the Consejo Mexiquense de Ciencia y Tecnología (COMECYT) under the program Cátedra de Investigación (CAT2024-0013). Y.S.PV. acknowledges the support provided by the Secretaría de Ciencia, Humanidades, Tecnología e Innovación (SECIHTI) under the program Estancias Postdoctoralespor México. All figures were created utilizing BioRender.com.

References

Abd El-Moaty HI, El-Dissouky A, Elhusseiny AF, Farag KM, Abu-Khudir R, Alkuwayti MA, Al Abdulsalam NK, Abdel Rahman SM (2024) Low-cost nano biochar: a sustainable approach for drought stress mitigation in faba bean (*Vicia faba* L.). Front Plant Sci 15. https://doi.org/10.3389/fpls.2024.1438893

Abd-Elsalam KA, Hashim AF, Abdelkhalek TE, Hassan RK (2024) Encapsulation of soil nutrients for plant growth and development. In: Abd-Elsalam KA, Hashim AF, Ahmed FK, Thomas S (eds) Biopolymeric nanoparticles for agricultural applications. Springer Nature, Cham, pp 145–174. https://doi.org/10.1007/978-3-031-68834-8_6

Ahmad I, Ahmad W, Nepal J, Junaid MB, Bukhari NA, Usman M, Ahmad N, Khan RN (2024) Synergistic enhancement of maize crop yield and nutrient assimilation via zinc oxide nanoparticles and phosphorus fertilization. J Sci Food Agric 104(11):6733–6745. https://doi.org/10.1002/jsfa.13500

Ali SM, Mohamed RA, Abdel-Khalek AA, Ahmed AM, Abukhadra M (2024) Physicochemical, steric, and energetic characterization of kaolinite based silicate nano-sheets as potential adsorbents for safranin basic dye: effect of exfoliation reagent and techniques. Front Chem 12. https://doi.org/10.3389/fchem.2024.1455838

Aloizou A-M, Siokas V, Vogiatzi C, Peristeri E, Docea AO, Petrakis D, Provatas A, Folia V, Chalkia C, Vinceti M, Wilks M, Izotov BN, Tsatsakis A, Bogdanos DP, Dardiotis E (2020) Pesticides, cognitive functions and dementia: a review. Toxicol Lett 326:31–51. https://doi.org/10.1016/j.toxlet.2020.03.005

Amutha C, Gopan A, Pushbalatatha I, Ragavi M, Reneese JA (2024) Nanotechnology and governance: regulatory framework for responsible innovation. In: George SC, Tawiah B (eds) Nanotechnology in societal development. Springer Nature, Singapore, pp 481–503. https://doi.org/10.1007/978-981-97-6184-5_14

Anand MA (2024) Global regulatory frameworks for hair care industry: an overview of key agencies and guidelines. In: Dwivedi A, Pant AB, Poojan S, Kotak M, Tripathi A (eds) Hair care products: efficacy, safety and global regulation. Springer Nature, Singapore, pp 201–221. https://doi.org/10.1007/978-981-97-6740-3_13

Antonio Chávez-Hernández J, Jimena Velarde-Salcedo A, Navarro-Tovar G, Gonzalez C (2024) Safe nanomaterials: from their use, application, and disposal to regulations. Nanoscale Adv 6(6):1583–1610. https://doi.org/10.1039/D3NA01097J

Atinafu DG, Wi S, Yun BY, Kim S (2021) Engineering biochar with multiwalled carbon nanotube for efficient phase change material encapsulation and thermal energy storage. Energy 216:119294. https://doi.org/10.1016/j.energy.2020.119294

Azmat A, Tanveer Y, Yasmin H, Hassan MN, Shahzad A, Reddy M, Ahmad A (2022) Coactive role of zinc oxide nanoparticles and plant growth promoting rhizobacteria for mitigation of synchronized effects of heat and drought stress in wheat plants. Chemosphere 297:133982. https://doi.org/10.1016/j.chemosphere.2022.133982

Bao L, Liu J, Mao T, Zhao L, Wang D, Zhai Y (2024) Nanobiotechnology-mediated regulation of reactive oxygen species homeostasis under heat and drought stress in plants. Front Plant Sci 15. https://doi.org/10.3389/fpls.2024.1418515

Ben Attia T, Nahdi A, Horchani M, Ben Ali R, Ben Jannet H, Galai S, Elmay MV, Mhamdi A (2024) Simultaneous exposure to noise and toluene induces oxidative and inflammatory damage in the heart of Wistar rats: therapeutic potential of *Oleaeuropaea* L. leaf extract. Cardiovasc Toxicol 24(7):667–686. https://doi.org/10.1007/s12012-024-09867-0

Caputo F, Favre G, Borchard G, Calzolai L, Fisicaro P, Frejafon E, Günday-Türeli N, Koltsov D, Minelli C, Nelson BC, Parot J, Prina-Mello A, Zou S, Ouf F-X (2024) Toward an international standardisation roadmap for nanomedicine. Drug Deliv Transl Res 14(9):2578–2588. https://doi.org/10.1007/s13346-024-01646-2

Channab B-E, Elhassani CE, Essamlali Y, Marrane SE, Chakir A, Zahouily M (2023) Urea-loaded hydroxyapatite–carboxylated cellulose composites as slow-release N fertilizer pellets for efficient delivery of nitrogen. Ind Eng Chem Res 62(37):14853–14865. https://doi.org/10.1021/acs.iecr.3c00737

Christudoss AC, Dimkpa CO, Mukherjee A (2023) Eco-corona formation diminishes the cytogenotoxicity of graphene oxide on *Allium cepa*: role of soil extracted–extracellular polymeric substances in combating oxidative stress. Plant Physiol Biochem 204:108123. https://doi.org/10.1016/j.plaphy.2023.108123

Chu Z-Y, Zi X-J (2024) Network toxicology and molecular docking for the toxicity analysis of food contaminants: a case of aflatoxin B1. Food Chem Toxicol 188:114687. https://doi.org/10.1016/j.fct.2024.114687

Chugh G, Siddique KHM, Solaiman ZM (2021) Nanobiotechnology for agriculture: smart technology for combating nutrient deficiencies with nanotoxicity challenges. Sustainability 13(4):1781. https://doi.org/10.3390/su13041781

Das KP, Satapathy BK (2024) Highly porous agro-augmenting urea-biochar/polylactic acid-based microfibrous electrospun mats as sustainable controlled release fertilizer carriers. ACS Appl Polym Mater 6(11):6262–6277. https://doi.org/10.1021/acsapm.4c00386

de la Nuez Pantoja EY, Iznaga IR, Fuentes GR, Petranovskii V, García AM, Gámez JJC, Jiménez DG, Cauqui MÁ, Rivero González LA, García OC (2024) Natural clinoptilolite materials enriched in nitrogen and phosphorous for agricultural purposes: a comprehensive study on their development. J Inorg Organomet Polym 34(6):2522–2542. https://doi.org/10.1007/s10904-023-02991-4

Di Ianni E, Erdem JS, Narui S, Wallin H, Lynch I, Vogel U, Jacobsen NR, Møller P (2024) Pro-inflammatory and genotoxic responses by metal oxide nanomaterials in alveolar epithelial cells and macrophages in submerged condition and air-liquid interface: an in vitro-in vivo correlation study. Toxicol In Vitro 100:105897. https://doi.org/10.1016/j.tiv.2024.105897

Elsabagh SS, Elkhatib EA, Rashad M (2024) Novel nano-fertilizers derived from drinking water industry waste for sustained release of macronutrients: performance, kinetics and sorption mechanisms. Sci Rep 14(1):5691. https://doi.org/10.1038/s41598-024-56274-0

Garg S, Rumjit NP, Roy S (2024) Smart agriculture and nanotechnology: technology, challenges, and new perspective. Adv Agrochem 3(2):115–125. https://doi.org/10.1016/j.aac.2023.11.001

Gobler CJ, Drinkwater RW, Anthony A, Goleski JA, Famularo-Pecora AME, Wallace MK, Straquadine NRW, Hem R (2024) Sewage-and fertilizer-derived nutrients alter the intensity, diversity, and toxicity of harmful cyanobacterial blooms in eutrophic lakes. Front Microbiol 15. https://doi.org/10.3389/fmicb.2024.1464686

Hao M, Jiang L, Lv M, Ding H, Zhou Y, Xu H (2024) Plant natural product-based pesticides in crop protection: semi-synthesis, mono-crystal structures and agrochemical activities of osthole ester derivatives, and study of their toxicology against *Tetranychus cinnabarinus* (Boisduval). Pest Manag Sci 80(12):6356–6365. https://doi.org/10.1002/ps.8364

Haydar MS, Ghosh D, Roy S (2024) Slow and controlled release nanofertilizers as an efficient tool for sustainable agriculture: recent understanding and concerns. Plant Nano Biol 7:100058. https://doi.org/10.1016/j.plana.2024.100058

Hermosillo-Abundis C, Méndez-Rojas MA, Arias-Carrión O (2024) Implications of environmental nanoparticles on neurodegeneration. J Neurosci Res 102(5):e25340. https://doi.org/10.1002/jnr.25340

Jakic K, Selc M, Razga F, Nemethova V, Mazancova P, Havel F, Sramek M, Zarska M, Proska J, Masanova V, Uhnakova I, Makovicky P, Novotova M, Vykoukal V, Babelova A (2024) Long-term accumulation, biological effects and toxicity of BSA-coated gold nanoparticles in the mouse liver, spleen, and kidneys. Int J Nanomedicine 19:4103–4120. https://doi.org/10.2147/IJN.S443168

Jarrar B, Almansour M, Al-Doaiss A, Lee SY, Melhem W, Jarrar Q, Sewelam A (2024) Metallic and metallic oxide nanoparticles toxicity primarily targets the mitochondria of hepatocytes and renal cells. Toxicol Ind Health 40(12):667–678. https://doi.org/10.1177/07482337241282860

Jumpapaeng P, Suwanakood P, Nanan S, Saengsuwan S (2023) Novel biodegradable nanocomposite hydrogels based on biopolymers and various montmorillonite contents as high-strength coating membranes for efficient slow-release fertilizers. J Ind Eng Chem 127:191–209. https://doi.org/10.1016/j.jiec.2023.07.005

Kandhol N, Singh VP, Pandey S, Sharma S, Zhao L, Corpas FJ, Chen Z-H, White JC, Tripathi DK (2024) Nanoscale materials and NO-ROS homeostasis in plants: trilateral dynamics. Trends Plant Sci. https://doi.org/10.1016/j.tplants.2024.06.009

Kang M, Bai X, Liu Y, Weng Y, Wang H, Ye Z (2024) Driving role of zinc oxide nanoparticles with different sizes and hydrophobicity in metabolic response and eco-corona formation in sprouts (*Vigna radiata*). Environ Sci Technol 58(22):9875–9886. https://doi.org/10.1021/acs.est.4c01731

Kazmi S, Farooqi H, Nawaz A, Afzal Z, Idress A, Yousafzai MO, Khan MU (2024) Rapid-onset methemoglobinemia from traditional-medicine-induced potassium nitrate poisoning: successful treatment with methylene blue—a case report. J Med Case Rep 18(1):439. https://doi.org/10.1186/s13256-024-04702-0

Kim Y (2024) The role of science in the policy subsystem: an application of the advocacy coalition framework to nanotechnology regulation policies. Int J Public Adm 47(6):397–414. https://doi.org/10.1080/01900692.2022.2123504

Le N, Chand A, Okafor O, Kim K (2023) The impact of cadmium selenide zinc sulfide quantum dots on the proteomic profile of *Saccharomyces cerevisiae*. Int J Mol Sci 24(22):16332. https://doi.org/10.3390/ijms242216332

Li M, Zhang P, Guo Z, Zhao W, Li Y, Yi T, Cao W, Gao L, Tian CF, Chen Q, Ren F, Rui Y, White JC, Lynch I (2024a) Dynamic transformation of nano-MoS_2 in a soil–plant system empowers its multifunctionality on soybean growth. Environ Sci Technol 58(2):1211–1222. https://doi.org/10.1021/acs.est.3c09004

Li K, Li H et al (2024b) Nanoparticle-mediated modulation of plant performance and microbiome dynamics: insights into interplay mechanisms. Biocatal Agric Biotechnol 61:103366. https://doi.org/10.1016/j.bcab.2024.103366

Liu W, Rose J, Plantevin S, Auffan M, Bottero J-Y, Vidaud C (2013) Protein corona formation for nanomaterials and proteins of a similar size: hard or soft corona? Nanoscale 5(4):1658–1668. https://doi.org/10.1039/C2NR33611A

Lodi LA, Borges R, Lopes MM, Graciano VA, Bortoletto-Santos R, Barud HS, de Oliveira-Paiva CA, Ribeiro C, Farinas CS (2024) Spray-drying microencapsulation of *Bacillus megaterium* in PVA/cationic starch/zinc oxide for promoting growth and zinc availability in soybean plants. ACS Agric Sci Technol. https://doi.org/10.1021/acsagscitech.4c00476

Lu J, Wu M, Luo L, Lu R, Zhu J, Li Y, Cai Y, Xiang H, Song C, Yu B (2025) Incorporating iron oxide nanoparticles in polyvinyl alcohol/starch hydrogel membrane with biochar for enhanced slow-release properties of compound fertilizers. Carbohydr Polym 348:122834. https://doi.org/10.1016/j.carbpol.2024.122834

Ma X, Zhang S, Yang Y, Tong Z, Shen T, Yu Z, Xie J, Yao Y, Gao B, Li YC, Helal MID (2023) Development of multifunctional copper alginate and bio-polyurethane bilayer coated fertilizer: controlled-release, selenium supply and antifungal. Int J Biol Macromol 224:256–265. https://doi.org/10.1016/j.ijbiomac.2022.10.121

Mahmoudi M, Landry MP, Moore A, Coreas R (2023) The protein corona from nanomedicine to environmental science. Nat Rev Mater 8(7):422–438. https://doi.org/10.1038/s41578-023-00552-2

Malheiro C, Prodana M, Cardoso DN, Morgado RG, Loureiro S (2024) Ageing influences the toxicity of two innovative nanofertilizers to the soil invertebrates *Enchytraeus crypticus* and *Folsomia candida*. Environ Pollut 350:123989. https://doi.org/10.1016/j.envpol.2024.123989

Mandal S, Yadav A, Panme FA, Devi KM, Kumar SMS (2024) Adaption of smart applications in agriculture to enhance production. Smart Agric Technol 7:100431. https://doi.org/10.1016/j.atech.2024.100431

Martins MAR, Kiirika LM, Schaffer N, Sajnóg A, Coutinho JAP, Franklin G, Mondal D (2024) Unveiling dissolution kinetics of CuO nanofertilizer using bio-based ionic liquids envisaging controlled use efficiency for sustainable agriculture. ACS Sustain Res Manag 1(6):1291–1301. https://doi.org/10.1021/acssusresmgt.4c00041

Mateen A, Khan A, Khan I, Ahmad L, Khan A, Salam A (2024) Formulation development, characterization, and evaluation of sorafenib-loaded PLGA–chitosan nanoparticles. Front Pharmacol 15. https://doi.org/10.3389/fphar.2024.1465363

Mei N, Hedberg J, Odnevall Wallinder I, Blomberg E (2019) Influence of biocorona formation on the transformation and dissolution of cobalt nanoparticles under physiological conditions. ACS Omega 4(26):21778–21791. https://doi.org/10.1021/acsomega.9b02641

Monopoli MP, Pitek AS, Lynch I, Dawson KA (2013) Formation and characterization of the nanoparticle-protein corona. Methods Mol Biol 1025:137–155. https://doi.org/10.1007/978-1-62703-462-3_11

Nam NN, Do HDK (2024) Ultrasonic sonication assisted in sol–gel stöber and reverse microemulsion method for fabrication of silica nanoparticles and the potential application in agriculture. Chem Pap 78(12):7089–7096. https://doi.org/10.1007/s11696-024-03588-y

Padhan SR, Kar I, Mohanty A, Panigrahi KK (2024) Nanofertilizers: types, synthesis, methods, and mechanisms. In: Abd-Elsalam KA, Alghuthaymi MA (eds) Nanofertilizers for sustainable agroecosystems: recent advances and future trends. Springer Nature Switzerland, Cham, pp 61–98

Parlak M, Ilban HI, Karsli K, Unal E, Demir HV (2024) Environmental tests and reliability characterization of pixel-sized colloidal QDs for next-generation display technologies. Next Nanotechnol 5:100059. https://doi.org/10.1016/j.nxnano.2024.100059

Pepe-Vargas P, Castro LR, Alves-de-Souza C, Llanos-Rivera A (2024) Effects of the harmful algal bloom toxin, okadaic acid, on the mechanoreceptors of larval anchoveta (*Engraulis ringens*) under varying environmental conditions. Front Mar Sci 11. https://doi.org/10.3389/fmars.2024.1446509

Perea-Velez YS et al (2022) Dissolution kinetics of citrate coated $CoFe_2O_4$ nanoparticles in soil solution. Environ Sci Nano. https://doi.org/10.1039/D2EN00330A

Prakash S, Gupta R, Deswal R (2023) Nanoparticle protein corona: understanding NP biomolecule interactions for safe and informed nanotechnological applications including stress alleviation in plants. J Plant Growth Regul 42(10):6377–6396. https://doi.org/10.1007/s00344-022-10857-1

Raza A, Charagh S, Salehi H, Abbas S, Saeed F, Poinern GEJ, Siddique KHM, Varshney RK (2023) Nano-enabled stress-smart agriculture: can nanotechnology deliver drought and salinity-smart crops? J Sustain Agric Environ 2(3):189–214. https://doi.org/10.1002/sae2.12061

Rodríguez-Gómez FD, Monferrer D, Penon O, Rivera-Gil P (2024) Implementing horizon scanning as a tool for the strategic development of regulatory guidelines for nanotechnology-enabled health products. Front Med 10. https://doi.org/10.3389/fmed.2023.1308047

Saini R, Manjaiah KM, Chobhe KA, Datta SP, Raju D, Jatav HS, Gawdiya S, Minkina T, Rajput VD (2024) Urea loaded nanoclay biopolymer composites: synthesis, characterization and evaluation for enhancing nitrogen use efficiency. Clean Techn Environ Policy. https://doi.org/10.1007/s10098-024-03040-5

Santás-Miguel V, Arias-Estévez M, Rodríguez-Seijo A, Arenas-Lago D (2023) Use of metal nanoparticles in agriculture. A review on the effects on plant germination. Environ Pollut 334:122222. https://doi.org/10.1016/j.envpol.2023.122222

Shanmugavel D, Solorza-Feria O, Kamaraj S-K (2024) Current trends in biopolymer-based hydrogels for use in agriculture. In: Abd-Elsalam KA, Hashim AF, Ahmed FK, Thomas S (eds) Biopolymeric nanoparticles for agricultural applications. Springer Nature Switzerland, Cham, pp 27–57. https://doi.org/10.1007/978-3-031-68834-8_2

Sitthisuwannakul K, Boonpavanitchakul K, Wirunmongkol T, Muthitamongkol P, Kangwansupamonkon W (2023) A tunable controlled-release urea fertilizer coated with a biodegradable polyurethane-nanoclay composite layer. J Coat Technol Res 20(2):635–646. https://doi.org/10.1007/s11998-022-00688-w

Song B, Cao X, Gao W, Aziz S, Gao S, Lam C-H, Lin R (2022) Preparation of nano-biochar from conventional biorefineries for high-value applications. Renew Sust Energ Rev 157:112057. https://doi.org/10.1016/j.rser.2021.112057

Stolte Bezerra Lisboa Oliveira L, Ristroph KD (2024) Critical review: uptake and translocation of organic nanodelivery vehicles in plants. Environ Sci Technol 58(13):5646–5669. https://doi.org/10.1021/acs.est.3c09757

Thangavelu RM, da Silva WL, Zuverza-Mena N, Dimkpa CO, White JC (2024) Nano-sized metal oxide fertilizers for sustainable agriculture: balancing benefits, risks, and risk management strategies. Nanoscale 16(43):19998–20026. https://doi.org/10.1039/D4NR01354A

Tourajzadeh O, Piri H, Naserin A, Cahri M (2024) Effect of nano biochar addition and deficit irrigation on growth, physiology and water productivity of quinoa plants under salinity conditions. Environ Exp Bot 217:105564. https://doi.org/10.1016/j.envexpbot.2023.105564

Vaudagna MV, Aiassa V, Marcotti A, Pince Beti MF, Constantín MF, Pérez MF, Zoppi A, Becerra MC et al (2023) Titanium dioxide nanoparticles in sunscreens and skin photo-damage. Development, synthesis and characterization of a novel biocompatible alternative based on their *in vitro* and *in vivo* study. J Photochem Photobiol 15:100173. https://doi.org/10.1016/j.jpap.2023.100173

Viswanadha LS, Arcot Y, Lin Y-T, Akbulut MES (2024) A comparative investigation of release kinetics of paclitaxel from natural protein and macromolecular nanocarriers in nanoscale drug delivery systems. JCIS Open 15:100120. https://doi.org/10.1016/j.jciso.2024.100120

Wang C, Miao Q, Wei Z, Guo Y, Li J, Fan Z, Hu Y, Zhang H, Sun J, Cui Z (2024a) Nutrient runoff and leaching under various fertilizer treatments and pedogeographic conditions: a case study in tobacco (*Nicotiana tabacum L.*) fields of the Erhai Lake basin, China. Eur J Agron 156:127170. https://doi.org/10.1016/j.eja.2024.127170

Wang C, Chen L, Xu J, Zhang L, Yang X, Zhang X, Zhang C, Gao P, Zhu L (2024b) Environmental behaviors and toxic mechanisms of engineered nanomaterials in soil. Environ Res 242:117820. https://doi.org/10.1016/j.envres.2023.117820

Wei L, Liu J, Jiang G (2024) Nanoparticle-specific transformations dictate nanoparticle effects associated with plants and implications for nanotechnology use in agriculture. Nat Commun 15(1):7389. https://doi.org/10.1038/s41467-024-51741-8

Wu H, Li Z (2022) Recent advances in nano-enabled agriculture for improving plant performance. Crop J 10(1):1–12. https://doi.org/10.1016/j.cj.2021.06.002

Wu Z, Chen X, Lu X, Zhu Y, Han X, Yan J, Yan L, Zou W (2024) Impact of combined organic amendments and chemical fertilizers on soil microbial limitations, soil quality, and soybean yield. Plant Soil. https://doi.org/10.1007/s11104-024-06733-4

Xin X, Judy JD, Sumerlin BB, He Z (2020) Nano-enabled agriculture: from nanoparticles to smart nanodelivery systems. Environ Chem 17(6):413–425. https://doi.org/10.1071/EN19254

Yılmaz B, Çilingir Tütüncü A, Saka AK, Demirkaya S, Yılmaz H, Ozer H (2024) Effects of different nutrient inputs in organic parsley cultivation. Biol Agric Hortic 40(3):161–172. https://doi.org/10.1080/01448765.2024.2351371

Yu Y, Dai W, Luan Y (2023) Bio- and eco-corona related to plants: understanding the formation and biological effects of plant protein coatings on nanoparticles. Environ Pollut 317:120784. https://doi.org/10.1016/j.envpol.2022.120784

Zhang H, Li X, Zhou J, Wang J, Wang L, Yuan J, Xu C, Dong Y, Chen Y, Ai Y, Zhang Y (2024) Combined application of chemical fertilizer and organic amendment improved soil quality in a wheat–sweet potato rotation system. Agronomy 14(9):2160. https://doi.org/10.3390/agronomy14092160

Nanotechnology in the Restoration of Degraded and Polluted Soil

2

Yazmín Stefani Perea-Vélez,
Ma. del Carmen A. González-Chávez,
Jorge L. Mejía-Méndez, and Rogelio Carrillo-González

Abstract

The degradation of soil quality due to industrial activities, agricultural practices, and urbanization poses significant challenges to environmental sustainability and food security. This chapter explores the use of nanotechnology in the restoration of degraded soils with special emphasis on contaminated soils. A general scenario of pollutants commonly found in soils was discussed, including metal(loid)s, pesticides, and persistent organic contaminants, and emerging ones were also included. The considered nanomaterials (NMs) to overcome the limitations of current remediation techniques encompass metal- and metal oxide-based nanoparticles (NPs), such as nano zero-valent iron (nZVI), carbon nanotubes, biochar-based NMs, and polymer-based nanostructures. The capacity of these NMs to be integrated with microorganisms to degrade, transform, or stabilize contaminants from soil was systematically addressed. The integration of NMs with conventional remediation technologies, such as phytoremediation and bioremediation, was analyzed to enhance the effectiveness of those technologies. The environmental fate of NMs used for in situ soil remediation technologies was also reviewed.

Keywords

Depleted nutrients · Nano-enabled remediation · Nano-fertilizers · Nanoparticles · Soil erosion · Soil salinity

---

Y. S. Perea-Vélez · Ma. del Carmen A. González-Chávez · J. L. Mejía-Méndez
R. Carrillo-González (✉)
Programa de Edafología, Colegio de Postgraduados, Montecillo, México
e-mail: yazmin.perea@colpos.mx; carmeng@colpos.mx; mejia.jorge@colpos.mx; crogelio@colpos.mx

S. S. Purewal et al. (eds.), *Agri-Nanotechnology: Innovations for Sustainable Agriculture and Environmental Restoration*,
https://doi.org/10.1007/978-981-96-9756-4_2

2.1 Introduction

Soil is a nonrenewable natural resource supporting us directly and indirectly (El-Ramady et al. 2024) through its multiple ecosystem services, such as nutrient cycling; water maintenance; and habitat, food, fiber, and fuel provision. Moreover, soils are crucial in regulating the climate by reducing atmospheric CO_2 (Arias-Navarro et al. 2024). In 2012, the term soil security was introduced, referring to the need for access to soil-ecosystem services to be on the same level as other human rights, meaning that soil is a common good, like water and air (Lehmann et al. 2020). In this respect, soil health impacts water quality, climate change dynamics, and human health, which is why soil has achieved prominence in the political agenda of several countries to address longer-term sustainability challenges, food security, and conservation of biodiversity (Lehmann et al. 2020; Panagos et al. 2024). On the other hand, soils can act as natural filters and transform the potentially harmful elements and compounds into their less toxic forms; unfortunately, human activities have reached critical limits, leading to soil degradation (FAO and ITPS 2015). According to the United Nations, 20% of the Earth's total land area was degraded in 2015 (UN 2024), and this soil threat has accelerated over time. It is estimated that each year, at least 100 million hectares of healthy and productive land is lost due to its degradation, and if the current trends persist, by 2030, the degraded soil will be 1.5 billion hectares (UNCCD 2023). In this perspective, soil degradation will not only impact the functionality and ecosystem services provided by soil but also amplify food-price volatility and potentially send millions of people into poverty (FAO and ITPS 2015).

Soil degradation is defined as a change in the soil health status, resulting in a diminished capacity of the ecosystem to provide goods and services for its beneficiaries (FAO 2024). It leads to a decline in the soil quality and functionality in terms of their physical, chemical, and biological properties. Thus, degraded soil presents declined soil fertility; loss of organic matter; erosion; changes in acidity, salinity, or alkalinity; excessive flooding; or the accumulation of toxic chemicals (such as pesticides, persistent organic contaminants, and dyes) and pollutants (such as metal(loid)s, nano-, and microplastics). This ultimately results in soil contamination (El-Ramady et al. 2024).

For agricultural productivity, soil degradation due to pollution constitutes a serious challenge since it can reduce soil fertility, decrease crop yields, and cause crop contamination leading to a risk of bioaccumulation of pollutants and water quality hazards (Moghimi Dehkordi et al. 2024). For human health, soil degradation is also considered a significant threat as it has been associated with nutritional deficiencies, respiratory disorders, and endocrine disruption cases (Angon et al. 2024). Economically, soil degradation by pollution has caused 20 billion USD in losses in agriculture productivity (Zhou et al. 2020), and it is expected to increase in the following years due to the drawbacks related to the use of current techniques considered to minimize it. Nanotechnology may improve the efficacy of conventional soil remediation techniques by addressing the limitations of traditional methods, such as the cost of their implementation, effectiveness, high energy consumption, and time constraints (Qaderi et al. 2023; Ningombam et al. 2024). Moreover, nanotechnology may offer more efficient, sustainable, and adaptable solutions (Dhanapal et al. 2024; Ningombam et al. 2024). It should be highlighted that sustainable remediation includes in its execution planning the possible outcomes across environmental, social, and economic factors (Bone et al. 2020; Visentin et al. 2024). For instance, Bone et al. (2020) evaluated the sustainability of the in situ nano-remediation process (nano-remediation and direct current electrokinetics) for soil polluted with chlorinated hydrocarbon (considered a contaminant source for groundwater). The sustainability of the nano-remediation was compared with conventional remediation technologies such as in situ chemical oxidation, excavation, and disposal. The authors reported that nano-remediation with nZVI integrated with direct current electrokinetics scored a low sustainability index (meaning that it is the most sustainable option) compared to the excavation and disposal treatment. When comparing nano-remediation to chemical oxidation, similar sustainability scores were found; however, for the environmental indicators, nano-remediation showed a positive impact in the longer term.

Considering the detrimental effects of soil degradation, this chapter describes nano-enabled technologies for agricultural soil remediation by addressing challenges associated with soil erosion, soil salinity, depleted soil fertility, soil pollution due to metal(loid)s and pesticides, and emerging contaminants (ECs). It also presents examples of contaminants and how nanostructures can be leveraged to remove, stabilize, or degrade them, assisted by plants or microorganisms. The environmental fate is covered, and the mechanisms by which NMs interact with soil components are illustrated and discussed. Thus, it is expected that the information exposed here may contribute to propitiating sustainable and safe nanotechnologies for soil in situ remediation, considering the gaps in knowledge discussed and the conscious use of NMs to prevent them from becoming an unwanted contaminant.

2.2 Nano-enabled Technologies for Agricultural Soil Remediation

Current agricultural practices have contributed to soil degradation (soil erosion, soil salinity, depleted soil fertility, and soil contamination) threatening soil ecosystem services, human health, and food/nutrition security. Innovations to address the above challenges are urgently needed, not only to alleviate the consequences of soil degradation but also to restore soil health. On this matter, nanotechnology approaches (Fig. 2.1) offer a revolutionary potential for the remediation of degraded soils (Dhanapal et al. 2024).

Both organic and inorganic NMs have been documented as efficient alternatives for removing contaminants from water sources that may be used for irrigation, or improving soil properties since their high surface area to volume ratio enables their reactivity, resulting in enhanced chemical reactions and degradation processes (Abbasi 2024). On this point, NMs designed for in situ soil remediation applications tend to exhibit improved transport and facilitated uptake, when NMs are

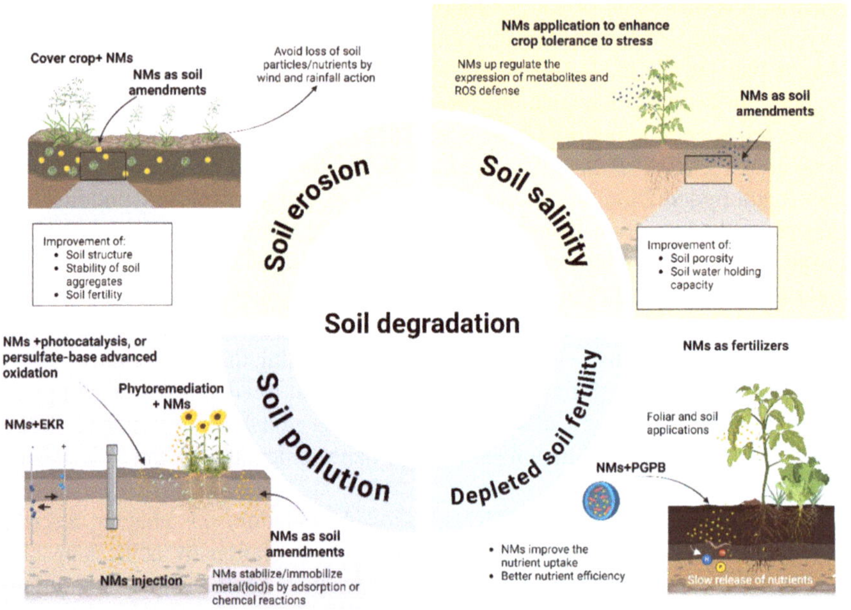

Fig. 2.1 Nano-enabled technologies for the remediation of degraded soils. *NMs* nanomaterials, *PGPB* plant growth promoting bacteria, *EKR* electrokinetic remediation

functionalized with synthetic (e.g., amines, crown ethers, or phosphonic acids) or natural (e.g., amino acids, bio-surfactants, or humic acids) ligands (Xue et al. 2024; Shirazian et al. 2025). This is important since it allows their movement through soil or water systems and promotes their arrival to confined contaminated sites.

The mechanisms by which NMs can remove contaminants from water sources or immobilize them in soils comprehend adsorption and catalytic events. The adherence of contaminants to the surface of NMs involves ionic interactions, hydrogen bonding, and Van der Waals forces, and it is driven by their zeta potential, presence of capping agents, or tailoring ligands. For example, it has been reported that negatively charged NMs, such as porous silicon NPs, are prone to attract cationic contaminants like cadmium (Cd) (Qu et al. 2024). In contrast, positively charged NMs such as ZrO_2 NPs immobilized into polyurethane sponges are predisposed to interact with anionic contaminants, such as fluorides (Lu et al. 2024). When amino, carboxyl, hydroxyl, phosphate, or sulfhydryl groups are distributed onto the surface of MnO_2-modified biochar NPs, they can remove contaminants by donating electron pairs and forming stable complexes with reduced mobility, bioavailability, and toxicity (Ojo et al. 2024; Ghandali et al. 2024).

The nanotechnology solutions that are currently available for the remediation of degraded soils use NMs to achieve several different goals. These include the detection, degradation, and stabilization of contaminants. In some cases, NMs are combined with the modern technologies to improve their effectiveness, for example, treatments based on photocatalysis, phytoremediation, and electrokinetics (Perea-Vélez et al. 2021; Falciglia et al. 2022; Karimi et al. 2023). Furthermore, in some other cases, the NMs are used to indirectly cope with soil degradation; for instance, NMs are used to enhance the tolerance of crops to salinity or to protect plants against the toxic effects of pollutants. The following sections present the nano-enabled technologies for soil erosion, soil salinity, declining soil fertility, and soil pollution due to metal(loid)s, pesticides, organic compounds, and ECs.

2.2.1 Soil Erosion

Soil erosion refers to the accelerated loss of soil due to anthropogenic activities or natural factors such as wind and rainfall. For agricultural soils, the principal drivers of soil erosion are soil management activities such as the use of poor or nonexistent soil cover and tillage (Quinton and Fiener 2024). Soil erosion poses a major risk to food security by causing the loss of topsoil and nutrients, which diminishes soil fertility. Additionally, it disrupts soil structure, diminishes ecosystem services, and heightens the severity of floods and droughts (Yang et al. 2024; Quinton and Fiener 2024; Sartori et al. 2024). On the other hand, the economic impacts of soil erosion are estimated at 8 billion USD per year to the global economy, equivalent to 33.7 million tons of agricultural and food losses (Sartori et al. 2019). Through the global bio-physical model, Sartori et al. (2024) estimated that the soil erosion rates will increase 30–66% from 2015 through 2070, resulting in a global economic contraction of up to 625 billion USD by 2070.

In agricultural land, the techniques to prevent soil erosion rates are reduced tillage, contour farming, terraces, afforestation of slopes, incorporation of plant residues, cover crops, grass margins, and brush layers (Sartori et al. 2019). However, the combination of these techniques with the use of NMs can be a new approach to control soil and nutrient loss. For instance, nano-biochar (40–200 nm) has been tested in combination with caragana and maize as vegetative covers to prevent soil erosion on slopes of 15° with loess soil under simulated rainfall events. The nano-biochar at rates of 0.1–1% reduced the runoff by 40–74% compared to bare soil without nano-biochar. Moreover, the vegetated plots reduced the cumulative nitrate loss by 14–62% compared with the bare plots (Chen et al. 2020).

The use of biopolymeric NMs and carbon nanotubes as soil amendments can control soil erosion indictment because these NMs can improve soil structure and stability of soil aggregates and retain nutrients (Hashim et al. 2024). For example, Huang et al. (2022) studied, at a laboratory scale, the effects of nano-aqueous adhesive (at a dose of 0.25–1%) on soil properties of silty clay soil. The authors found that nano-aqueous adhesive (polyvinyl acetate) can increase the number of aggregates with a particle size up to 0.25 mm by 98% compared to the untreated soil. Moreover, the stability coefficient of these soil aggregates increased by 0.2–8% compared to natural soil, leading to enhanced water stability of the soil.

2.2.2 Soil Salinity

Soil salinity is associated with the high accumulation of ions (e.g., Na^+, K^+, Mg^{2+}, or Ca^{2+}) in the soil and can arise from natural and anthropogenic causes. The former includes climate change, high evaporation rates in arid and semi-arid regions, weathering of rocks, and the upward movement of saline groundwater. The latter encompasses excessive irrigation, inadequate drainage systems, overuse of fertilizers, poor crop rotation, and industrial activities such as mining and oil extraction. Geographically, soil salinity has affected 7% of land area and 33% of irrigated lands over the last decades. Economically, soil salinity has been related to the loss of 27.3 billion USD in recent years (Chele et al. 2021).

Comparably to other current agricultural challenges, soil salinity is related to reduced crop yields (by up to 50%), increased soil erosion, biodiversity loss, poor water quality, and significant economic losses for agricultural, food security, and industrial sectors. Current approaches to minimizing the impact of soil salinity include improved irrigation techniques that reduce evaporation and minimize salt accumulation, application of soil amendments to improve soil structure while enhancing water retention, installation of drainage systems to remove water excess and salts from the soil, and rotation of crops to improve soil health by varying nutrient uptake (Li et al. 2024a, b, c). Such approaches have resulted in significant advantages; however, they have also been correlated to high initial costs for installation, nutrient imbalances, environmental challenges, labor-intensive practices, and inconsistent effectiveness (Qaderi et al. 2023).

The number of research studies focused especially on NMs used as soil amendments for soil salinity is limited; however, it has been reported that pH-responsive gel spheres based on sodium alginate and charcoal-activated carbon and loaded with humic acid remediated soil salinity after 50 d by improving its porosity and water-holding capacity together with enhancing the presence of available nitrogen, phosphorus, and fast-acting potassium (Wang et al. 2021). In another study, it has been documented that amendments with iron oxide NPs can alleviate salinity stress and promote the growth of *Eucalyptus tereticornis* by modifying the activities of the superoxide dismutase and malondialdehyde enzymes, while up-regulating the expression of salt-responsive genes such as HKT1, SOS1, and NHX1 (Singh et al. 2021). On this detail, most of the research conducted on the use of NMs as soil conditioners/soil amendments has been developed at a laboratory scale. The main drawback of NMs as soil amendments is the lack of information about the amounts of NMs required at the field scale, which is necessary to ascertain the feasibility and efficacy of NMs on a large scale (Hofmann et al. 2020).

On the other hand, NMs can be used as fertilizers to enhance crop tolerance to salinity. Nanotechnology is a tool for developing salinity-tolerant and adaptive agriculture; for instance, in a two-year field experiment conducted with William banana plants irrigated with saline groundwater (EC = 4.12 dS m^{-1}). The foliar spray of green nanosilica (20 nm size) at doses of 150 and 300 mg L^{-1} increased fruit yield. In the first year, the banana yield increased by 30% and 45% compared to banana plants without foliar fertilization, respectively. In the second year, the yield increased by 21% and 38%, correspondingly. Furthermore, the Na^+ concentration in banana leaves was reduced by 32–36% compared to the control plants (Ding et al. 2022). Other examples of NMs used to boost crop tolerance to salt stress are NPs of CeO_2, TiO_2, ZnO, $ZnSO_4$, and carbon nanotubes (Ahmed et al. 2023; Hassan et al. 2023; Singh et al. 2024a, b). The precise molecular interplays between NMs and plants that promote plant tolerance to salt stress remain unclear. However, it has been observed that NMs promote ROS clearance by mediating the plant antioxidant repose or by the antioxidant activity of NMs, lowering osmotic and ionic stress by the regulation of ionic homeostasis. Furthermore, NMs can improve the uptake of other nutrients and regulate the function of genetic material and the synthesis of metabolites (such as proline, glycine betaine, and soluble sugars) Singh et al. 2024a, b; Mustafa et al. 2024). In Chap. 1, a detailed review of the molecular interactions between NMs and plants to improve their growth and functionality is presented.

2.2.3 Declining Soil Fertility

Soil mineral depletion is the progressive decrease of vital nutrients and minerals in the soil, leading to reduced crop yield (Musa et al. 2024). According to the FAO, 33% of global soils are affected by reduced fertility due to the degradation process (FAO 2022). The main drivers for soil nutrient depletion are intensive farming methods, uneven fertilization, insufficient inputs of replacement of nutrients, and poor soil management (such as the lack of crop rotation, cover crops, soil testing, excessive tillage, monocropping, etc.) (FAO 2022; Musa et al. 2024). Moreover, nutrient-depleted soils are more prone to erosion due to less plant presence (Musa et al. 2024).

On the other hand, the consumption of edible plant parts cultivated in nutrient-deficient soils leads to malnutrition and health complications (such as anemia, low birth weight, child and maternal deaths, and learning impairment) due to the inappropriate composition of essential elements for human well-being (Musa et al. 2024; Weffort and Lamounier 2024). From a socioeconomic point of view, undernutrition impacts the health burden and economy of countries. Jain et al. (2024) estimated the cost of inaction regarding global nutrition targets, resulting that preventable undernutrition costs the world at least 761 billion USD per year, equivalent to 2.1 billion USD per day. While human nutrition problems are multifactorial and dependent on the distribution of and access to food, as well as lifestyle, it is evident that soils play a pivotal role in the production of nutritious food and the economy of countries. In this regard, soil mineral depletion poses a potential threat to global food and nutritious security (Tan et al. 2005). Thus, preserving soil integrity and restoring essential nutrients are imperative for the sustained production of nutritious food and the improvement of human well-being (FAO 2022; Musa et al. 2024).

There is no single solution to address soil fertility decline; however, the application of fertilizers is one of them. Conventional fertilizers do, nevertheless, present certain challenges, including low absorption efficiency, fixation of nutrients following their application, and the underuse of fertilizers (in the case of some micronutrients) (FAO 2022). Nano-fertilizers and their use with precision farming offer an alternative by circumventing the limitations of conventional fertilizers. To know more about nanotechnology solutions related to fertilizers, see Chap. 1.

In general terms, nano-fertilizers intend to supply nutrients to crops in a more efficient form due to their unique properties such as small size, volume ratio, and surface area (Kale et al. 2024). Both soil and foliar applications have been demonstrated to improve plant nutrition. The NMs mainly used as nano-fertilizers are carbon-based NMs, ZnO, CuO, Fe_2O_3, $CoFe_2O_4$, SiO_2 NPs, SeNPs, hydroxyapatite, and polymeric NMs for the encapsulation of nutrients (Jakhar et al. 2022; Perea-Vélez et al. 2023; Khaliq et al. 2023). Furthermore, NMs combined with biotechnology have been reported to increase crop growth and nutrition. For instance, the encapsulation of plant-growth-promoting bacteria such as *Pseudomonas* sp. DN18 into alginate beads supplemented with salicylic acid and ZnO-NPs demonstrated to be effective in increasing the growth of *Oryza sativa* seedlings and inhibiting the growth of *Sclerotium rolfsii* at 0.5–2 mM (Panichikkal et al. 2021).

The technological readiness level (TLR) for macronutrient fertilizers is relatively high. At the field scale, these NMs have been shown to deliver nutrients in a slow, or prolonged form, resulting in diminishing the nutrient lost due to runoff or infiltration. In the case of micronutrients, their efficiency has been proved at laboratory and greenhouse scale, and their TRL level is medium (Hofmann et al. 2020). However, it is important to emphasize that the proper use of nano-fertilizers, i.e., applying them in the right amount, time, and form can contribute to restoring depleted nutrients of soil in a safe and sustainable form. In addition, the fertilization rates should be based on soil characteristics, soil management, and the cultivars to be grown as well as the previous crops grown (FAO 2019).

2.2.4 Soil Pollution Due to Metal(loid)s, Pesticides and Organic Compounds

Some of the major risks of agricultural soil pollution due to metal(loid)s are the accumulation of these elements on the edible parts of the crops and their subsequent transmission to the food chain and the negative impact on the crop yield (Chowardhara et al. 2024). The main inputs of metal(loid) to agricultural soils are atmospheric deposition, sewage irrigation, sewage sludge, animal manure, mineral fertilizers, and pesticides (Wan et al. 2024). On the other hand, solvents such as perchloroethylene (PCE), trichloroethylene (TCE), carbon tetrachloride (CCl_4), and dichloroethylene (DCE) are highly problematic due to their persistence, mobility, toxicity, and complex remediation. In agriculture, TCE, CCl_4, and DCE have been utilized as fumigants to control pests and prevent the contamination of crops since they can act as cleaning agents (Cano-Ortiz et al. 2024). Given their lipophilicity, volatility, low water solubility, or existence in various isomeric forms, such solvents can persist in soils, leach into groundwater, affect water retention, and decrease crop yield and quality (Pathak et al. 2024). Another group of pollutants that must receive attention for remediation is those mentioned by the Stockholm Convention and named the

Dirty Dozen (EPA 2014). These are those organic pollutants produced by men, also called persistent organic pollutants, such as aldrin, chlordane, dieldrin, dichlorodiphenyltrichloroethane, endrin, heptachlor, hexachlorobenzene, mirex, polychlorinated biphenyls, polychlorinated dibenzo-p-dioxins (dioxins), polychlorinated dibenzofurans, and toxaphene. Most of them include herbicides, fungicides, and insecticides, which are synthetic, toxic, hazardous, persistent, and cause health human problems. Moreover, they may accumulate in human and animal tissues over time. Other nasty pollutants are hydrocarbon-derived chlorinated and polycyclic aromatic (Jin et al. 2020; Li et al. 2022).

For agricultural land, some of the strategies used to minimize or reduce the toxic effects of metal(loid)s and other contaminants in crops encompass the management of soil moisture, and the application of soil amendments, such as biochar (Chowardhara et al. 2024). Meanwhile, in situ remediation techniques for polluted soils include surface capping, encapsulation, electrokinetic extraction, soil flushing, chemical immobilization, phytoremediation, and bioremediation. Every one of the above-mentioned techniques has its advantages and limitations, but some of the major drawbacks of remediation technologies are the high cost of implementation, the large clean-up times required (for some of the techniques), and the final disposal of polluted substances after their remotion (Zhao et al. 2022; Ghazaryan et al. 2024). On this point, nano-enabled technologies may offer economic and sustainable advantages over conventional remediation methods.

The potential use of NMs as soil amendments to immobilize/stabilize metal(loid)s in soils has been highlighted in several studies (Michálková et al. 2017; Wang et al. 2019, 2020a). Most of them have been performed at the laboratory scale, but rare examples are at field scale (Wang et al. 2020b). The NMs present advantages over bulk materials due to their high surface reactivity, and high area/volume ratio (Perea-Vélez et al. 2021). Another form to use NMs for soil remediation is the injection of NPs into the subsurface near where the contamination plume is located. The advantage of using NPs in this form is that the transport of NPs is dominated by Brownian movement. This allows the NPs to remain suspended in the groundwater/soil for a longer time during the remediation period. Moreover, NPs can travel longer distances reaching all the extensions of the contamination plume (Puiatti 2023).

To date, there are a few studies at the pilot/field scale that reported the effectiveness and the sustainability of NMs injection for soil and groundwater remediation (Bone et al. 2020; Visentin et al. 2024). For instance, Visentin et al. (2024) evaluated the sustainability of five case studies of nZVI used on a pilot scale through the life cycle analysis tool. The selected case studies were located in Brazil, China, Spain, Hungary, and the United States. While the type of contaminant and concentration were Cr(VI) (Brazil and China cases, 100 and 800 mg kg^{-1}, respectively), As (Spain case, 5527 μg L^{-1}), chlorinated hydrocarbons (Hungary case, 15–20,000 μg L^{-1}), and solvents such as PCE,TCE, DCE (United States case, PCE = 1–12 mg L^{-1}, TCE = 2–24 mg L^{-1}, and DCE = 9–20 mg L^{-1}). The amount of nZVI used for the nano-remediation for the case study of Brazil was 40 g nZVI used per kg of soil, while for the Chinese case study was 11 mg per kg of soil. For the Spain, Hungary, and US case studies, the concentration of nZVI injected was 10 g per L of groundwater treated. The monitoring time in the cases was 120 d for Brazil and China, 6 months for Spain, 360 d for Hungary, and 600 d for the United States. The remediation efficiency in all the cases was over 80–95%, except for Hungary, where the remotion efficiency was >60%. According to the authors, the most sustainable nano-remediation process was in the US, followed by Hungary>Spain>China>Brazil. The sustainability of nano-remediation with nZVI was related to the concentration of NPs injected, which explains why the case studies of Brazil and China scored the lowest sustainability index. It should be barred in mind that the sustainability and the cost of nano-remediation will be related to different factors such as the amount of NM used, the synthesis method of NMs, the type of contaminant and matrix to be treated, the characteristics of the soil, the degree of pollution, as well as the materials used in the remediation process (Perea-Vélez et al. 2021; Visentin et al. 2024).

On the other hand, the most widely used NMs for soil remediation is nZVI due to their low toxicity, high reactivity, and contaminant reduction capacity (Komárek 2024; Visentin et al. 2024). The mechanism described through nZVI immobilizes metal(loid)s or organic pollutants on its surface including adsorption, and reduction. nZVI can adsorb simultaneously various metal(loid)s; however, the type of adsorption reaction involved depends on the redox potential of the metal species (Komárek 2024). Recently, nZVI supported on biochar has been proposed to improve the stability, mobility, and stabilization efficiency of nZVI. Biochar helps to limit the aggregation of nZVI in the soil and thus possibly reduces their leaching through the soil profile and maintains the surface reactivity of NPs (Song et al. 2024; Komárek 2024). For instance, the use of biochar loaded with nZVI (nZVI@BC) was explored to remediate soil polluted with Cd. nZVI@BC at the dose of 5% after 40 d significantly reduced the Cd fraction of weak acid extractable by 37%, while the residual Cd increased by 46% compared to the control treatment (Song et al. 2024).

The indirect application of photocatalysis based on nanotechnology represents a further potential solution for soil remediation, given its principal role in water treatment, a review of the nanotechnology-base photocatalysis for pesticides-contaminated water was done by Kajitvichyanukul et al. (2022). In batch studies with aqueous solutions, green synthesized Cu-Zn bimetallic nanoparticles demonstrated photodegraded 53%–95% of pesticides such as monocroptophos, chlorpyrifos, and carvendazim after 24–72 h of incubation. Nevertheless, a few studies at the laboratory scale have shown the potential of TiO_2 NPs to be used in soil for photocatalytic nano-remediation. For example, Dong et al. (2025) evaluated the degradation process of fluoranthene in soil suspension (10:1 water:soil) by photocatalysis using a GQDs/TiO_2/α-FeOOH composite (50 mg g^{-1}). The authors reported that 64% of fluoranthene was degraded after 24 h of exposure to simulated sunlight irradiation. In another research, bare TiO_2 NPs and Fe-doped TiO_2 NPs were used to degrade polycyclic aromatic hydrocarbons (PAHs, phenanthrene, anthracene, or fluoranthene extracted from polluted soil) by visible light-induced photocatalysis. When it was used 1% (m/v) TiO_2 NPs, the mineralization of PAHs was about 80% and the main intermediates of PAHs photodegradation were 9-octadecanoic acid, heptadecane, octadecane, cyclohexane, and 15-hydroxypentadecanoic acid (Theerakarunwong and Phanichphant 2018). EL-Saeid et al. (2022) evaluated the photodegradation of organic pesticides in samples of agricultural soils at a laboratory scale. Soils were amended with 1% (m/m) of TiO_2 and ZnO NPs. Contrary to the former studies, the authors UV irradiated the soil and not soil suspension. The pesticides evaluated (organochlorines, organophosphorus, carbamates, and pyrethroids) were completely degraded by photocatalysis within 20–24 h of irradiation. TiO_2 NPs showed higher photocatalytic activity compared to ZnO NPs. Among the photocatalyst NMs (ZnO, Fe_2O_3, CuO, and CdS NPs), TiO_2 NPs is one of the most widely used due to its low cost, nontoxicity, and high photocatalytic performance (Dong et al. 2025).

NMs can be combined with the current technologies for soil remediation to enhance their efficiency. For instance, sulfidated nZVI (S-nZVI) was used with electrokinetic remediation to treat Cr-contaminated soil. The S-nZVI was used as a third electrode. Compared with other electrodes, the S-nZVI electrode led to higher currents and a more uniform voltage distribution. The leaching toxic residues were reduced by 50–67%, and the removal efficiencies of total Cr and Cr(VI) were increased by 15% and 23%, respectively (He et al. 2021). The use of NMs in combination with microorganisms (bioremediation) or plants (phytoremediation) has been documented as well (this topic will be discussed in Sect. 2.3: Nano-remediation assisted by plants and microorganisms). Rajput et al. (2023) summarized reports on nano-remediation assisted by microorganisms of soil contaminated by PAHs. For example, NPs such as Fe_2O_3 and $Zn_5(OH)_8Cl_2$ improved the degradation by *Bacillus licheniformis*. The functionalization of silica with lipid,1,2-dimyristoylsn-glycero-3-phosphocholine NPs, and two layers of *Pseudomonas aeruginosa* were effective for adsorption of PAHs. The degradation of derivates of pyrene improved by the combination of yeast (*Candida tropicalis* NN4) and Fe NPs. Similarly, two yeast strains (YC01 and YC04) and NPs of ZnO significantly enhance the degradation of PAHs of five or six fused benzene rings in a short time of 6 d.

Nano-coated enzymes are an example of amalgamated remediation technology breaking down organic compounds (Ramezani et al. 2021). Fang et al. (2012) observed increased enzymatic activity by adding iron NPs, which resulted in higher efficiency of degradation of 2,4-diclorophenoxyacetic acid (2,4-D) than compared with the use of only microorganisms. Iron NPs potentially influence the diversity and structure of soil microbial communities with high and rapid degradation activity, therefore being advantageous for soil remediation.

2.2.5 Emerging Contaminants

ECs are synthetic or natural compounds and microorganisms produced and used by humans that tend to have ecological and human health effects once they reach the environment (Wang et al. 2024a). Examples of ECs in agricultural soils are pharmaceuticals and veterinary drugs (e.g., antibiotics or animal husbandry), hormones, per- and polyfluorinated chemicals, NMs (TiO_2 NPs, carbon NPs, and hydroxyapatite), nano- and micro-plastics, disinfectants, pesticides, and endocrine-disrupting substances (Naeem et al. 2023; Rajput et al. 2024). These compounds enter agricultural systems through the use of polluted water or treated wastewater for crop irrigation (Denora et al. 2023) and the application of manures, organic fertilizers, biosolids, or pesticides (Liu et al. 2024). For example, in the case of nano- and micro-plastic, poor plastic waste management is the entry pathway to agricultural soils (Wang et al. 2024b), as well as through atmospheric deposition, and waterways (Zhu et al. 2024). Nanoplastics are plastic particles less than 100 nm whereas microplastics are less than 5 mm. Current classifications of nano- and micro-plastics categorize them according to their origin: primary or secondary (Hoang et al. 2024). The former category is related to plastics intentionally manufactured at small sizes, whereas the latter classification refers to the breakdown of larger plastic items into their nano- or microplastic form.

The environmental threat of ECs is due to their complex chemical composition, persistence, and potential for bioaccumulation, leading to their introduction into the food chain (Liu et al. 2024). For instance, Denora et al. (2023) evaluated the fate of pharmaceutical contaminants in soil and tomato plants in a pilot-scale study. For this study, tomato plants were irrigated with tertiary treated wastewater effluent spiked with ECs (antibiotics, antifungal, and anti-inflammatory; at one or three times the European average dose). Fluconazole and carbamazepine were found in soil, plant tissue, and drain water. In fruits, the concentrations of fluconazole, carbamazepine, and metoprolol were 110, 83, and 1.2 ng g^{-1}, respectively. Meanwhile, Conde-Cid et al. (2018) assessed the occurrence of antibiotics (tetracycline, and sulfonamides) in samples of cattle, pig and poultry manure used as soil amendments, as well as, soil and vegetation samples. The authors reported the detection of antibiotics in 42% of manures tested. In the case of soils, 17% presented the occurrence of antibiotics, and the maximum concentration detected was 0.6 mg kg^{-1}. Antibiotics were detected in 44% and 33% of tested crop and wheat samples, respectively.

On the other hand, the disruption of ecosystems, biodiversity loss, alteration of physiological systems, and negative human health effects (such as endocrine disruptions, reproductive disorders, cardiopulmonary diseases, and neurotoxicity) are other concerns related to ECs in agricultural soils (Liu et al. 2024). Wang et al. (2024a) highlighted that the ECs can disrupt the physiological function of microbial soil communities, affecting the abundance and diversity of key microorganisms involved in nutrient cycling and organic matter decomposition.

The migration of ECs from agricultural soils to groundwater and sources of drinking water is another issue related to the ECs. For example, it has been estimated that about 22,700 tons of nano- and micro-plastic enter the Chinese riverine system due to agricultural water-driven soil erosion (Wang et al. 2024b). In the case of nano- and micro-plastics, they can impact the structure of the soil altering size distribution, aeration, and hindering root growth (Winiarska et al. 2024). Moreover, they can potentially reduce the water and nutrient retention in soils (Wang et al. 2024a). On the other hand, soil erosion, spray drift, or the volatilization of pesticides after their application poses a potential risk for the ECs to enter the atmosphere (Wang et al. 2024b).

Nowadays, technologies such as oxidation processes or incineration have been adopted to address ECs in the environment. However, they can be expensive and less sustainable. Innovative treatment techniques are still in development, and their adoption depends on regulations and the knowledge associated with the toxicology of ECs (Bell et al. 2019). In this regard, nanotechnology encompasses one of the most innovative solutions for the treatment of ECs. The vast majority of the nano-enabled technologies are designed for wastewater treatment. For instance, carbon nanotubes, graphene oxide, and nanocomposites have been used to enhance the efficiency of filtration membranes that have been shown to remove metal(loids)s, organic pollutants, pharmaceuticals, and microplastics from water (Liu et al. 2024). Another nano-enabled solution is the use of NMs for the advanced oxidation process in wastewater treatments (Ambaye et al. 2024). The TiO_2 NPs are an example of NMs used in photocatalysis, e.g., film-coated with TiO_2 NPs can degrade completely the pesticide alpha-hexachlorocyclohexane after 20 min (Yu et al. 2007).

There are scarce examples of nano-enabled solutions for the treatment of soils polluted with ECs. However, the use of NMs to enhance the removal of herbicides like atrazine has been tested on a laboratory scale. Biochar-supported Fe NPs (FeNPs@BC) were used as catalysts for persulfate-base advanced oxidation technology. Li et al. (2024c) conducted a series of batch experiments to determine the atrazine degradation in soils by the persulfate-base advanced oxidation assisted with FeNPs@BC. The authors reported that the combination of 2% (w/w) of FeNPs@BC with 3 mM of persulfate can remove 100% of atrazine in soils after 10 min. Furthermore, the FeNPs@BC was able to maintain their catalytic activity after four cycles of recycling. Another example is the use of nZVI as catalysts for persulfate-base advanced oxidation technology. In four agricultural soil, batch experiments were conducted to assess the degradation of sulfamethoxazole (an antibiotic, initial concentration of 20 mg kg^{-1}). When nZVI at 30 mg g^{-1} was added to the oxidation system, the degradation of the antibiotic was from 81% to 96%. Moreover, after soil treatment soil enzyme activity was observed after 21 d (Zhou et al. 2020). On the other hand, soil amendment with S-ZVI or nZVI at dose of 5% (w/w) showed 94% and 45% degradation of decabromodiphenyl ether (a flame retardant) after 48 h, respectively (Zhang et al. 2023).

Dyes are another emerging contaminant; in particular, azo dyes are toxic, persistent, and difficult to degrade by conventional technologies (Tripathi et al. 2021; Sathish et al. 2024). They are carcinogenic due to the release of aromatic amines during dye degradation. Moreover, their remediation is more difficult because some dyes also contain toxic metals such as lead, cadmium, and chromium. On this point, Khan et al. (2022) suggested the involvement of NMs for dye remediation. Some examples presented were NPs of CuS_3, Fe_3O_4, carbon-based NMs such as multi-walled and single-walled nanotubes, and MgO NPs, which remove dyes by adsorption, while TiO_2 nanotubes remove dyes by photocatalysis. Nano and microparticles of nickel polyvinyl pyrrolidone accelerated nearly 99% of dye removal after 2 h treatment. Other NMs were NPs of Au and Ag, FeO, Cu, NZO, Al-Zn-O, and Pa-Ag. Some dyes studied were methylene blue, crystal violet, eosin, blue CP, yellow 3RS, malachite green, Evans blue, Congo red, rhodamine, orange II and G, acid, and reactive red. These authors also summarized that those biologically synthesized NMs were efficient for the degradation of dyes. Although soil contamination by dyes occurs, many examples of remediation are available for water remediation. An exceptional case is a review of the bioremediation of soils contaminated with dyes (Saravanan et al. 2024). The authors presented that bacterial remediation was superior in efficiency for dye degradation. Moreover, they mentioned other emerging strategies to complete the degradation of dyes: desalination, electrolysis, microbial oil cells, and bioelectrochemical systems; however, they do not include the involvement of NMs. Therefore, concerning dye soil pollution, nano-remediation assisted by microorganisms is an unexplored field of research that requires attention because of the toxicity and risk of dyes to human beings.

2.3 Nano-remediation Assisted by Plants and Microorganisms

Soil pollution is a concern that requires cost-effective and environmentally friendly solutions. Among the nano-remediation alternatives for restoring polluted soils, nanotechnology with microorganism combination is a potentially effective option to improve NMs and microbial efficiency. This combination may leverage unique properties to enhance the breakdown or modification of contaminants in comparison to conventional technologies (Hidangmayum et al. 2023; Asghar et al. 2024; Ningombam et al. 2024; Mokrani et al. 2024).

Nano-remediation assisted by microorganisms is an innovative approach to increase environmental remediation in a more sustainable, cost-effective, and eco-friendly manner (Singh and Saxena 2022; Chaudhary et al. 2023). Different names have received this alternative: nano-enhancement bioremediation, nano-enhanced microbial remediation, NPs-mediated bioaugmentation, nano-bioremediation, and microbial nano-bioremediation. Still, a more specific denomination nano-bioremediation mediated by fungi and NPs (Singh and Saxena 2022; Rajput et al. 2023; Chaudhary et al. 2023); however, all involved in the general and more accepted denomination nano-remediation assisted by microorganisms.

Similar to other alternatives to environmental contamination, nano-remediation assisted by microorganisms may be ex-situ or in situ. NMs and microorganisms are used for remediation because both have several similar action modes or complementary forms to tackle soil pollution. The bioremediation option involves the participation of bacteria, filamentous fungi, arbuscular mycorrhizal fungi, and archaea, among the most common to decrease environmental pollution problems (Perea-Vélez et al. 2021; Hidangmayum et al. 2023). The combined use of microorganisms and NMs, either sequentially or simultaneously, has several successful examples showing the increase in the effectiveness of bioremediation and soil health restoration (Hidangmayum et al. 2023; Ningombam et al. 2024).

Although NMs and pollutants may influence microorganisms for biocidal or stimulation effects (Hidangmayum et al. 2023), several examples show that some NMs stimulate microbial pathways that enhance pollutants´ biodegradation. The type, concentration, size, form, dispersion, NMs, and pollutant concentrations are relevant to direct microbial behavior (Asghar et al. 2024). The interaction of microorganisms in nano-remediation first occurs in green nanomaterial production to decrease the use of toxic/hazardous chemicals or their aggregation in solutions. Biological NMs synthesis uses bacterial/fungal extracts, exopolysaccharides produced by these microorganisms or algae, their enzymes, and biosurfactants to produce diverse NMs. These microbial sources improve NMs' biodegradability, performance, stability, and effectiveness (Singh and Saxena 2022; Asghar et al. 2024). Both NMs and microorganisms work together to strengthen their effect on remediation.

Theoretically, inorganic and organic may be treated by nano-remediation assisted by microorganisms. The most hazardous and persistent pollutants in the environment are metal(oils)s (Rajput et al. 2023). For example, Cd, Hg, Pb, As, Cr, Ni, and Zn are components of formulations of several agrochemicals and mining activities, among some pollution sources. These pollutants are of concern because they are not degraded and may cause biomagnification. It is metal(loid) accumulation in the body of plants, animals, and microorganisms and increases their concentration through the food chain (McMullen et al. 2024; Gao et al. 2024). Perea-Vélez et al. (2021) reviewed that some plant-beneficial microorganisms, by extracellular polymeric substances, hormones, and organic acids, may interact with NMs, influencing the surface and bioavailability of metal(loids). These authors resumed examples of the use of TiO_2 and nZVI (bare or starch-stabilized) NPs shown synergic effect with arbuscular mycorrhizal fungi on *Vigna unguiculata* and *Sorghum bicolor* plants established in soil contaminated with $CdCl_2$ or multipolluted soil, respectively. Similarly, plant growth-promoting bacteria such as *Bacillus fortis* or *Pseudomonas fluorescens* may positively interact with ZnO NPs of nZVI to protect *Cucumis melo* or *Salix alba*, respectively, plants against the toxic effects of Cd soil contamination.

The use of NMs as soil amendments to improve the phytoextraction of some metal(loid)s has been also analyzed. In this context, it has been noted that amendments with MoO_3 NPs increased the tolerance and Cd phytoextraction capability of *Solanum nigrum* by modifying its abscisic acid signaling pathway, which can contribute to its carbohydrate biosynthesis and antioxidant system potential (Qin et al. 2024). Similarly, it has been documented that amendments with 50 and 100 $mgkg^{-1}$ of green-synthesized ZnO-NPs can aid the phytoextraction potential of *Pennisetum giganteum* in Cd contaminated soils and enhance the levels of soil enzyme activities such as urease, catalase, saccharase, and acid phosphatase (Li et al. 2024b). It is necessary to emphasize that there are many environmental studies suggesting several solutions through nano-remediation assisted by microorganisms for remediation of water contamination by several toxic compounds; however, there is limited information on the remediation of polluted soils, a foundation resource of community concern. The successful integration of nano-remediation assisted by microorganisms into soil remediation practices, and the risk assessment into regulatory agendas are vital.

2.4 Nanomaterials and Soil Components Interactions

Predicting the behavior over time of the different NMs used for in situ soil remediation techniques is essential to avoid the migration of these NMs beyond the contamination plume. For this reason, understanding the interactions between NMs with soil components allows to design of better nano-remediation approaches. Detailed reviews on this subject are offered by Boccardo et al. (2020), Perea-Vélez et al. (2021), Nguyen et al. (2024), and Yusuf et al. (2024). However, from those works, it could be highlighted that some of the main NMs-soil interactions are the following:

- Transport and deposition of NMs in soil. The transport and deposition of NMs through porous media depend on the properties of the NMs themselves, of the soil and soil solution. Regarding NMs their composition, surface coating, and surface charge are the main features affecting their transport. On the other hand, chemical soil properties such as pH, ionic strength, ionic composition, and content of organic matter can influence the NMs' transport (Boccardo et al. 2020; Hwang and Kim 2024). Moreover, the type of porous media is also relevant because NMs can be immobilized by physical straining (Perea-Vélez et al. 2021).
- NMs move through soil porous as individual or agglomerates/aggregates. For individual particles, diffusion and mass flow are the main transport mechanisms (Perea-Vélez et al. 2021). Natural colloids such as bacteria, organic matter, metal oxides, and clay can modify the surface charge of NPs, as well as immobilize the NMs by adsorption onto the surface of those soil components (Hwang and Kim 2024).
- NMs aggregation/agglomeration. The aggregation of NMs depends on the soil properties (ionic strength, pH, organic matter) and NMs properties (shape, surface charge, surface bare or surface coating, and NMs concentration) (Perea-Vélez et al. 2021). On the other hand, van der Waals forces and the formation of an electric double layer of counter ions are the main forces involved in the aggregation of NMs. The agglomeration/aggregation could be homo-aggregation (NMs-NMs are aggregated) or hetero-aggregation; it occurs when different components interact with the NMs and promote their aggregation (Ahmed et al. 2021). For in situ remediation nano-enabled technologies, avoiding the aggregation of NMs is important because it generally decreases the chemical reactivity of NMs or their bioavailability, or uptake in the case of nano-fertilizers (Ahmed et al. 2021).
- NMs transformations. The NMs transformations in soils include sulfidation, dissolution, oxidation/reduction, corona formation, and agglomeration/aggregation (Ahmed et al. 2021; Wei et al. 2024).

2.5 Environmental Fate of Nanomaterials and Their Ecological Impact of Nano-enabled Technologies for Soil Remediation

The use of NMs for in-situ soil remediation is raising concerns about the potential impact of NMs as a secondary source of contamination in soils and aquifers (European Commission 2017; Yusuf et al. 2024). For instance, the migration of NMs to other locations, such as the atmosphere or groundwaters, or the uncontrolled distribution of NMs beyond the application zones leads to disrupted nutrient cycling. Moreover, NMs may induce changes in the microbial soil communities, and their accumulation may cause toxicity to organisms (Boccardo et al. 2020; Marcon et al. 2021; Yusuf et al. 2024; Umair et al. 2024). Hence, before the widespread use of NMs for in situ soil remediation, the risk of NMs to the environment and living organisms must be evaluated (Yusuf et al. 2024). To this end, understanding the transportation dynamics of NMs in the environment may enable the design and implementation of an effective nano-remediation strategy, preventing the unintended consequences of the release of NMs in soil (Yusuf et al. 2024). The transport of NMs in porous media can be elucidated by two principal models: continuous models and discrete models (Nguyen et al. 2024; Yusuf et al. 2024) (Table 2.1).

The continuous models are based on the advection-dispersion-reaction equation modified (Viotti et al. 2022) to include terms that describe the interactions between NMs and the porous medium. These interactions may include the transport of NMs by fluid flow, attachment/detachment of NMs on the solid phase, agglomeration of NMs, physical straining, site blocking, and spread of NMs due to the diffusion mixing (Viotti et al. 2022; Yusuf et al. 2024), In contrast, the discrete models are based on the tracking of individual particles (Boccardo et al. 2020). A detailed review of the transport of NMs in porous media is provided by Yusuf et al. (2024). The fate of NMs in the environment is another concern, which involves their transformations,

Table 2.1 A summary of the models used for describing NMs transport in porous media

Model	Principle	References
Filtration theory	The retention of NMs in porous media can be attributed to the interception of NMs with solid particles of the media, sedimentation due to gravity, and diffusion due to Brownian motion.	Yusuf et al. (2024)
Advection-dispersion-adhesion	The model considers NMs collisions, leading to aggregated structures with varying mobility and reactivity. The NMs may be suspended or attached to the solid matrix reversibly or irreversibly.	Yusuf et al. (2024)
Lagrangianparticle tracking method with force equilibrium	Particle movement is a result of convection and molecular diffusion caused by the fluid velocity and the Browinan motion of NMs, respectively. This model ignores the particle interactions.	Nguyen et al. (2024)
Monte Carlo method	Probability-based method for estimating the movement and transport of particles	Nguyen et al. (2024)

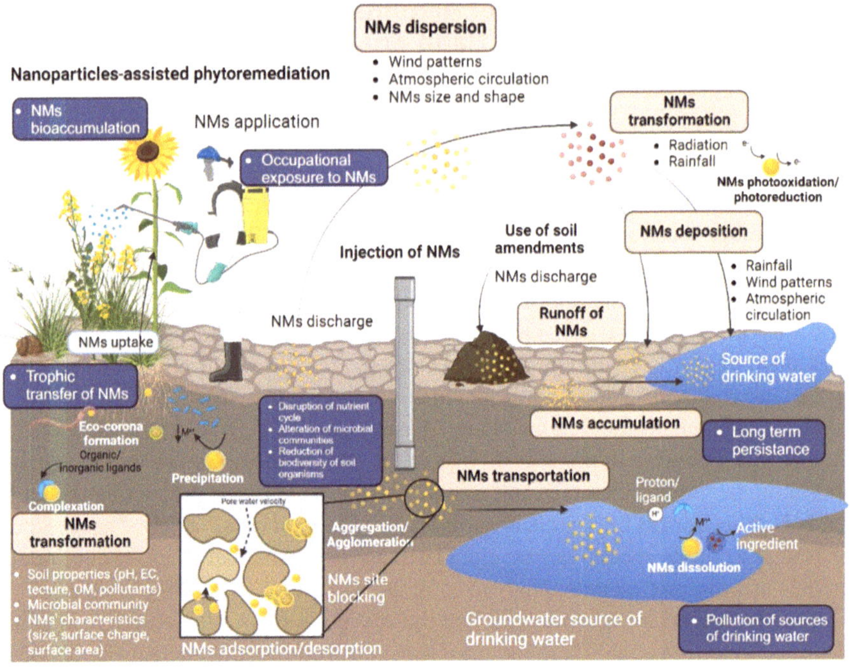

Fig. 2.2 Diagram of factors that influence the fate of nanomaterials (NMs) in the environment following their utilization in soil remediation in situ processes

transport, bioaccumulation, and interactions with the environment. It is a complex process that depends on the interaction of multiple factors (Fig. 2.2).

The brown boxes describe the interaction between NMs and the environment. After the NMs discharge, the NMs can interact with the soil components and suffer sedimentation due to gravity or be diffused due to the Brownian motion. Moreover, NMs can be subjected to transformations, such as eco-corona formation; to be dissolved by the action of microorganisms, root activity, or protons and ligands present in the rhizosphere zone; or the NMs oxidation, reduction, photooxidation, and photoreduction can occur. The organic matter in the soil, as well as the electrical conductivity (EC) and ionic strength, influence the NMs aggregation/agglomeration. The transformation of NMs affects their transportation through the soil leading to their mobility in other spheres (water bodies or atmosphere). The blue boxes highlight the NMs' ecological impacts. These factors include the medium into which the NMs are discharged (soil, water, air) and their characteristics, the properties of the NMs themselves, and the transformations that occur in the environment following their release (Keller et al. 2024). Nevertheless, despite the complexity of environmental processes that affect the NMs and their interactions with abiotic and biotic mediums (Fig. 2.2), computational tools have been developed to predict the mobility/transportation, distribution, fate, and toxicity of NMs (Tang et al. 2024). The environmental fate models (EFMs) are computational tools that allow assessing the impact, exposure, and hazards of NMs on the environment; thus, they evaluate the potential danger of NMs. These models have emerged as a rapid, effective, and low-time-consuming alternative compared to the field monitoring methods or the in vitro/in vivo assays (Keller et al. 2024; Tang et al. 2024).

The models that determine the flow of NMs in the environment from their production to their end-of-life encompass the exposure assessment. For instance, the material flow analysis (MFA) was one of the first tools to provide information on the magnitude of NMs release (from the NMs' production, use, and end-of-life) by tracking the flows of NMs through the technosphere into the environment (Adam et al. 2018). However, one of the disadvantages of these models is that they do not account for the fate process (such as homo and hetero-aggregation, transformations, dissolution, or corona formation) providing unrealistic predicted environmental concentrations (Keller et al. 2024). Adam et al. (2021) proposed an update of the MFAs taking into account the form of release of NMs through their lifecycle and the flows of NMs out of recycling. These updates resulted in the integrated dynamic probabilistic material flow analysis.

Multimedia Environmental Models (MEMs) are useful for understanding the distribution of NMs among the different environmental compartments (i.e., soil, water, air) over time. The MEMs consider the process that determines the transport process of NMs between the compartments, for instance, atmospheric deposition, runoff, and infiltration (Keller et al. 2024; Tang et al. 2024). Examples of these multimedia environmental models are the MendNano model (Liu and Cohen 2014), SimpleBox4nano (Meesters et al. 2014), and the Nanofatemodel (Garner et al. 2017).

The models that assess the bioaccumulation and toxicokinetic of NMs in living organisms allow us to assess the hazards of NMs. These models take into account the anatomy and physiological structures of living systems, and the mass transport process to predict the absorption, distribution, metabolism, and excretion of NMs (Tang et al. 2024). The nanoBio model is one example of the physiologically based toxicokinetics model. While the nano-QSAR model predicts the toxicological effects of NMs on biological systems considering the NMs' properties. On the other hand, the nano-QSAR model assesses the fate of NMs in the environment by the identification of physicochemical descriptors that are most relevant for predicting toxicity. These include surface area, surface charge, shape, and composition. Once the descriptors have been identified, computational descriptors can be created and associated with toxicity. This process improves the accessibility of nanomaterial models that are capable of generating assumptions without the need to incorporate hypothetical data (Singh et al. 2024b).

From a regulatory perspective, data on the environmental release and exposure of NMs are vital to estimating their environmental risk, therefore creating safe nano-enabled technologies for soil remediation (Gayathiri et al. 2024; Keller et al. 2024). However, the current analytical methods are limited to the detection of NMs in environmental samples. Another limitation of the EFMs is that they need a large amount of data, in the case of MFAs largely depend on the market data to estimate the production of NMs (Baun et al. 2017). Moreover, the models need to capture the complexity of the interactions that determine the fate and risk of NMs to the environment and living organisms. Validation of models with field data is also needed to ensure their reliability and accuracy.

2.6 Conclusions

Soil pollution is a serious challenge to agricultural productivity, human health, water quality, and biodiversity. Nano-enabled technologies have been developed to provide novel solutions to challenges associated with soil degradation such as soil erosion, depleted soil fertility, soil salinity, and soil pollution. Even though the interest in using nanotechnology for agricultural productivity has exponentially increased in recent years, few efforts have been performed and reported through research articles, which are also missing to extensively characterize the physical and chemical features of the developed NMs. This is problematic since the fate of NMs in the environment, especially in soil, is a complex issue that requires a thorough understanding of the physical, chemical, biological, and ecological processes involved. Pollutants with a major impact on agricultural productivity and human health encompass solvents and nano- and microplastics, for which, the use of nano-remediation approaches to mitigate its consequences in agricultural productivity and human health has been scarcely reported. Nano-remediation approaches assisted by plants have been explored with certain species and the pathways involved in phytostabilization or phytoextraction have been merely investigated. It is crucial to assess the effects of NMs at various levels of biological organization, from individual cells and organisms to populations and communities, to understand their ecological impact. Access to reliable data on the properties of nanomaterials (NMs), their release pathways, and environmental conditions is essential for accurate modeling. Models need to reflect the complexity of environmental processes that influence the fate of NMs, including aggregation, transformation, and interactions with organic matter and living organisms. Validating these models with field data is critical for ensuring their reliability and accuracy.

Acknowledgments Y.S.PV. recognizes the support provided by the Secretaría de Ciencia, Humanidades, Tecnología e Innovación (SECIHTI) under the program Estancias Postdoctorales por México, J.L.MM. acknowledges the support provided by the Consejo Mexiquense de Ciencia y Tecnología (COMECYT) under the program Cátedra de Investigación (CAT2024-0013). All figures were created utilizing BioRender.com.

References

Abbasi S (2024) Investigation the changes in the concentration of dye organic pollutant using magnetic titania nanoparticles incorporated on surface of graphene oxide. Appl Water Sci 14:18. https://doi.org/10.1007/s13201-023-02074-2

Adam V, Caballero-Guzman A, Nowack B (2018) Considering the forms of released engineered nanomaterials in probabilistic material flow analysis. Environ Pollut 243:17–27. https://doi.org/10.1016/j.envpol.2018.07.108

Adam V, Wu Q, Nowack B (2021) Integrated dynamic probabilistic material flow analysis of engineered materials in all European countries. NanoImpact 22:100312. https://doi.org/10.1016/j.impact.2021.100312

Ahmed B, Rizvi A, Ali K, Lee J, Zaidi A, Khan MS, Musarrat J (2021) Nanoparticles in the soil–plant system: a review. Environ Chem Lett 19:1545–1609. https://doi.org/10.1007/s10311-020-01138-y

Ahmed R, Zia-ur-Rehman M, Sabir M, Usman M, Rizwan M, Ahmad Z, Alharby HF, Al-Zahrani HS, Alsamadany H, Aldhebiani AY, Alzahrani YM, Bamagoos AA (2023) Differential response of nano zinc sulphate with other conventional sources of Zn in mitigating salinity stress in rice grown on saline-sodic soil. Chemosphere 327:138479. https://doi.org/10.1016/j.chemosphere.2023.138479

Ambaye TG, Hassani A, Vaccari M, Franzetti A, Prasad S, Formicola F, Rosatelli A, Rehman MZ et al (2024) Emerging technologies for the removal of pesticides from contaminated soils and their reuse in agriculture. Chemosphere 362:142433. https://doi.org/10.1016/j.chemosphere.2024.142433

Angon PB, Islam MS, Das A, Anjum N, Poudel A, Suchi SA (2024) Sources, effects and present perspectives of heavy metals contamination: soil, plants and human food chain. Heliyon 10:e28357. https://doi.org/10.1016/j.heliyon.2024.e28357

Arias-Navarro C, Baritz R, Jones A (2024) The state of soils in Europe. Publications Office of the European Union, Luxembourg. https://data.europa.eu/doi/10.2760/7007291

Asghar N, Hussain A, Nguyen DA, Ali S, Hussain I, Junejo A, Ali A (2024) Advancement in nanomaterials for environmental pollutants remediation: a systematic review on bibliometrics analysis, material types, synthesis pathways, and related mechanisms. J Nanobiotechnol 22:26. https://doi.org/10.1186/s12951-023-02151-3

Baun A, Sayre P, Steinhäuser KG, Rose J (2017) Regulatory relevant and reliable methods and data for determining the environmental fate of manufactured nanomaterials. NanoImpact 8:1–10. https://doi.org/10.1016/j.impact.2017.06.004

Bell CH, Horst J, Anderson P, Forsberg N (2019) Considerations for future contaminants of emerging concern. In: Bell CH, Gentile M, Kalve E, Ross I, Horst J, Suthersan S (eds) Emerging contaminants handbook. CRC Press, Boca Raton

Boccardo G, Tosco T, Fujisaki A, Messina F, Raoof A, Aguilera DR, Crevacore E, Marchisio DL, Sethi R (2020) A review of transport of nanoparticles in porous media. In: Bonelli B, Freyria FS, Rossetti I, Sethi R (eds) . Elsevier, Nanomaterials for the Detection and Removal of Wastewater Pollutants, pp 351–381. https://doi.org/10.1016/B978-0-12-818489-9.00013-X

Bone B, Bardos P, Edgar S, Kvapil P (2020) The sustainability of nanoremediation—two initial case studies from Europe. In: Hou D (ed) Sustainable remediation of contaminated soil and groundwater. Elsevier, Oxford, pp 367–404. https://doi.org/10.1016/B978-0-12-817982-6.00014-8

Cano-Ortiz A, Piñar Fuentes JC, Cano E (2024) Analysis of toxic contaminants in agriculture: educational strategies to avoid their influence on food. Res J Ecol Environ Sci 4:1–15. https://doi.org/10.31586/rjees.2024.718

Chaudhary P, Ahamad L, Chaudhary A, Kumar G, Chen W-J, Chen S (2023) Nanoparticle-mediated bioremediation as a powerful weapon in the removal of environmental pollutants. J Environ Chem Eng 11:109591. https://doi.org/10.1016/j.jece.2023.109591

Chele KH, Tinte MM, Piater LA, Dubery IA, Tugizimana F (2021) Soil salinity, a serious environmental issue and plant responses: a metabolomics perspective. Meta 11:724. https://doi.org/10.3390/metabo11110724

Chen X, Zhou B, Wang Q, Tao W, Lin H (2020) Nano-biochar reduced soil erosion and nitrate loss in sloping fields on the loess plateau of China. Catena 187:104346. https://doi.org/10.1016/j.catena.2019.104346

Chowardhara B, Saha B, Awasthi JP, Deori BB, Nath R, Roy S, Sarkar S, Santra SC, Hossain A, Moulick D (2024) An assessment of nanotechnology-based interventions for cleaning up toxic heavy metal/metalloid-contaminated agroecosystems: potentials and issues. Chemosphere 359:142178. https://doi.org/10.1016/j.chemosphere.2024.142178

Conde-Cid M, Álvarez-Esmorís C, Paradelo-Núñez R, Nóvoa-Muñoz JC, Arias-Estévez M, Álvarez-Rodríguez E, Fernández-Sanjurjo MJ, Núñez-Delgado A (2018) Occurrence of tetracyclines and sulfonamides in manures, agricultural soils and crops from different areas in Galicia (NW Spain). J Clean Prod 197:491–500. https://doi.org/10.1016/j.jclepro.2018.06.217

Denora M, Candido V, Brunetti G, De Mastro F, Murgolo S, De Ceglie C, Salerno C, Gatta G, Giuliani MM, Mehmeti A, Bartholomeus RP, Perniola M (2023) Uptake and accumulation of emerging contaminants in processing tomato irrigated with tertiary treated wastewater effluent: a pilot-scale study. Front Plant Sci 14. https://doi.org/10.3389/fpls.2023.1238163

Dhanapal AR, Thiruvengadam M, Vairavanathan J, Venkidasamy B, Easwaran M, Ghorbanpour M (2024) Nanotechnology approaches for the remediation of agricultural polluted soils. ACS Omega 9:13522–13533. https://doi.org/10.1021/acsomega.3c09776

Ding Z, Zhao F, Zhu Z, Ali EF, Shaheen SM, Rinklebe J, Eissa MA (2022) Green nanosilica enhanced the salt-tolerance defenses and yield of Williams banana: a field trial for using saline water in low fertile arid soil. Environ Exp Bot 197:104843. https://doi.org/10.1016/j.envexpbot.2022.104843

Dong H, Yin B, Li J, Guo W, Meng D, Zhu X, Zhang G, Zhang G, Xin Y, Chen Q (2025) Photocatalytic remediation of fluoranthene contaminated soil by eco-friendly GQDs/TiO_2/α-FeOOH composite photocatalyst: efficiency, influence factors, mechanism, and toxicity analysis. Sep Purif Technol 357:130045. https://doi.org/10.1016/j.seppur.2024.130045

El-Ramady H, Brevik E, Abowaly M, Ali R, Saad Moghanm F, Gharib M, Mansour H, Fawzy Z, Prokisch J (2024) Soil degradation under a changing climate: management from traditional to nano-approaches. Egypt J Soil Sci 64. https://doi.org/10.21608/ejss.2023.248610.1686

EL-Saeid MH, BaQais A, Alshabanat M (2022) Study of the photocatalytic degradation of highly abundant pesticides in agricultural soils. Molecules 27:634. https://doi.org/10.3390/molecules27030634

EPA (2014) Persistent organic pollutants: a global issue, a global response [Overviews and Factsheets]. https://www.epa.gov/international-cooperation/persistent-organic-pollutants-global-issue-global-response. Accessed on 18 Dec 2024

European Commission (2017) Talking nanotechnological remediation processes from lab scale to end user applications for the restoration of a clean environment. https://cordis.europa.eu/project/id/309517/reporting. Accessed on 17 De 2024

Falciglia PP, Gagliano E, Scandura P, Bianco C, Tosco T, Sethi R, Varvaro G, Agostinelli E, Bongiorno C, Russo A, Romano S, Malandrino G, Roccaro P, Vagliasindi FGA (2022) Physico-magnetic properties and dynamics of magnetite (Fe_3O_4) nanoparticles (MNPs) under the effect of permanent magnetic fields in contaminated water treatment applications. Sep Purif Technol 296:121342. https://doi.org/10.1016/j.seppur.2022.121342

Fang G, Si Y, Tian C, Zhang G, Zhou D (2012) Degradation of 2,4-D in soils by Fe_3O_4 nanoparticles combined with stimulating indigenous microbes. Environ Sci Pol 19(3):784–793. https://doi.org/10.1007/s11356-011-0597-y

FAO (2019) The international code of conduct for the sustainable use and management of fertilizers. Food and Agriculture Organization of the United Nations, Roma

FAO (2022) Soils for nutrition: state of the art. Food and Agriculture Organization of the United Nations, Roma

FAO (2024) Degradation/Restoration | FAO SOILS PORTAL | Food and Agriculture Organization of the United Nations. https://www.fao.org/soils-portal/soil-degradation-restoration/en/. Accessed on 14 Dec 2024

FAO, ITPS (2015) Status of the world's soil resources (SWSR)-technical summary. Food and Agriculture Organization of the United Nations, Roma

Gao S, Li Z, Zhang S (2024) Trophic transfer and biomagnification of microplastics through food webs in coastal waters: a new perspective from a mass balance model. Mar Pollut Bull 200:116082. https://doi.org/10.1016/j.marpolbul.2024.116082

Garner KL, Suh S, Keller AA (2017) Assessing the risk of engineered nanomaterials in the environment: development and application of the nanofate model. Environ Sci Technol 51:5541–5551. https://doi.org/10.1021/acs.est.6b05279

Gayathiri E, Prakash P, Pandiaraj S, Ramasubburayan R, Gaur A, Sekar M, Viswanathan D, Govindasamy R (2024) Investigating the ecological implications of nanomaterials: unveiling plants' notable responses to nano-pollution. Plant Physiol Biochem 206:108261. https://doi.org/10.1016/j.plaphy.2023.108261

Ghandali MV, Safarzadeh S, Ghasemi-Fasaei R, Zeinali S (2024) Heavy metals immobilization and bioavailability in multi-metal contaminated soil under ryegrass cultivation as affected by ZnO and MnO_2 nanoparticle-modified biochar. Sci Rep 14:10684. https://doi.org/10.1038/s41598-024-61270-5

Ghazaryan K, Agrawal S, Margaryan G, Harutyunyan A, Rajput P, Movsesyan H, Rajput VD, Singh RK, Minkina T, Elshikh MS, Alwahibi MS, Alexiou A, Papadakis M, Sousa JR, Singh A (2024) Soil pollution: an agricultural and environmental problem with nanotechnological remediation opportunities and challenges. Discov Sustain 5:453. https://doi.org/10.1007/s43621-024-00666-9

Hashim AF, Ahmed FK, Abd-Elsalam KA, Thomas S (2024) Biopolymeric nanoparticles for agricultural applications: note from editors. In: Abd-Elsalam KA, Hashim AF, Ahmed FK, Thomas S (eds) Biopolymeric nanoparticles for agricultural applications. Springer Nature, Cham, pp 1–26. https://doi.org/10.1007/978-3-031-68834-8_1

Hassan MU, Kareem HA, Hussain S, Guo Z, Niu J, Roy M, Saleem S, Wang Q (2023) Enhanced salinity tolerance in alfalfa through foliar nano-zinc oxide application: mechanistic insights and potential agricultural applications. Rhizosphere 28:100792. https://doi.org/10.1016/j.rhisph.2023.100792

He X, Cui G, Zhang Q, Wang Z, Tang T, Liu Y (2021) Application of sulfide-modified nanoscale zerovalent iron electrodes for electrokinetic remediation of chromium-contaminated soil in a three-dimensional electrode system. J Environ Chem Eng 9:106791. https://doi.org/10.1016/j.jece.2021.106791

Hidangmayum A, Debnath A, Guru A, Singh BN, Upadhyay SK, Dwivedi P (2023) Mechanistic and recent updates in nano-bioremediation for developing green technology to alleviate agricultural contaminants. Int J Environ Sci 20:11693–11718. https://doi.org/10.1007/s13762-022-04560-7

Hoang V-H, Nguyen M-K, Hoang T-D, Ha MC, Huyen NTT, Bui VKH, Pham M-T, Nguyen C-M, Chang SW, Nguyen DD (2024) Sources, environmental fate, and impacts of microplastic contamination in agricultural soils: a comprehensive review. Sci Total Environ 950:175276. https://doi.org/10.1016/j.scitotenv.2024.175276

Hofmann T, Lowry GV, Ghoshal S, Tufenkji N, Brambilla D, Dutcher JR, Gilbertson LM, Giraldo JP, Kinsella JM, Landry MP, Lovell W, Naccache R, Paret M, Pedersen JA, Unrine JM, White JC, Wilkinson KJ (2020) Technology readiness and overcoming barriers to sustainably implement nanotechnology-enabled plant agriculture. Nat Food 1:416–425. https://doi.org/10.1038/s43016-020-0110-1

Huang W, Du J, Lai H, Zhang Q, Zhou C, Liu Z (2022) Mechanical properties and mechanism of soil treated with nano-aqueous adhesive (NAA). Sci Rep 12:14711. https://doi.org/10.1038/s41598-022-19108-5

Hwang G, Kim D (2024) Transport and retention of positively charged zinc oxide nanoparticles in saturated porous media: effects of metal oxides and clays. Environ Pollut 351:124007. https://doi.org/10.1016/j.envpol.2024.124007

Jain S, Ahsan S, Robb Z, Crowley B, Walters D (2024) The cost of inaction: a global tool to inform nutrition policy and investment decisions on global nutrition targets. Health Policy Plan 39:819–830. https://doi.org/10.1093/heapol/czae056

Jakhar AM, Aziz I, Kaleri AR, Hasnain M, Haider G, Ma J, Abideen Z (2022) Nano-fertilizers: a sustainable technology for improving crop nutrition and food security. NanoImpact 27:100411. https://doi.org/10.1016/j.impact.2022.100411

Jin R, Zheng M, Lammel G, Bandowe BAM, Liu G (2020) Chlorinated and brominated polycyclic aromatic hydrocarbons: Sources, formation mechanisms, and occurrence in the environment. Prog Energy Combust Sci 76:100803. https://doi.org/10.1016/j.pecs.2019.100803

Kajitvichyanukul P, Nguyen V-H, Boonupara T, Phan Thi L-A, Watcharenwong A, Sumitsawan S, Udomkun P (2022) Challenges and effectiveness of nanotechnology-based photocatalysis for pesticides-contaminated water: a review. Environ Res 212:113336. https://doi.org/10.1016/j.envres.2022.113336

Kale SS, Chauhan R, Nigam B, Gosavi S, Chaudhary IJ (2024) Effectiveness of nanoparticles in improving soil fertility and eco-friendly crop resistance: a comprehensive review. Biocatal Agric Biotechnol 56:103066. https://doi.org/10.1016/j.bcab.2024.103066

Karimi N, Mirbagheri SA, Nouri R, Bazargan A (2023) Sequential application of aerated electrocoagulation and γ-Fe_2O_3 nanoparticle adsorption for COD removal: consuming the least amount of energy and economic evaluation. Results Eng 17:100770. https://doi.org/10.1016/j.rineng.2022.100770

Keller AA, Zheng Y, Praetorius A, Quik JTK, Nowack B (2024) Predicting environmental concentrations of nanomaterials for exposure assessment—a review. NanoImpact 33:100496. https://doi.org/10.1016/j.impact.2024.100496

Khaliq M, Hanif MA, Bhatti IA, Mushtaq Z (2023) A novel study for producing complexed and encapsulated nutrients at nanometric scale to enhance plant growth. Sci Rep 13(1):11100. https://doi.org/10.1038/s41598-023-37607-x

Khan A, Roy A, Bhasin S, Bin ET, Khusro A, Eftekhari A, Moradi O, Rokni H, Karimi F (2022) Nanomaterials: an alternative source for biodegradation of toxic dyes. Food Chem Toxicol 164:112996. https://doi.org/10.1016/j.fct.2022.112996

Komárek M (2024) Perspectives of soil nanoremediation: the case of nano zerovalent iron and metal(loid) contaminants. NPJ Mater Sustain 2:8. https://doi.org/10.1038/s44296-024-00013-z

Lehmann J, Bossio DA, Kögel-Knabner I, Rillig MC (2020) The concept and future prospects of soil health. Nat Rev Earth Environ 1:544–553. https://doi.org/10.1038/s43017-020-0080-8

Li R, Cai J, Li J, Wang Z, Pei P, Zhang J, Krebs P (2022) Characterizing the long-term occurrence of polycyclic aromatic hydrocarbons and their driving forces in surface waters. J Hazard Mater 423:127065. https://doi.org/10.1016/j.jhazmat.2021.127065

Li D, Yang Y, Zhao Y, Zhou X, Han Q, Liu H, Li M (2024a) Optimizing cotton yield and soil salinity management: integrating brackish water leaching and freshwater drip irrigation with subsurface drainage. Field Crop Res 314:109454. https://doi.org/10.1016/j.fcr.2024.109454

Li H, Rehman A et al (2024b) Biosynthesized zinc oxide nanoparticles modulate the phytoremediation potential of *Pennisetum giganteum* and its rhizocompartments associated microbial community structure. J Clean Prod 434:140346. https://doi.org/10.1016/j.jclepro.2023.140346

Li R, Shen X, Zhang J, Jiang Q, Wang L, Zhang Y (2024c) Tailoring biochar supported iron nanoparticles to activate persulfate for atrazine degradation in soil. J Environ Chem Eng 12(2):111967. https://doi.org/10.1016/j.jece.2024.111967

Liu HH, Cohen Y (2014) Multimedia environmental distribution of engineered nanomaterials. Environ Sci Technol 48:3281–3292. https://doi.org/10.1021/es405132z

Liu X, Sathishkumar K, Zhang H, Saxena KK, Zhang F, Naraginti S, Anbarasu K, Rajendiran R, Rajasekar A, Guo X (2024) Frontiers in environmental cleanup: recent advances in remediation of emerging pollutants from soil and water. J Hazard Mater Adv 16:100461. https://doi.org/10.1016/j.hazadv.2024.100461

Lu W, Zhang C, Lin Z, Su P, Qin Z, Li X, Wang R, Liu Y, Luo P, Qiu K (2024) Soy protein isolate amyloid fibril-coated ZrO_2 nanoparticles anchored in the polyurethane sponge for rapid and effective fluoride adsorption in the wide pH range. J Environ Chem Eng 12:111591. https://doi.org/10.1016/j.jece.2023.111591

Marcon L, Oliveras J, Puntes VF (2021) In situ nanoremediation of soils and groundwaters from the nanoparticle's standpoint: a review. Sci Total Environ 791:148324. https://doi.org/10.1016/j.scitotenv.2021.148324

McMullen K, Vargas FH, Calle P, Alavarado-Cadena O, Pakhomov EA, Alava JJ (2024) Modelling microplastic bioaccumulation and biomagnification potential in the Galápagos penguin ecosystem using ecopath and ecosim (EwE) with ecotracer. PLoS One 19:e0296788. https://doi.org/10.1371/journal.pone.0296788

Meesters JAJ, Koelmans AA, Quik JTK, Hendriks AJ, van de Meent D (2014) Multimedia modeling of engineered nanoparticles with simplebox4nano: model definition and evaluation. Environ Sci Technol 48:5726–5736. https://doi.org/10.1021/es500548h

Michálková Z, Martínez-Fernández D, Komárek M (2017) Interactions of two novel stabilizing amendments with sunflower plants grown in a contaminated soil. Chemosphere 186:374–380. https://doi.org/10.1016/j.chemosphere.2017.08.009

Moghimi Dehkordi M, Pournuroz Nodeh Z, Soleimani Dehkordi K, Salmanvandi H, Rasouli Khorjestan R, Ghaffarzadeh M (2024) Soil, air, and water pollution from mining and industrial activities: sources of pollution, environmental impacts, and prevention and control methods. Results Eng 23:102729. https://doi.org/10.1016/j.rineng.2024.102729

Mokrani S, Houali K, Yadav KK, Arabi AIA, Eltayeb LB, AwjanAlreshidi M, Benguerba Y, Cabral-Pinto MMS, Nabti E (2024) Bioremediation techniques for soil organic pollution: mechanisms, microorganisms, and technologies—a comprehensive review. Ecol Eng 207:107338. https://doi.org/10.1016/j.ecoleng.2024.107338

Musa IO, Samuel JO, Adams M, Abdulsalam M, Nathaniel V, Maude AM, Adedayo OA, Tiamiyu AT (2024) Soil erosion, mineral depletion and regeneration. In: Aransiola SA, Babaniyi BR, Aransiola AB, Maddela NR (eds) Prospects for soil regeneration and its impact on environmental protection. Springer Nature, Cham, pp 159–172. https://doi.org/10.1007/978-3-031-53270-2_7

Mustafa G, Chaudhari SK, Manzoor M, Batool S, Hatami M, Hasan M (2024) Zinc oxide nanoparticles mediated salinity stress mitigation in *Pisum sativum*: a physio-biochemical perspective. BMC Plant Biol 24:835. https://doi.org/10.1186/s12870-024-05554-y

Naeem M, Gill R, Gill SS, Singh K, Sofo A, Tuteja N (2023) Editorial: emerging contaminants and their effect on agricultural crops. Front Plant Sci 14. https://doi.org/10.3389/fpls.2023.1296252

Nguyen VT, Pham NH, Papavassiliou DV (2024) Prediction of the aggregation rate of nanoparticles in porous media in the diffusion-controlled regime. Sci Rep 14:1916. https://doi.org/10.1038/s41598-023-50643-x

Ningombam L, Mana T, Apum G, Ningthoujam R, Disco Singh Y (2024) Nano-bioremediation: a prospective approach for environmental decontamination in focus to soil, water and heavy metals. Environ Nanotechnol Monit Manag 21:100931. https://doi.org/10.1016/j.enmm.2024.100931

Ojo O, Vaňková Z, Beesley L, Wickramasinghe N, Komárek M (2024) Evaluating the effectiveness of sulfidated nano zerovalent iron and sludge co-application for reducing metal mobility in contaminated soil. Sci Rep 14:8322. https://doi.org/10.1038/s41598-024-59059-7

Panagos P, Borrelli P, Jones A, Robinson DA (2024) A 1 billion euro mission: a soil deal for Europe. Eur J Soil Sci 75:e13466. https://doi.org/10.1111/ejss.13466

Panichikkal J, Prathap G, Nair RA, Krishnankutty RE (2021) Evaluation of plant probiotic performance of *Pseudomonas* sp. encapsulated in alginate supplemented with salicylic acid and zinc oxide nanoparticles. Int J Biol Macromol 166:138–143. https://doi.org/10.1016/j.ijbiomac.2020.10.110

Pathak HK, Chauhan PK, Seth CS, Dubey G, Upadhyay SK (2024) Mechanistic and future prospects in rhizospheric engineering for agricultural contaminants removal, soil health restoration, and management of climate change stress. Sci Total Environ 927:172116. https://doi.org/10.1016/j.scitotenv.2024.172116

Perea-Vélez YS, Carrillo-González R et al (2021) Interaction of metal nanoparticles–plants–microorganisms in agriculture and soil remediation. J Nanopart Res 23:206. https://doi.org/10.1007/s11051-021-05269-3

Perea-Vélez YS, Carrillo-González R et al (2023) Citrate-coated cobalt ferrite nanoparticles for the nano-enabled biofortification of wheat. Food Funct 14:4017–4035. https://doi.org/10.1039/D2FO03835H

Puiatti GA (2023) Using nanoremediation strategies: cost–benefit analysis. In: Policarpo Tonelli FM, Roy A, Ananda Murthy HC (eds) Green nanoremediation: sustainable management of environmental pollution. Springer International Publishing, Cham, pp 357–374. https://doi.org/10.1007/978-3-031-30558-0_16

Qaderi F, Tamadoni A, Banisheikholeslami A (2023) Cost estimation for application of ultrasonication–ozonation hybrid process in remediation of PAH-contaminated soil. Environ Dev Sustain 26:12441–12466. https://doi.org/10.1007/s10668-023-03828-3

Qin M, Gong J, Zeng G, Cao W, Song B, Zhang Y, Fang S, Xu F, Wu Y (2024) Molybdenum trioxide nanoparticles as promising stimulators for facilitating phytoremediation by Cd hyperaccumulator *Solanum nigrum* L. ACS Sustain Chem Eng 12:4767–4778. https://doi.org/10.1021/acssuschemeng.4c00385

Qu X, Zhang H, Kong H, Chen D, Yang Z, Mäkilä E, Salonen J, Santos HA, Hai M, Weitz DA (2024) Porous silicon nanoparticles conjugated magnetite-chitosan graphene oxide nanoparticles for effective removal of complex pollutants. Adv Sustain Syst 8:2300471. https://doi.org/10.1002/adsu.202300471

Quinton JN, Fiener P (2024) Soil erosion on arable land: an unresolved global environmental threat. Prog Phys Geogr Earth Environ 48:136–161. https://doi.org/10.1177/03091333231216595

Rajput VD, Kumari S, Minkina T, Sushkova S, Mandzhieva S (2023) Nano-enhanced microbial remediation of PAHs contaminated soil. Air Soil Water Res 16. https://doi.org/10.1177/11786221231170099

Rajput P, Kumar P, Priya AK, Kumari S, Shiade SRG, Rajput VD, Fathi A, Pradhan A, Sarfraz R, Sushkova S, Mandzhieva S, Minkina T, Soldatov A, Wong MH, Rensing C (2024) Nanomaterials and biochar mediated remediation of emerging contaminants. Sci Total Environ 916:170064. https://doi.org/10.1016/j.scitotenv.2024.170064

Ramezani M, Rad FA, Ghahari S, Ghahari S, Ramezani M (2021) Nano-bioremediation application for environment contamination by microorganism. In: Panpatte DG, Jhala YK (eds) Microbial rejuvenation of polluted environment, vol 2. Springer, Singapore, pp 349–378. https://doi.org/10.1007/978-981-15-7455-9_14

Saravanan A, Yaashikaa PR, Ramesh B, Shaji A, Deivayanai VC (2024) Microorganism-mediated bioremediation of dyes from contaminated soil: mechanisms, recent advances, and future perspectives. Food Chem Toxicol 185:114491. https://doi.org/10.1016/j.fct.2024.114491

Sartori M, Philippidis G, Ferrari E, Borrelli P, Lugato E, Montanarella L, Panagos P (2019) A linkage between the biophysical and the economic: assessing the global market impacts of soil erosion. Land Use Policy 86:299–312. https://doi.org/10.1016/j.landusepol.2019.05.014

Sartori M, Ferrari E, M'Barek R, Philippidis G, Boysen-Urban K, Borrelli P, Montanarella L, Panagos P (2024) Remaining loyal to our soil: a prospective integrated assessment of soil erosion on global food security. Ecol Econ 219:108103. https://doi.org/10.1016/j.ecolecon.2023.108103

Sathish T, Ahalya N, Thirunavukkarasu M, Senthil TS, Hussain Z, Haque Siddiqui MI, Panchal H, Kumar Sadasivuni K (2024) A comprehensive review on the novel approaches using nanomaterials for the remediation of soil and water pollution. Alex Eng J 86:373–385. https://doi.org/10.1016/j.aej.2023.10.038

Shirazian S, Alshehri S, Almalki AH, Alzhrani RM, Pirestani N, Soltani R (2025) Eco-friendly synthesis of crown-ether functionalized silica nanospheres from sorghum waste for thallium adsorption. J Taiwan Inst Chem Eng 167:105851. https://doi.org/10.1016/j.jtice.2024.105851

Singh Y, Saxena MK (2022) Insights into the recent advances in nano-bioremediation of pesticides from the contaminated soil. Front Microbiol 13:982611. https://doi.org/10.3389/fmicb.2022.982611

Singh D, Sillu D, Kumar A, Agnihotri S (2021) Dual nanozyme characteristics of iron oxide nanoparticles alleviate salinity stress and promote the growth of an agroforestry tree, *Eucalyptus tereticornis* Sm. Environ Sci Nano 8:1308–1325. https://doi.org/10.1039/D1EN00040C

Singh A, Agrawal S, Rajput VD, Ghazaryan K, Yesayan A, Minkina T, Zhao Y, Petropoulos D, Kriemadis A, Papadakis M, Alexiou A (2024a) Nanoparticles in revolutionizing crop production and agriculture to address salinity stress challenges for a sustainable future. Discov Appl Sci 6:317. https://doi.org/10.1007/s42452-024-06009-7

Singh AV, Shelar A, Rai M, Laux P, Thakur M, Dosnkyi I, Santomauro G, Singh AK, Luch A, Patil R, Bill J (2024b) Harmonization risks and rewards: nano-QSAR for agricultural nanomaterials. J Agric Food Chem 72:2835–2852. https://doi.org/10.1021/acs.jafc.3c06466

Song P, Liu J, Ma W, Gao X (2024) Remediation of cadmium-contaminated soil by biochar-loaded nano-zero-valent iron and its microbial community responses. J Environ Chem Eng 12:112311. https://doi.org/10.1016/j.jece.2024.112311

Tan ZX, Lal R, Wiebe KD (2005) Global soil nutrient depletion and yield reduction. J Sustain Agric 26:123–146. https://doi.org/10.1300/J064v26n01_10

Tang W, Zhang X, Hong H, Chen J, Zhao Q, Wu F (2024) Computational nanotoxicology models for environmental risk assessment of engineered nanomaterials. Nano 14:155. https://doi.org/10.3390/nano14020155

Theerakarunwong CD, Phanichphant S (2018) Visible-light-induced photocatalytic degradation of PAH-contaminated soil and their pathways by Fe-doped TiO_2 nanocatalyst. Water Air Soil Pollut 229:291. https://doi.org/10.1007/s11270-018-3951-6

Tripathi M, Kumar S, Singh DN, Pandey R, Pathak N, Fatima H (2021) Bioremediation of dye contaminated soil. In: Parray JA, Mahmoud AHAE, Sayyed R (eds) Soil bioremediation. Wiley, pp 115–142. https://doi.org/10.1002/9781119547976.ch6

Umair M, Huma Zafar S, Cheema M, Usman M (2024) New insights into the environmental application of hybrid nanoparticles in metal contaminated agroecosystem: a review. J Environ Manag 349:119553. https://doi.org/10.1016/j.jenvman.2023.119553

UNCCD (2023) At least 100 million hectares of healthy land now lost each year. https://www.unccd.int/news-stories/press-releases/least-100-million-hectares-healthy-land-now-lost-each-year. Accessed on 2 Dec 2024

United Nations (2024) Protec, restore and promote sustainable use of terrestrial ecosystems, sustainably management forests, combat desertification, and halt reverse land degradation and halt biodiversity loss. https://unstats.un.org/sdgs/report/2019/goal-15/#. Accessed on 18 Dec 2024

Viotti P, Sappa G, Tatti F, Andrei F (2022) nZVI mobility and transport: laboratory test and numerical model. Hydrology 9:196. https://doi.org/10.3390/hydrology9110196

Visentin C, Braun AB, Reginatto C, Cecchin I, Vanzetto GV, Thomé A (2024) Are contaminated soil and groundwater remediation with nanoscale zero-valent iron sustainable? An analysis of case studies. Environ Pollut 352:124167. https://doi.org/10.1016/j.envpol.2024.124167

Wan Y, Liu J, Zhuang Z, Wang Q, Li H (2024) Heavy metals in agricultural soils: sources, influencing factors, and remediation strategies. Toxics 12:63. https://doi.org/10.3390/toxics12010063

Wang Y, Ma X, Wang J et al (2019) Effects of mercapto-functionalized nanosilica on cd stabilization and uptake by wheat seedling (*Triticum aestivum* L.) in an agricultural soil. Bull Environ Contam Toxicol 103:860–864. https://doi.org/10.1007/s00128-019-02729-4

Wang Y, Zhan W, Zheng K et al (2020a) Effect of surface-modified nano-silica on the mobility and fraction of cd in contaminated agricultural soils. Soil Sediment Contam An Int J 29:96–106. https://doi.org/10.1080/15320383.2019.1679086

Wang Y, Liu Y, Zhan W et al (2020b) A field experiment on stabilization of cd in contaminated soils by surface-modified nano-silica (SMNS) and its phyto-availability to corn and wheat. J Soils Sediments 20:91–98. https://doi.org/10.1007/s11368-019-02416-1

Wang W, Qu K, Zhang X, Teng M, Huang Z (2021) Integrated instillation technology for the synthesis of a pH-responsive sodium alginate/biomass charcoal soil conditioner for controlled release of humic acid and soil remediation. J Agric Food Chem 69:13386–13397. https://doi.org/10.1021/acs.jafc.1c04121

Wang F, Xiang L, Sze-Yin Leung K, Elsner M et al (2024a) Emerging contaminants: a one health perspective. Innovation 5:100612. https://doi.org/10.1016/j.xinn.2024.100612

Wang Y, Jing S, Hou P, Ni R, Niu L, Wanger TC, Liu W, Liu K (2024b) Soil erosion is a major drive for nano and micro-plastics to enter riverine systems from cultivated land. Water Res 256:121597. https://doi.org/10.1016/j.watres.2024.121597

Weffort VRS, Lamounier JA (2024) Hidden hunger—a narrative review. J Pediatr 100:S10–S17. https://doi.org/10.1016/j.jped.2023.08.009

Wei L, Liu J, Jiang G (2024) Nanoparticle-specific transformations dictate nanoparticle effects associated with plants and implications for nanotechnology use in agriculture. Nat Commun 15:7389. https://doi.org/10.1038/s41467-024-51741-8

Winiarska E, Jutel M, Zemelka-Wiacek M (2024) The potential impact of nano- and microplastics on human health: understanding human health risks. Environ Res 251:118535. https://doi.org/10.1016/j.envres.2024.118535

Xue S, Wang Y, Bo W, Wan K, Miao Z (2024) Calcium-doped magnetic humic acid nano particles for the efficient removal of heavy metals from wastewater: the role of ca. Environ Technol 45:3228–3243. https://doi.org/10.1080/09593330.2023.2213832

Yang Q, Peng J, Ni S, Zhang C, Wang J, Cai C (2024) Erosion and deposition significantly affect the microbial diversity, co-occurrence network, and multifunctionality in agricultural soils of Northeast China. J Soils Sediments 24:888–900. https://doi.org/10.1007/s11368-023-03687-5

Yu B, Zeng J, Gong L, Zhang M, Zhang L, Chen X (2007) Investigation of the photocatalytic degradation of organochlorine pesticides on a nano-TiO_2 coated film. Talanta 72:1667–1674. https://doi.org/10.1016/j.talanta.2007.03.013

Yusuf EO, Amber I, Officer S, Oluyemi GF (2024) Transport of nanoparticles in porous media and associated environmental impact: a review. J Eng Res 12:275–284. https://doi.org/10.1016/j.jer.2024.01.006

Zhang X, Chen R, Li Z et al (2023) The influence of various microplastics on PBDEs contaminated soil remediation by nZVI and sulfide-nZVI: impedance, electron-accepting/-donating capacity and aging. Sci Total Environ 880:163233. https://doi.org/10.1016/j.scitotenv.2023.163233

Zhao H, Wu Y, Lan X, Yang Y, Wu X, Du L (2022) Comprehensive assessment of harmful heavy metals in contaminated soil in order to score pollution level. Sci Rep 12:3552. https://doi.org/10.1038/s41598-022-07602-9

Zhou Z, Liu J, Zeng H, Zhang T, Chen X (2020) How does soil pollution risk perception affect farmers' pro-environmental behavior? The role of income level. J Environ Manag 270:110806. https://doi.org/10.1016/j.jenvman.2020.110806

Zhu J, Xu A, Shi M, Su Y, Liu W, Zhang Y, She Z, Xing X, Qi S (2024) Atmospheric deposition is an important pathway for inputting microplastics: insight into the spatiotemporal distribution and deposition flux in a mega city. Environ Pollut 341:123012. https://doi.org/10.1016/j.envpol.2023.123012

Nanotechnology as a New Perspective in Precision Agriculture

3

Muhammad Adeel, Noman Shakoor, Mughees Mustafa, and Xu Ming

Abstract

Nanotechnology has emerged as a transformative tool for precision agriculture, addressing challenges in sustainable food production, resource efficiency, and environmental protection. This chapter explores the intersection of nanotechnology and precision agriculture, emphasizing the use of nanomaterials such as nanoparticles, nanotubes, and quantum dots for real-time monitoring of crops, soil health, and environmental changes. Applications include improved pest management, maximized soil nutrition, and enhanced crop productivity through the integration of digital technologies. While nanotechnology offers reduced material costs, increased efficiency, and enhanced resilience against climate variability, its widespread adoption remains constrained by insufficient understanding of nanoparticle behavior in complex agricultural ecosystems. Additionally, potential risks of nanoparticle accumulation in the food chain and the environment necessitate detailed investigation. The research gap lies in the lack of comprehensive studies evaluating the long-term effects of nanomaterials on soil microbiota, plant health, and human safety. This chapter highlights the need for a multidisciplinary approach to develop sustainable nano-enabled technologies while minimizing unintended consequences. Finally, we outline the necessary policy options and actions for enhancing sustainable agricultural productivity,

M. Adeel (✉) · X. Ming
BNU-HKUST Laboratory of Green Innovation, Advanced Institute of Natural Sciences, Beijing Normal University at Zhuhai, Guangdong, China

N. Shakoor
College of life science and oceanography, Shenzhen University, Shenzhen, China

M. Mustafa
Department of Agronomy, Pir Mehr Ali Shah Arid Agriculture University, Rawalpindi, Pakistan

S. S. Purewal et al. (eds.), *Agri-Nanotechnology: Innovations for Sustainable Agriculture and Environmental Restoration*,
https://doi.org/10.1007/978-981-96-9756-4_3

proposing research pathways to address the challenges associated with the practical implementation of nanotechnology. Future research should focus on eco-friendly nanomaterials, scalable technologies, and robust regulatory frameworks to ensure safe integration into agricultural systems. Bridging nanotechnology with precision agriculture promises innovative solutions for global food security, offering a roadmap for sustainable agricultural development.

Keywords

Nanotechnology · Precision agriculture · Nano-sensors · Sustainable agriculture · Food security

3.1 Introduction

Agriculture, the basis of human civilization, has always changed to fulfill people's demands and solve environmental concerns. As we navigate the challenges of the twenty-first century, agriculture faces enormous pressure to fulfill the difficulties of a worldwide population estimated to exceed 9 billion by 2050. These difficulties impose a change from traditional farming to innovative agriculture methods as an alternative that provides both food security and environmental sustainability. By exploiting these attributes, nanotechnology delivers novel things to address the complex challenges coping with agriculture today, problem-solving solutions that enhance crop productivity, improve sustainability, and protect the world (Scrinis and Lyons 2007). The polygonal discipline of nanotechnology provides novel ways to enhance agricultural yield. This tactic has played an important role in the Green Revolution's extreme use of synthetic fertilizers and pesticides, which declined soil biodiversity and uplifted disease and insect resistance. Innovative instruments like nanoparticles and nanochips can be used to produce multifaceted biosensors for precision agriculture and permit planned nutrition delivery. Besides, accurate application is made possible by nano-encapsulated insecticides, herbicides, and fertilizers, enhancing crop productivity. Nanotechnology is also improving crop management and promoting more effective and sustainable agricultural techniques (Scrinis and Lyons 2007). An innovative field that provides cheap, long-lasting, and protected solutions is nanotechnology, which emphasizes ingredients at the nanoscale (1–100 nm). They have many uses contain communication technologies, domestic things, food manufacturing, medicine, and agriculture(Mahmood, Ansari, and Rizvi 2024).

The approach that is environmentally friendly and practical for agrifarming is primarily supported by nanotechnology. The creation of innovative technologies, such as nanomaterials, propels agricultural progress by addressing the challenges of both traditional and precision agriculture (He, Deng, and Hwang 2019). Its influence is crucial for the development of sustainable agricultural systems, primarily in terms of inclined specific surface area. Nanotechnology presents a possible alternative to traditional agricultural systems by decreasing uncertainty and refining management plans in agricultural output. In many cases, agro-nanotechnology offers immediate and creative solutions to problems in modern industrial farming. The applications of nanotechnology in agriculture stress its potential to promote sustainable farming techniques and environmental protection. This chapter focuses on exploring recent literature on the application of nanotechnology in sustainable agriculture. The global importance of nanotechnology in agricultural sector is discussed, and future challenges in the use of nanotechnology are highlighted. Further, we also pointed out that how computation approaches like machine learning modeling, nano-sensors, and artificial intelligence are playing their role in nano-enabled agriculture.

3.2 Nanotechnology in Precision Agriculture: Innovations and Applications

Agriculture is the basis for creating important assets such as food, feed, fiber, fuel, and firewood. Food demand is expected to surge dramatically as the world's population expands. However, the natural resources like water, land, and soil fertility are limited. The input cost of crop production such as chemical fertilizers and pesticides is estimated to rise due to the reduction of natural resources. Precision farming gives a feasible solution to these challenges by decreasing inputs while increasing agricultural outputs. Precision farming involves paying modern monitoring schemes to maximize crop production while lowering the usage of herbicides, pesticides, and fertilizer. One emerging field in agriculture is the use of nanotechnology, which is still comparatively new and needs more investigation. Nanotechnology has the capacity to allow novelty such as controlled pesticide release, custom-made nutrition, and macromolecule supply. These innovations can help crops survive against disease, use nutrients more efficiently, and grow quicker. Nanotechnology-based precision agriculture products contain nanopesticides, nanoherbicides, nanofertilizers, and also nanosensors that are used to detect nutrient deficiency symptoms. This speedily developing technology is now having a significant effect on the agriculture sector and is expected to play a larger role in the future (Goyal et al. 2023).

Agriculture remains an important and established sector, providing raw resources to the food and feed sectors. With limited natural resources like arable land, water, soil, and a growing global population, agricultural production must become more economically viable, environmentally supportable, and effective. These developments are critical for handling the problems of recent years and creating progress (Mukhopadhyay 2014). Agriculture's nutrient balances move significantly with economic growth, emphasizing the importance of restoring soil fertility, mainly in rising markets (Campbell et al. 2014). Nanoparticles have strained a lot of scientific interest because they are essential in bridging the gap between bulk materials and atomic or molecular structures. Many studies conducted in the last 20 years have illustrated the extensive range of uses of nanotechnology in agriculture. While fertilizers are important for increasing agricultural yield, overuse of them has changed the chemistry of the soil permanently and declined the amount of area available for crops. Lowering pesticide use is necessary for sustainable agriculture in order to defend biodiversity and the environment (Prasad, Bhattacharyya, and Nguyen 2017). Likewise, nanomaterials provide a solution for the effective use of agricultural inputs by increasing crop yield. They make it possible for nutrients to be carried in a site-specific, controlled manner, lowering the need for agrochemicals while protecting environmental sustainability.

The pre-Green Revolution era was marked by widespread starvation, while the post-Green Revolution period significantly enhanced the use of fertilizers and pesticides to meet the growing food demands of an increasing population. However, this progress has led to detrimental effects, including the irreversible damage to microorganisms, depletion of soil humus, the emergence of pathogen resistance, and a decline in organic matter, which now pose serious challenges. Nanotechnology, with its diverse applications in nutrient delivery, remediation, and advanced nanosensor tools, offers transformative potential for sustainable agriculture. Sensors embedded in nanoparticles can be designed to monitor microbiological activity, assess soil health, and enhance detoxification processes with greater precision. Furthermore, nano-biofertilizers enriched with essential nutrients, combined with field-deployable nanosensors, can support smart agricultural practices, boosting productivity and ensuring sustainable crop management (Duhan et al. 2017). The innovative approach of nano-enabled precision agriculture utilizes nano-biofertilizers integrated with nano-chips and neuro-receptors to improve nutrient recycling, conserve resources, and optimize crop yields. Field deployment of nanosensors allows for early detection of disease outbreaks, providing farmers and stakeholders with sufficient time to make informed decisions. Studies on the use of nano-biosensors (e.g., Cu, Ag, Au) in agriculture have shown their effectiveness in accurately monitoring fertilizer distribution, pH balance, moisture variations, pathogen tracking, nutrient losses, and elemental concentrations, all contributing to enhanced agricultural productivity (Fatima et al. 2018).

3.2.1 Exploring Nanosensor Types and Roles in Precision Farming

Each nano-based material possesses distinct physicochemical properties that make it well-suited for various sensing applications (Elizabath et al. 2019). These materials function as analytical tools equipped with highly advanced and enhanced biosensors, capable of detecting biological components through signals generated by transducers (Kaushal and Wani 2017). This technology provides critical data, enabling plant scientists to assess crops' nutritional and water requirements and predict disease onset (Mariano, El Kissi, and Dufresne 2014). A well-designed nano-scale system incorporating these materials can facilitate the implementation of effective agronomic practices, such as composting and insecticide application. This review highlights the applications of nanosensors and nano-biosensors, emphasizing their advantages in advancing precision agriculture. In precision agriculture, several categories of nano-biosensors are practiced. Optical nanosensors based on recognition of variation in optical-signal make them appropriate for several spectroscopic measurements, like absorption, phosphorescence, fluorescence, Raman effect, refraction, and Raman scattering by sensing the variations in wavelength, intensity, polarity, time, and phase of the light (Dar, Qazi, and Pirzadah 2020). Altogether, optical biosensors rely on fluorescence spectroscopy, interferometry, surface plasmon resonance, and spectroscopy to function (Srivastava, Dev, and Karmakar 2018). Further, near-infrared fluorescent H_2O_2 nanosensors report early signs of stress and are considered a novel tool in precision agriculture (Wu et al. 2020). Furthermore, Ahmed et al. (2020) documented that red-colored selenium nanoparticles obtained from bio-reduction could be used as optical biosensor for heavy-metal toxicity assays.

The nanosensor-based electronic nose (e-Nose) and electronic tongue (e-Tongue) are highly accurate, sensitive, redundant, selective, and reliable devices with wider applications in precision agricultural system (Srivastava, Dev, and Karmakar 2018). Their functions are similar to human sensory organs nose and tongue and are used to detect a group of gases, taste, aromas, and their variable concentrations (Tan and Xu 2020). They have attained critical rank to regulate the quality as well as quantity of material, e.g., foods, beverages, agriculture, pharmacology, personal care product engineering, and processing (Tan and Xu 2020; Baldwin, Hagenmaier, and Bai 2011). For example, e-Nose can be applied to evaluate rice aroma and thus prevent their deprivation through timely intercession (Hu et al. 2020). Furthermore, e-Noses are used to monitor the manufacturing processes and to measure the plant diseases, insect invasions, and soil or water contaminants (Hu et al. 2019).

Wireless sensor networks (WSNs) play a pivotal role in advancing precision agriculture by facilitating automation and data-driven decision-making. Wireless nanosensors deployed in fields collect real-time data on crop health, soil conditions, and weather, ensuring both productivity and sustainability (Shafi et al. 2019). Key applications of WSNs in precision farming include smart irrigation, precise fertilization, pest management, and greenhouse monitoring. Accurate fertilization is particularly critical for enhancing productivity. Automated, wireless sensor-based systems enable real-time monitoring of soil fertility, ensuring the precise application of fertilizers to optimize crop yields.

Electrochemical sensors are also used for different purposes in agricultural industry. Electrochemical sensors gather chemical data of the soil by detecting specific ions in the soil. This data is obtained in the form of pH and soil nutrient concentrations (Pudake, Jain, and Kole 2021). Excluding spectroscopic detection, electrochemical sensors are used for the detection of nitrogen, phosphorus and potassium (NPK) (Chauhan and Urooj 2019), pesticide residues (Umapathi et al. 2022), pathogenic *Staphylococcus aureus* (Braiek et al. 2012), chloramphenicol (Yang and Zhao 2015), and nitrogen species (Ryu et al. 2020).

3.2.2 Integration of Variable Rate Technology and Nanotechnology in Precision Agriculture

Variable rate technology (VRT) is a cornerstone of precision agriculture, enabling the targeted application of inputs such as herbicides, fertilizers, and pesticides based on soil variability and crop health. By ensuring inputs are applied only where and when needed, VRT helps farmers reduce costs, minimize waste, and maximize crop productivity. Nanoparticles further enhance VRT by serving as efficient carriers for these inputs, enabling precise delivery to specific areas, particularly in fields with varying soil types (Dar, Qazi, and Pirzadah 2020) and crop conditions. Additionally, VRT facilitates optimal seeding rates tailored to specific field types (Šarauskis et al. 2022). The integration of nano-carriers with VRT offers numerous benefits: precise resource application, reduced environmental pollution through targeted nutrient delivery, and enhanced crop growth and productivity, making it a transformative approach in sustainable agriculture.

3.2.3 Enhancing Automated Machinery with Nanotechnology

Automated machinery plays a vital role in precision agriculture by executing accurate and efficient farm management tasks. Integrating nanosensors with automated systems significantly enhances agricultural operations. For instance, nanosensors can monitor soil moisture levels and trigger automatic irrigation systems, enabling real-time water delivery to crops while preventing overwatering, which could otherwise negatively impact crop growth. Beyond soil moisture, nanosensors also track environmental parameters such as humidity, temperature, and soil nutrient levels (Mahmoud and Fawzy 2021). This data can be integrated with artificial intelligence (AI) to address global food security challenges effectively. Moreover, the integration of nanosensors with automated machinery offers numerous benefits, including reducing labor dependency, minimizing human intervention, and improving overall efficiency in precision agriculture.

3.2.4 Nanosensors for Heavy Metals Detection

Natural resources, agriculture, and the food chain are increasingly challenged by issues of vulnerability, sustainability, human health, and quality of life. Agricultural productivity is particularly constrained by various biotic and abiotic factors, including weeds, diseases, and insect pests (Giraldo and Kruss 2023). The presence of heavy metal ions such as Pb^{2+}, Li^{+}, Hg^{2+}, Cd^{2+}, Ag^{+}, and Cu^{2+} from various sources poses significant risks to humans, animals, and plants (Adeel et al. 2023; Sharma et al. 2021). These metals primarily originate from chemical fertilizers, lead-acid batteries, and vehicle exhaust emissions (Gammoudi et al. 2014). While soils naturally contain essential nutrient ions like K^{+}, H^{+}, and Na^{+} necessary for optimal plant growth, heavy metal contamination severely limits crop productivity (Goyal et al.

2020). Furthermore, heavy metal exposure induces oxidative stress in plants, causing both visible and invisible damage to their physiological and biochemical processes (Wang et al. 2019; Yadav 2010).

Nanotechnology plays a crucial role in advancing diagnostic methodologies through modern tools such as nanosensors. These nanosensors simplify the detection of food-borne microorganisms, pesticides, drug residues, heavy metal ions, and toxic contaminants (Acharya et al. 2020; Shabaninejad et al. 2019). In order to create nanosensors and nano-biosensors with potent detecting capabilities, scientists are integrating configurations of biology, nanoscience, computers, and electronics for proper crop management (Amjad and Serajuddin 2022). The precision agricultural techniques may be significantly impacted by the nanosensors or nano-biosensors (Sekhon 2014). Nano-biosensors detect physiological signals and convert them into measurable, regularly spaced signals for visualization (Mondal et al. 2022).

A wide range of nanosensors has been developed for the detection of heavy metal ions. Among these, field-effect transistor sensors show significant potential for rapid and real-time heavy metal detection (Falina et al. 2021). Graphene-based optical nanosensors have also demonstrated superior performance in detecting heavy metal ions (Zhang et al. (2016). Soil heavy metal levels can be effectively measured using electrochemical nanosensors (Zhang et al. 2016; Roy et al. 2014). For in-situ detection, DNA optical biosensors combined with evanescent wave assessment have proven successful in identifying heavy metal ions (Long et al. 2013). Additionally, beta-carotene derived from palm kernels has been identified as a biological reporter for biosensors made of biopolymers, particularly for detecting metals such as aluminum (Wong and Wong 2014). Gold nanorods (Au NRs) have also been evaluated as nanosensors for Fe^{3+} detection, with a sensitivity of 100 ppb, where Fe^{3+} adsorption on Au NRs led to their oxidation, enabling accurate detection Thatai et al. (2014).

3.3 Boosting Agricultural Productivity with Nanotechnology

3.3.1 Advancing Seed Germination and Plant Growth

Seed is an essential input that affects crop productivity. Usually, seed quality has been found by germination percentage. But seeds usually show lower germination rates in the field when compared to higher rates (80–90%) documented in laboratory conditions. The researchers have explored the effect of nanomaterials on seed germination and plant growth with the goal of improving their agricultural applications (Zheng et al. 2011). Seed germination is an essential stage in the life cycle of plants determining seedling development, survival, and population dynamics. Several factors, including environmental conditions, genetic features, moisture levels, and soil fertility, have a significant impact on seed germination (Manjaiah et al. 2019). Nanotechnology is the affection of matter at the atomic, molecular, and supramolecular levels. Nanoparticles have created extensive scientific attention because of their various uses in sectors such as biological sciences, medicine, electronics, and energy. Over the last few years, scientists have progressively investigated the application of nanoscience in agriculture. The effective use of agrochemicals and the rule of numerous physiological processes in plants are two major areas of study. The current global research on plant nanotechnology emphasizes its benefits and problems in agricultural applications (Banerjee and Kole 2016).

These findings exposed the objectives of nanomaterials to improve crop yields and product quality. While the exact mechanism behind the promotion of crop growth and enriched quality is not clear, it may be at least partially explained by the ability of nanomaterials to absorb more nutrients and water that in turn helps to enhance the vigor of root systems with increased enzymatic activity (Shojaei et al. 2019). Nanobiotechnology uses live organisms to create nanostructures, and plants have arisen as a crucial contributor to the synthesis of metal nanoparticles. Plants provide a simpler alternative to other eco-friendly solutions because they do not require the maintenance of sophisticated cell cultures. Using plants or their extracts to produce nanoparticles is especially advantageous when these particles can be formed outside of cells and in a controlled manner, with specific sizes, shapes, and distributions. Furthermore, plants facilitate the scaling of the process for large-scale productivity. The investigation shows how plants and their extracts are utilized to produce silver and gold nanoparticles, emphasizing their potential for a variety of human applications (Kumar and Yadav 2009).

Manjaiah et al. (2019) outlined how ZnO nanoparticles affected the development of Cicer arietinum and Vigna radiata seedlings. In order to prevent cohesion caused by the water-insoluble nanoparticles, tests were conducted using plant agar media. Using agar techniques, the researcher examined the application effect of ZnO nanoparticles in roots and shoots. The nanoparticle adhering to the rhizosphere was found using scanning electron microscopy. The research using atomic emission spectroscopy clarified how the roots absorbed these nanoparticles. The findings demonstrated that seedling growth performed best at the ideal dosage compared to the control but was hampered at higher concentrations.

Likewise, pea crops were treated with iron nanoparticles. It was discovered that 50 mg/L of iron nanoparticles was the ideal dosage, whereas noticeable growth response was displayed by the peas plant. In comparison to the control, it raised the number of pods per plant by 47%, the iron content in the leaves by 34%, the thousand-grain weight by 7%, and the chlorophyll content by 10% through increases of 4%, 45%, 28%, and 12%, in that order. It demonstrated that the group treated with iron nanoparticles outperformed the group treated with iron salts (Delfani et al. 2014). Furthermore, iron nanoparticles were applied to boost vigor and seed germination, which enhances crop development and growth. Nanomaterials have special qualities that allow them to become nanoscale by reducing their weight.

3.3.2 Optimizing Fertilization for Higher Yields

In order to develop a receptor layer that can effectively sense soil moisture, a variety of materials were tried in this work, including graphite oxide, molybdenum disulphide, vanadium oxide, and molybdenum oxide. Unlike both graphite oxide, molybdenum oxide materials, and molybdenum disulfide showed the highest sensitivity, nearly three times higher (1200 pF) than vanadium pentoxide across all soil moisture levels. Future autonomous agriculture depends on the development of in-situ soil moisture sensors using new technologies. One of the main challenges for these sensors is controlling variations in capacitance rather than resistance while maintaining stability, high performance, and reliability. At temperatures as high as 45 °C, the sensitivities for MoO_3, GO, MoS_2, and V_2O_5 surge approximately 13%, 11%, 30%, and 9%, respectively. For all policies, there was a small change in sensitivity when the temperature was altered from 25 °C to 50 °C. MoS_2 sensors were an effective option, achieving a record sensitivity of 540% in black soil with response and recovery times of 65 and 27 seconds, respectively (Surya et al. 2020). Due to unsustainable farming techniques and environmental contamination, rigorous farming plays a significant role in the depletion of essential resources like land and water. These difficulties are further compounded by the effects of climate change and the surging world population. Numerous cutting-edge technologies, such as nanostructured biosensors, have been created to solve these problems. By consuming the special qualities of functional materials at the nanoscale, these nanotechnology-based biosensors allow farmers to do quick, precise, economical, and field-based assessments of (i) soil humidity, (ii) water and soil nutrients/pesticides, and (iii) plant diseases. This study highlights recent developments in nanostructured biosensors for smart agriculture (Antonacci et al. 2018). Demands for food security, resource limitations, and climate change are increasing difficulties for conventional farming. Cleaner production emphasizes knowledge-based and ecologically friendly farming methods that minimize chemical inputs and improve natural recycling. There are many different sensing technologies available for identifying chemical compounds; their unequal procedures, operational needs, and skill requirements may make it difficult to make wise decisions in a single field. On the other hand, a combined detection platform that is readily adaptable to evaluate many factors is a viable soil management option. By adapting macro-parameters, this work presents a flexible lipid nano-platform that makes it simple to customize sensor sensitivity and selectivity. The development of a soil nitrite biosensor is described in detail as a case study, including its design, optimization of physical and chemical variable, and first validation on both simulated and real soil samples. The sensor has a minimum detectable nitrite concentration of 1 ppm in the field, a detection limit of 2.1 mg/L for soil extracts and a high tolerance to interfering ions (Siontorou and Georgopoulos 2016).

In general, adding basic nutrients is unavoidable for enhancing crop productivity and soil fertility. Field research in the semi-arid Loess Plateau of China discovered that the ridge-furrow (M2) planting style, paired with a N: P fertilizer rate of 200:100 kg ha^{-1} (F2), considerably improved maize production, water usage efficiency (WUE), and economic returns. The M2F2 therapy surged WUE by 37.7% and economic benefit by 54% when compared to controls. Beyond this acceptable rate, agronomic efficiency decreased. The M2F2 model is shown for high-productivity and efficient maize growing in semi-arid areas (Li et al. 2018). Excessive use of fertilizer can reduce agricultural yields, deplete nutrient efficiency, and leave nitrate-N residues in soils, generating environmental and economic issues. This study created optimal fertilizer recommendations based on soil testing following the previous crop's harvest. From 2013 to 2016, field tests were carried out on the dryland winter wheat production area of China's Loess Plateau. Improved fertilizer application (OFA) decreased fertilizer use by 96 kg N ha^{-1} and 43 kg P_2O_5 ha^{-1}, Nitrate-N residues by 120 kg N ha^{-1} in the 0–100 cm soil layer, and enhanced winter wheat yield by 386 kg ha^{-1} and economic returns by \$230 ha^{-1} on average.

The optimized fertilizer application (OFA) approach retained the same nitrogen (N), phosphorous (P), and potassium (K) uptake as conventional fertilizer application (CFA) while increasing the harvest index for N, P, and K by 4%, 3%, and 7%, respectively. It also increased nutrient internal efficiency rates (N, P, and K) by 9%, 4%, and 8%, respectively, as well as partial factor productivity by 85%, 61%, and 31%. This strategy strikes a negotiation between environmental protection, increased crop yields, enhanced nutrient usage efficiency, and economic benefits, making it a possible alternative for areas where fertilizer misuse is prevalent (Huang et al. 2017). Fertilizers account for about 50% of agricultural production, but excessive use does not always result in higher yields and can lead to soil deterioration and water pollution. The use efficiency of N, P, and K fertilizers remains poor at 30–35%, 18–20%, and 35–40%, respectively, with much of the fertilizer leaching down into soils and aquatic systems, resulting in eutrophication. Nano fertilizers can improve plant nutrient absorption, whereas non-toxic fertilizer additives such as titanium dioxide (TiO_2) can improve water retention and increase crop yield (Emadian 1988).

Nanotechnology has significant potential for enhancing plant nutrient usage in agriculture. In this study, zinc oxide nanoparticles (ZnO-NPs) were produced from $ZnSO_4$. $7H_2O$ as an originator. The ZnO-NPs had an average particle size of 63.60 nm, were stable as shown by the zeta potential, and had a semi-spherical agglomerated morphology under scanning electron microscopy (SEM). Their purity was authenticated using energy-dispersive X-ray spectroscopy (EDAX). When groundnut seeds were primed with ZnO-NPs, zinc adsorption and distribution were improved compared to bulk $ZnSO_4$, with no physical damage and outstanding biocompatibility. The nanoparticles' small size and high surface area permitted greater zinc uptake, as seen by increased zinc levels in treated seedlings. Priming with ZnO-NPs also enhanced germination and seedling vigor 10 days after sowing. Biochemical research exposed that ZnO-NP-treated plants had greater amounts of chlorophyll, soluble protein, total soluble sugar, and free amino acids in their leaves. Besides these seedlings, they indicated higher DPPH radical scavenging activity, indicating increased antioxidant capacity. Also, the increased activity of antioxidant enzymes (superoxide dismutase-SOD, catalase-CAT, ascorbate peroxidase-APX, guaiacol peroxidase-GPX, and polyphenol oxidase-PPO) in ZnO-NP-primed seeds helps with seed hydration, enzyme activity for reactive oxygen species neutralization, cellular integrity, and overall plant health. These findings emphasize the potential of ZnO-NPs as an environmentally friendly method of improving seed germination, seedling vigor, and stress resilience, resulting in increased crop growth and yield (Ashwini et al. 2024).

Fertilizers are important for increasing crop yield, soil productivity, and nutrient availability. While excessive use of traditional fertilizers can cause soil pollution, deplete soil fertility, and health hazards. The increased demand for sustainable crop production has prompted the development of nano fertilizer processes. Controlled-release fertilizers (CRFs) decrease nutrient loss and offer timely nutrient delivery, supporting optimal plant growth. However, CRFs can be costly and are often coated with petroleum-based synthetic polymers. These non-renewable, fossil fuel-derived products can aggravate environmental problems such as soil pollution and resource depletion. To overcome these problems, researchers are working to create biodegradable, cost-effective, and environmentally acceptable coating materials utilizing nanotechnology. The use of nanotechnology can improve the effectiveness of CRFs (Asghar and Kazerooni).

Drought stress is a significant abiotic factor lowering agricultural production, especially in semi-arid environments. This study investigates the impact of drought on wheat growth phases (D0: control, D1: drought at tillering, D2: drought at anthesis) and the use of Cu nanoparticles at different concentrations (T0: 0 mg/L, T1: 300 mg/L, T2: 700 mg/L, T3: 950 mg/L) to improve drought resilience. The findings exposed that dryness drastically affected wheat growth and production during critical growth stages. However, adding Cu nanoparticles aided in buffering the harmful impacts of drought by enhancing plant height, spike length, 1000-grain weight, stomatal conductance, chlorophyll content, water consumption efficiency, leaf turgor potential, and relative water content, resulting in increased grain yield. Primary component analysis was used to incorporate and explain these findings, which confirmed that drought stress during the anthesis stage had the greatest impact on yield. A 300 mg/L application of copper nanoparticles significantly increased wheat productivity by relieving drought stress (Raza et al. 2024).

Globally, food supply depends on fertilizers to reach production goals, but the harmful environmental effects of traditional fertilization processes have encouraged the mission for new options, such as smart fertilizer technologies (SFTs). These SFTs are classified into three categories: delayed- and controlled-release fertilizers (SCRFs), nano fertilizers, and biological fertilizers, and explain how each works. It also measures the environmental disadvantages of traditional fertilizers and discusses how SFTs address these concerns. A comparison of various technologies underlines the importance of a balanced strategy that takes into account efficiency, environmental safety, and cost-effectiveness. While no one SFT attains the ideal balance, combining several technologies can help mitigate individual constraints. Financial and cost-benefit calculations are critical for determining their possibility in various agricultural situations (Singh et al. 2024b).

To address these difficulties, novel solutions are immediately required. Nanotechnology has played an important role in the agro-technological revolution, with the capacity to alter agriculture and assure food security. Nanoparticles (NPs)

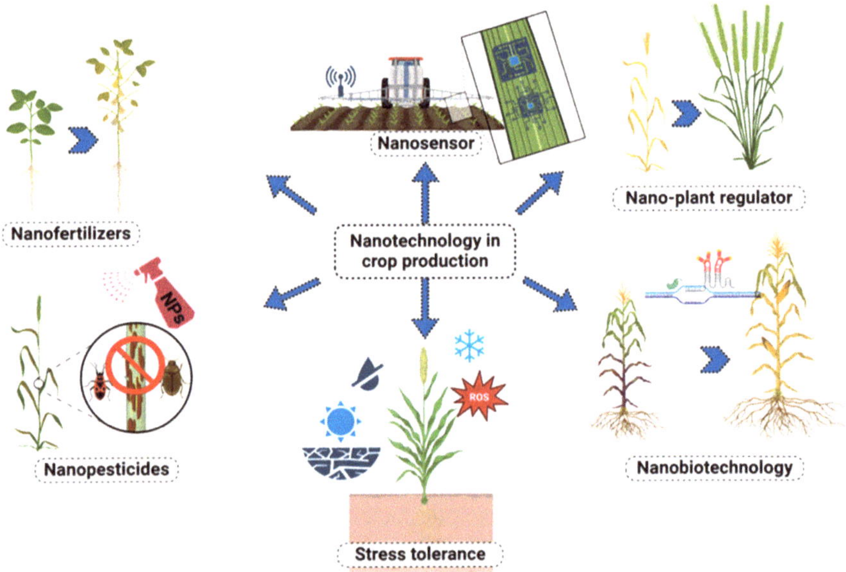

Fig. 3.1 The pictorial representation of the role of nanotechnology in boosting crop production by multidisciplinary approaches

are emerging as a possible tool for enhancing existing agriculture techniques. Various NP-based formulations, including nano-sized insecticides, herbicides, fungicides, fertilizers, and sensors, have been examined for their ability to regulate plant health and improve soil fertility. NPs have illustrated usefulness as macro- and micronutrient sources, increasing soil fertility, crop nutrient absorption, and overall productivity. The emphasis should be on synthesizing NPs from biodegradable organic or natural sources and improving their application dosages to increase healthy crop growth and yield. Nanofertilizers play a vital role in balanced crop nutrition, nutrient management, and enhancing fertilizer usage efficiency. These benefits contribute to increased soil fertility and agricultural yields, while also addressing potential limitations and emphasizing the need to protect soil health and crop productivity (Chandra et al. 2024; Shakoor et al. 2023b) (Fig. 3.1).

The controlled-release nitrogen (CRN) fertilizer application (150 kg ha^{-1} for early rice and 180 kg ha^{-1} for late rice) boosted yields by 7% and 11%, respectively, while increasing nitrogen use efficiency (NUE) from 23% and 24% to 33% and 37% compared to traditional urea. CRN fertilizer also minimized nitrogen losses, as evidenced by reduced runoff (23.1% and 19.4%), leaching (12.7% and 12.1%), ammonia volatilization (28% and 30%), and N_2O emissions (10.4% and 16.1%). Importantly, a 10% reduction in CRN fertilizer input maintained crop yield levels. Reduction of 10%, 20%, and 30% in CRN application improved NUE by 7.0–7.6%, 7.3–7.4%, and 11.6–12.6%, respectively, while decreasing runoff by 16.1–17.9%, 27.9–30.7%, and 35.0–37.2%; leaching by 7.6–12.8%, 18.1–22.6%, and 26.5–31.4%; ammonia volatilization by 9.9–12.3%, 16.3–22.7%, and 23.2–29.3%;

and N_2O emissions by 7.8–13.3%, 12.8–32.8%, and 20.3–36.9%. CRN-treated soils exhibited higher total and inorganic nitrogen levels compared to urea-treated soils, with nitrogen concentrations increasing in proportion to CRN application rates. Additionally, soil nitrogen concentration and runoff were closely associated with yield and nitrogen absorption, while runoff and leaching had a direct correlation with NUE (Su et al. 2024).

The ease of application and availability of urea is the most used nitrogen fertilizer in commercial agriculture due to its high nitrogen concentration. However, its nutrient use efficiency (NUE) is limited to approximately 30%, due to environmental losses, primarily through leaching into the soil and volatilization via air oxidation. This not only causes economic losses but also contributes to environmental challenges such as global warming, water pollution, and human and animal health threats. Furthermore, the Haber-Bosch process, the traditional method for urea manufacturing, is both energy-intensive and unsustainable. With the world population growing quickly and the desire for higher agricultural yields increasing, there is a need to find solutions that can increase the efficiency of urea supply and nitrogen release in order to line with plant demand. Nano-delivery technology has been established as an efficient solution to the problems related to urea. Engineered nanoparticles are being examined as transporters for delayed and regulated nitrogen delivery to crops, later enhancing NUE and lowering negative environmental impacts. The various forms of nano-carriers, including nanoparticles, layered nanostructures, and nanotubes, have been used to develop urea-based slow-release nano fertilizers (Dissanayake et al. 2024).

3.3.3 Managing Plant Diseases with Nanotechnology

Global food losses are caused by a variety of biotic, abiotic, and human factors. Increased food production is essential to feed the world's growing population, while natural resources are becoming more limited and unobtainable. These losses are mostly caused by weeds, pests, illnesses, infections, drought, salt, climate changes, and the loss of arable land. Another important factor in worldwide food losses is post-harvest losses. Arable land is further exploited and declined as a result of resource mismanagement. Currently, weeds account for 34% of agricultural losses, and insect pests account for 38%. As a result of industrialization and urbanization, arable land is gradually declining (Junaid and Gokce 2024). Interest in practical replacement has grown as it becomes gradually clear that conventional farming techniques cannot satisfy the enhancing requirement for higher yield. The restrictions and prohibitions of traditional chemical nematodes because of the hazards to the environment and public health are further causes of anxiety. Around the world, plant-parasitic nematodes (PPNs) extremely harm and destroy a large number of commercially valuable crops. Overcoming PPNs with nematocidal nanomaterials (NNMs) offers a practical, safe, and economical substitute. NNMs are produced, formulated, and used in a way that is ideal for modern, sustainable farming. While novel formulations can lower doses, restrict human exposure, and encapsulate nematicides to improve stability, permeability, solubility, gradual release, and resistance to premature degradation, proper synthesis can provide nematicides that are both effective and reasonably priced. Synergistic effects could result from combining NNMs with other agricultural inputs, increasing their ability to inhibit PPNs and enhance crop yields. The benefits and drawbacks of NNMs, as well as their mechansims of action, are investigated. The effective use of nanomaterials is to manage the main PPN species (Abd-Elgawad 2024). Effective healthcare management requires fast decisions based on informatics, accurate data analysis, and quick diagnostic instruments. When balancing with nanotechnology and biotechnology, microfluidics allows for quick analysis and detection at the point of care. Nanobiosensors use nanotechnology to check, monitor, diagnose, and treat diseases quickly. With the use of a transducer, nano biosensors transform physio-chemical signals into optical or electrical outputs in order to identify biomolecules using bioreceptors. These sensors are essential for early disease detection because of their brilliant sensitivity and selectivity. The advantages of mixing microfluidics and nanobiosensing techniques are demonstrated through a number of examples in this chapter, which discusses detection techniques used with microfluidic systems for nanobiosensors. Along with the latest developments in the creation of microfluidics-based nanobiosensors for point-of-care detection and remedial monitoring, it inspects their basic features, uses, and developments (Kumari et al. 2024).

Because identification must take place exactly during the preventive stage, such as viral DNA replication or first viral protein production, managing plant diseases is difficult (Trippa et al. 2024). This reduces environmental systems and raises the cost of producing crops. For plants to be secured from pests and crop loss to be avoided, active substances must be present at target places at the lowest possible effective concentration. For many years, agricultural research has placed a high priority on creating novel plant protection formulations. The development of innovative plant protection compounds has been a significant field of agricultural research. The encapsulation or nanoformulation of pesticides is one such innovation that has completely changed the plant protection sector. Biocompatible materials were used in the production of a dimethoate nanoformulation in order to decrease the dangers involved with conventional formulations. Using glycerol, adipic acid, maleic anhydride acid, and polyethylene glycol 1000, the best copolymers for encapsulating dimethoate were created by a polycondensation process. These copolymers were combined with ZnO nanoparticles (ZnO-NPs), which were generated by a chemical precipitation process, to add photodegradable qualities (Fig. 3.2). Over time, nanoformulations (7%) significantly increased the effectiveness of pest control as compared to conventional formulations (EC, 40%). ZnO nanoparticle-based formulations presented the best efficiency. Solvents and active ingredients leak and leach significantly into the environment from traditional EC formulations. A sustainable way to decrease these threats is to improve dimethoate by creating water-based, pH-sensitive, and photodegradable nanoformulations (Jahanbakhshi et al. 2024).

The vital elements of nanotechnology, which provides effective and long-lasting ways to improve agricultural yields while protecting the environment, are nanoparticles. Nanotechnology in agriculture makes it possible to detect and diagnose pests,

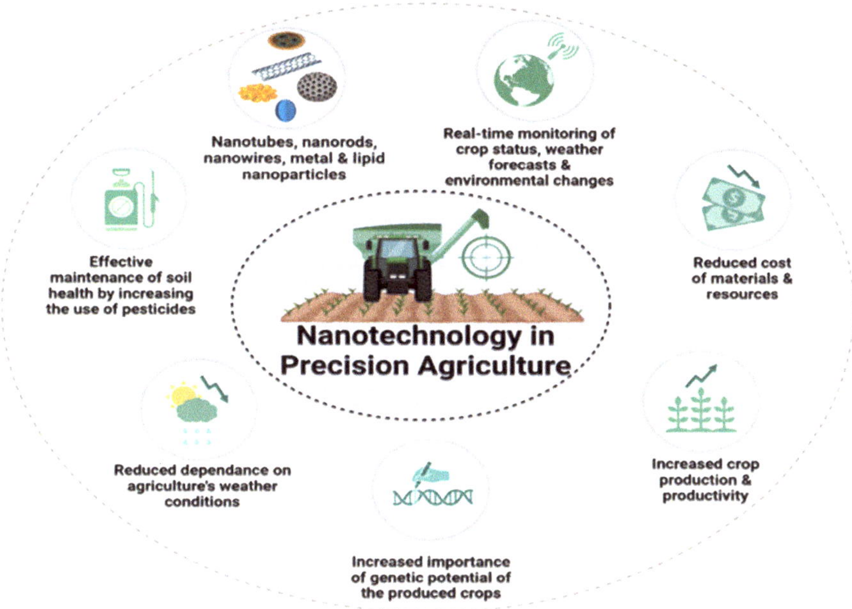

Fig. 3.2 Schematic diagram highlighting the application of nanotechnology in precision agriculture

illnesses, disorders, nutrient deficiencies, and pesticide residues early on, allowing for rapid and correct remedial treatment. It also assists with soil health and environmental stress monitoring. As antifungal, antibacterial, and antiviral agents, nanoparticles can be active in drug supply systems to transmission genetic materials, plant hormones, and agrochemicals to specific areas. Green-engineered nanoparticles like Ag, Cu, Au, and ZnO have been confirmed in studies to efficiently regulate plant health by raising the concentration of important secondary metabolites in treated plants (DUTTA). Although silver (Ag) has been used since ancient time, biogenic AgNPs were just produced in the last few decades and are mostly used as antioxidants and antimicrobials in the food and pharmaceutical industries (Fig. 3.2). AgNPs can inhibit bacteria like *Salmonella typhi*, *Klebsiella pneumonia*, *Vibrio parahaemolyticus*, *Citrobacter koseri*, *Bacillus cereus*, *Escherichia coli*, *Staphylococcus aureus*, and *Pseudomonas aeruginosa* through mechanisms like cell diffusion, free radical production, ROS generation, and modulation of transduction pathways. Biogenic AgNPs are very influential antibacterial agents because of the new potential for fighting germs brought about by improvements in nano-biotechnology-based treatments. Significant physio-chemical characteristics of AgNPs, such as surface charge, colloidal state, shape, concentration, and size, influence their antibacterial ability (Rodrigues et al. 2024). Both gram-positive and gram-negative bacteria can be effectively opposed by the 1:3 and 3:1 ZnO:MgO compounds, which exhibit best performance at lower MIC values (12500–15,000 μg/mL). Their antibacterial activity is increased by the smaller particle size and higher density. These ratios indicate superior surface contact and ROS formation when compared to pure ZnO, MgO, or 1:1 mixture. The 1:3 ratio works very well against gram-positive bacteria at a much lower MIC (~5000 μg/mL) (Fig. 3.2). Because these ratios are less harmful and less cost-effective, they can be used in a broader range of applications. Particle size and structural qualities can be improved to further increase their efficiency (Amir et al. 2021).

3.3.4 Controlling Weeds Using Nano Herbicides

The use of nanotechnology in plant protection products has grown significantly, offering encouraging opportunities to boost agricultural productivity. Plant protection and environmental resistance adaptation are the primary priorities in agriculture. Aside from climate change, a variety of additional factors, including soil salinity, drought stress, temperature fluctuations, and heavy metal use, might interfere with plant growth and development. Therefore, one of the hottest answers to these problems is nanotechnology, with the primary goal being to help plants in ecosystems become more adaptable (Goyal et al. 2023).

Crop monitoring has been made possible by the advancement and integration of new technologies in agricultural sensors. These nanosensors are employed in the administration of fertilizer and pesticides, as well as in the detection of plant diseases, crop growth, soil health, and environmental contamination. The development and application of nanosensors in precision farming have allowed for the verification and monitoring of all these factors. This innovation in the agricultural sector continuously aids farmers in managing plant development and soil health while also enhancing quality and safety measures (Chen et al. 2016).

The importance of the agriculture sector to developing nations' economies highlights the need for technology that increases agricultural output in an economical and environmentally responsible way. Through harmless mechanisms, environment-friendly techniques like biopesticides and bioinsecticides are already being utilized to control pests. Using nanoparticles, nanofibers, nano mixtures, and nanocapsules, nanobiotechnology provides innovative instruments to increase crop yields. These nanoparticles make it easier to transfer macromolecules and agrochemicals that sponsor plant development and strengthen tolerance to stress. Agrochemicals are supplied precisely, which maximizes the use of water and nutrients and increases yields. Agrochemical ability is further enhanced by site-specific supply and controlled release. By altering plant gene expression, nanoparticle-mediated gene transformation and macromolecule management can further increase agricultural production and quality. Numerous nanomaterials have been the subject of optimistic research aimed at improving crop quality and quantity (Mali, Raj, and Trivedi 2020). Nanobioherbicides can protect plants in two ways: either as carriers for traditional herbicides or as crop protection nanoparticles (NPs). Herbicide misuse can be minimized while maintaining safe and effective release by using NPs for plant protection. Herbicide use can be reduced and crop production increased with nanostructured herbicides, improving farming methods (Igiebor et al. 2024). Crop production depends on the effective management of agricultural systems, including pest control, irrigation, and fertilizers. Significant yield losses are caused by weeds, and excessive pesticide use has resulted in resistance, water pollution, environmental damage, and severe health effects. Plants naturally release allele chemicals, which may be used as bioherbicides. However, their efficacy is disadvantage by their poor soil tenacity. These allele chemicals' stability, duration, and availability are enhanced when they are encapsulated in nano-carriers such nanocapsules, nanospheres, and nanogels, which improves their ability to suppress weeds. The possibility of covering allele chemicals in nano-carriers for improved weed control is inspected in this chapter (Marimuthu et al. 2023). Plant diseases, insect damage, and crop loss due to weed competition all have an impact on global agricultural productivity. 34% of crop loss worldwide is caused by weeds, and another 34% is caused by pests and microbiological diseases. In addition to lowering crop loss, the use of synthetic pesticides and herbicides that contain N, P, and K improves food quality by improving nutrient uptake and retention (Bindraban et al. 2015).

3.3.5 Enhancing Pest Control with Nano Pesticides

In order to ensure food security and prevent biological disasters, pesticides are essential. A capable field for developing effective and eco-friendly pesticide formulations is the creation of biocompatible, biodegradable, and responsive materials. Through molecular and chemical changes, nanotechnology has a great deal to offer for creating targeted insecticides with controlled-release, environmentally sensitive qualities. Nano pesticides with controlled release mechanisms that respond to light, temperature, humidity, soil pH, and enzyme activity. Additionally, it highlights nano pesticide release methods based on nanomaterials, which improve targeting efficiency, dispersibility, and stability of active ingredients (Huang et al. 2018). Tactics based on nanotechnology have demonstrated promising outcomes in sustainable agriculture, especially in the development of new fertilizers and pesticides. It is essential to assess these nano formulations' advantages for the economy and the environment (Table 3.1). This study evaluates the impact of nanofertilizers on greenhouse gas emissions, finds practical applications and important challenges, and estimates the possible income changes related to the usage of nanopesticides and nanofertilizers. The cost-benefit analysis shows that although existing nanoformulations enhance agricultural yields and lessen their impact on the environment, more efficiency improvements are required before they can be widely used. Economic and environmental results can be greatly enhanced by innovations in targeted distribution, lowering greenhouse gas emissions from nanomaterials, and decreasing their concentration in nanopesticides and fertilizers (Su et al. 2022). overuse of harmful pesticides and mineral fertilizers has led to pollution and serious health hazards. Nanoscience provides high-performance nanomaterials as possible remedies. The elements N, P, K, Fe, Mn, Zn, Cu, Mo, and carbon nanotubes are among the elements found in nanofertilizers that show enhanced release and targeted delivery efficiency (Table 3.1). Comparatively speaking, nanopesticides like Ag, Cu, SiO_2, ZnO, and nanoformulations offer better broad-spectrum pest control than traditional pesticides (Chhipa 2017; Shakoor et al. 2023a).

Table 3.1 Application of NFs to different plants under various concentrations and their responses

Plant species	Duration (days)	NPs (types)	Concentration (mg kg-1, mg L-1)	Plant response	Reference
Rice	55	Si & TiO_2	0, 5, 10, 15, & 20 (Foliar application)	chlorophyll content ↑, shoot biomass ↑, Ed bioaccumulation ↓ &improve the antioxidant activity.	(Rizwan et al. 2019)
Peanut	30	Fe_2O_3	0, 500, &4000	protein content ↑, Glycoprotein ↑,	(Suresh, Karthikeyan, and Jayamoorthy 2016)
Cucumis sativus L. cv. Dominus	30	Bulk Zn $(CH_3COO)_2{\cdot}2H_2O$ &nZnO	0, 10, 100, &1000	Zn bioaccumulation ↑&shoot and root dry biomass ↑	(Moghaddasi et al. 2017)
Triticum Aestivum L.	—	$nTiO_2$ and nZO	0, 50, 500, 1000, &2000	Chlorophyll content ↑ & protein content ↑. Non-significant impact on germination.	(Ramesh, Palanisamy, and Kumar 2014)
Black-eyed pea	120	Fe & Mg	0, 250, &500	Seed yield↑,Chlorophyll content ↑, seed protein ↑, bioaccumulation of Fe and Mg ↑,	(Delfani et al. 2014)
Catharanthus roseus	70	Fe_3O_4	0, 5, 10, 20, 30, &40 mM	Bioaccumulation of K, P, and Fe ↑. Growth parameters ↑, photosynthetic pigments ↑, and total protein contents ↑.	(Askary, Amirjani, and Saberi 2017)
Cherry radish	30	Fe_2O_3, and γ-Fe_3O_4 nZVI, α-Fe_2O_3, β-FeO(OH), Fe_3O_4	0, 50, 100, 200, &500	, Zn ↑, vitamin C ↑, crude protein contents ↑, essential and non-essential amino acid ↑, Fresh fruit weight ↑, bioaccumulation Fe content ↑, daily intake of Fe ↑	(Shakoor et al. 2022)

3.4 Sustainability and Environmental Advantages of Nanotechnology

Finding and assessing environmentally hazardous threats is morally right, especially when taking into account the concepts of justice (equitably allocating risks), autonomy (respecting self-determination), and non-maleficence (avoidance of harm) (Schulte and Salamanca-Buentello 2007). Materials science, electronics, medicine, agriculture, and energy are just a few of the sectors that have seen revolutionary breakthroughs because of nanotechnology. The wide-ranging effects of nanomaterials (NMs) in several fields are examined in this chapter. It discusses the advancements in healthcare, environmental cleanup, and material science, showing how nanotechnology is increasing the land of what is possible. The effects of nanotechnology on society are also examined, including the necessity for regulatory frameworks, ethical issues, and environmental consequences. Addressing the opportunities and difficulties presented by nanotechnology is becoming more and more crucial as it develops in order to reach its full potential. This study offers guidance for future research and development while giving a thorough overview of the dynamic and quickly evolving subject of NMs (Singh et al. 2024a). Plant disease management, stress monitoring, and improved nutrient usage efficiency are all aided by these nanoscale materials. Biotechnology was employed to produce nanoparticles for a particular application in precision agriculture systems. Nanoparticles are created by combining physical, chemical, and biological processes (Fig. 3.3). When silver and zinc oxide nanoparticles interacted with bacterial membranes, their unique antifungal and antibacterial properties were triggered (Thakur, Thakur, and Kumar 2018).

For example, silica zinc titanium dioxide nanofertilizers can enhance plant growth characteristics and nutrient uptake. To use nanoscale technology to detect food pathogens and enhance food quality. The use of nanoparticles in food packaging to prevent bacterial growth and penetration can also extend the shelf life of food. A fundamental component of biosecurity is biotechnology, which is used to provide novel approaches to food production and their safety components. However, it's crucial to take into consideration the issues that nanotechnology causes. Therefore, ethical considerations about the use of nanotechnology are necessary (Saritha, Anju, and Kumar 2022).

3.4.1 Mitigating Environmental Impacts

New materials with the best properties and the least environmental impact were created using nanotechnology, a significant shift in the business by developing the best nanomaterials possible, which are biodegradable, environmentally benign, and low in toxicity. Waste management, packaging, renewable energy, and water purification all made use of these nanoparticles. Because of the unique properties of nanoparticles, such as their high surface area to volume ratio and catalytic activity, nanotechnology has been able to alleviate environmental issues. Solomon et al. (2024)

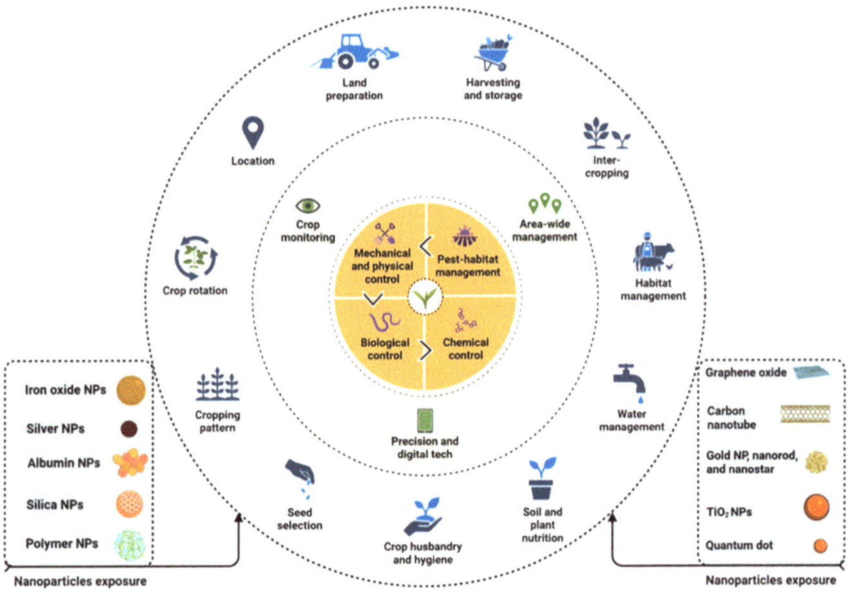

Fig. 3.3 Schematic representation of nanotechnology applications focusing on pest management strategies (mechanical, physical, biological, chemical control, and pest-habitat management) integrated with precision and digital technologies

demonstrated how these nanoparticles were effectively linked to a decrease in waste, resource consumption, and environmental contamination. The findings brought to light ethical issues and how nanotechnology can be used to address them by the use of nanomaterials that can guard against pollution in the environment. These creative and long-lasting applications to the green supply chain were made possible. Nwokediegwu et al. (2024) carried out a study on the management and policy of water in the United States and Africa. The study's primary goals were to assess many facets of climate change, water scarcity, and arrangements, as well as institutional structure and major issues in the field. While Africa's governance lacked these advantages, the United States has an advantage in terms of established institutional structure, guidelines, and best practices; however, they need environmental solutions in the local areas of African government. This study aims to increase climate change resilience, improve the governance structure, and attract investors through advertising. It emphasizes information exchange and capacity enhancement via global cooperation (Fig. 3.3).

A collaborative vision for sustainable water management on a global scale is proposed, providing valuable insights to policymakers, stakeholders, and communities. In conclusion, vertical farming offers a ground-breaking approach to sustainable urban agriculture when paired with architectural design. This partnership addresses critical challenges such as food security, environmental sustainability, and urbanization. They highlight the innovative ideas that result from this lively exchange (Akintuyi 2024).

3.4.2 Contributions of Nanoparticles to Reducing Greenhouse Gases and Carbon Footprint

Through their ability to retain heat in the atmosphere, greenhouse gases are essential to keeping the Earth habitable. However, global warming and climate change are the result of their excessive build-up, which is mostly caused by industrial processes, the burning of fossil fuels, and agricultural practices. Reducing existing emissions and bringing greenhouse gas levels back to pre-industrial levels must be the main goals of mitigation initiatives. This calls for the broad availability of carbon-free energy options, the implementation of cutting-edge emission capture technologies, and a shift to renewable energy sources. Effective tactics at manageable scales can be tailored with the aid of regional assessments, but innovation and regulatory changes are also necessary to provide lasting solutions. To highlight the significance of cutting greenhouse gas emissions and switching to sustainable energy sources, public involvement is crucial (Sarika et al. 2023).

Because of their special optical, magnetic, and electrical qualities as well as their affordability, nanomaterials are being used more and more to improve the functionality of technology goods and services. Although a lot of research has been done to confirm the technical viability of using nanoparticles, nothing is known about how sustainable they are. In order to identify gaps, obstacles, and environmental performance measures, it critically evaluates life cycle assessment (LCA) studies on the use of nanomaterials in technology. In conclusions, energy-intensive operations have a major impact on both cumulative energy demand and global warming. 60% and 80% of cumulative energy impacts are attributed to the manufacture of raw materials and the embodied energy in solvents and precursors, respectively. Using the most effective methods is essential to reducing these responsibilities. Environmental impacts could also be decreased by 20% and 73%, respectively, by scaling up bench and pilot procedures, as well as by recycling and reusing resources within a circular economy framework (Alves et al. 2022).

Nanotechnology involves designing and using functional structures at the atomic or molecular scale, with at least one dimension in nanometers. Among various nanomaterials, magnetic nanoparticles (mNPs) offer unique advantages, such as easy separation using magnetic fields. The environmental impact of different synthesis routes for mNPs, focusing on production schemes that involve coating preformed mNPs with a shell for use as immobilization support in biotechnological applications. From a LCA perspective, energy and chemical usage during the manufacturing stage emerged as significant contributors to environmental impacts. The study found that polyethylenimine-coated mNPs achieved an effective balance between environmental impact and enzyme immobilization efficiency. However, applying LCA to nano-products presents challenges, including limited comparable reports and data availability, which increase uncertainty in the analysis (Feijoo et al. 2018).

Agriculture-related ammonia (NH_3) emissions are a major cause of air pollution, soil acidification, water eutrophication, biodiversity loss, and health hazards. Although there are many mitigation methods available, there is a lack of an integrated assessment that takes into account both agricultural and livestock systems, and the efficiency of these strategies varies greatly. To develop solutions for ammonia emissions in agricultural systems, a global meta-analysis is required. Compared to controls, dietary supplements, urease inhibitors, manure acidification, and deep manure disposal showed the highest decrease potential in animal production, with reductions of 35.1–54.2%, 24.3–68.7%, 88.8–95.0%, and 93.8–99.7%, respectively. Reductions ranging from 70.0 to 82.1% further revealed effective management of manure storage. For agricultural systems, fertilizer application and selection techniques were essential. Ammonium nitrate, controlled-release fertilizers, and deep placement all reduced NH_3 emissions by 48.0%, 56.8%, and 88.3%, respectively (Ti et al. 2019). Fertilizer selection and application techniques must be carefully considered in order to reduce emissions. Viable strategies for lowering NH_3 emissions globally are presented by well-thought-out, geographically focused initiatives. Future research should assess the impact of these initiatives on greenhouse gas emissions in order to ensure overall environmental benefits.

The ammonia selective catalytic reduction system influences both the amount and size distribution of solid and total exhaust particles downstream of a diesel particulate filter. In order to assess the impact of ammonia during active diesel particulate filter regeneration, fuel post-injection studies were included. Due to possible particle formation, the adoption of selective catalytic reduction systems to fulfill rigorous diesel no emission limits have raised exhaust ammonia levels, which raises health and air quality issues. When 100 ppm ammonia was supplied downstream of the selective catalytic reduction catalyst, post-diesel particulate solid particles >23 nm rose by 129% (range 80–193%) and total particles <23 nm increased by 67% (range 26–136%) for efficient nitrous oxide control (Amanatidis et al. 2014). However, ammonia did not affect the already high particle concentrations seen during diesel particulate renewal. Sulfate, nitrate, and chloride salts with ammonium are potential sources of new particle generation, depending on species availability and formation conditions. The environmental issue caused by this ammonia-induced particle formation is not adequately addressed by current regulations, highlighting the need for increased attention to reduce its consequences.

This study looked at the emissions of ammonia (NH_3) and greenhouse gases (N_2O, CH_4, and CO_2) from an aerobic composting plant that uses municipal sewage sludge. It focused on seasonal variations in ambient temperature and aeration frequency. The study found that whereas ammonia and carbon dioxide emissions peaked at 86% and 74%, respectively, during the thermophilic phase, nitrous oxide and methane emissions peaked at 56% and 95% during the mesophilic phase (Han et al. 2018). Raising the temperature increased both ammonia and greenhouse gas emissions, while increasing the frequency of aeration reduced greenhouse gas emissions and raised ammonia emissions. Of the GHG gases, carbon dioxide and nitrous oxide had the biggest impact on CO_2 equivalent. The results indicated that keeping temperatures between −3 and 5 °C and increasing aeration frequencies were beneficial. It reduces ammonia emissions from composition and greenhouse gas emissions from sewage sludge. Carbon emissions can be significantly reduced by the electrocatalytic CO_2 reduction reaction. The calculation of catalysts with high activity and good stability has generated attention in the substantial gap between nanotechnology and industrial usage. The metal-organic structure serves as the electro-catalytic CO_2 reduction reaction, while the porous crystalline substance provides pyrolysates. They have regulated release charges, porosity, and skeletons that can be altered. The structural activity of these materials was thoroughly examined by theoretical computation and nanotechnology innovation. The element type, local environment, and microstructure are all included in the materials accountability factor (Wang et al. 2023).

3.4.3 Nanotechnology Strategies for Soil Pollution Remediation

Urbanization has numerous benefits, but there are also significant risks to the environment and public health from the release of heavy metals, organic and inorganic pollutants, personal care products, and pharmaceuticals. Effective solutions are desperately needed to protect the environment, even in the face of continuous efforts to limit the emission of dangerous compounds. With its many uses, nanotechnology has demonstrated tremendous potential in the treatment of water and soil contamination through nanoremediation. For example, iron nanoparticles have demonstrated exceptional efficacy in removing all hexavalent chromium. Numerous studies have examined the use of metal-, carbon-, and polymer-based nanomaterials for cleanup; however, most of them have not thoroughly examined the advantages and disadvantages of each material. With an emphasis on creative, eco-friendly techniques that employ green-synthesized nanomaterials, the benefits and drawbacks of various nanomaterials are thoroughly evaluated. Their main focus is to monitor the relationship between the activity and structure of covalent and organic metals and also pyro lysates. Furthermore, they described the essential factors the affect electrocatalytic CO_2 reduction reaction performance by designing functional porosity. Pollution was reduced because of the environmentally friendly green nanoparticles. However, there are obstacles and problems in using green nanotechnology widely (Sathish et al. 2024).

The researchers emphasize the limits of nanomaterials and concentrate on their applications. The long-term usage of nanomaterials in soil and water remediation, as well as their safety for the environment, requires further study. Numerous environmental issues are brought on by human activities, including contaminated soil and water, as well as air pollution that affects human health. The majority of such issues are brought on by traditional farming methods. The advancement of nanotechnology and its potential to detect and address environmental hazards with less energy and resources while handling these harmful substances. It learns about the various uses of nanomaterials that contain various kinds of nanoparticles, such as oxide-based and carbon-based nanoparticles (Ibrahim et al. 2016). Additionally, they recommended the advantages of nanotechnology above conventional therapeutic approaches The following techniques for applying nanomaterials to ecosystems are explained: (i) Soil as the transformative agent enhances the effectiveness of the process and phytoremediation. (ii) Using air filtration, degradation, and adsorption, nanoproducts were utilized to clean GHG gases, biofertilizer, and organic waste, causing volatilization. (iii) They used adsorption and photocatalytic membranes to eliminate pathogens and heavy metals from water.

Nanotechnology has become a vital tool in soil and groundwater remediation because it offers quick, simple, and efficient solutions to heavy metal and petroleum pollution contamination. This work investigates the application of various nanomaterials, such as carbon nanotubes, metallic and magnetic nanoparticles, and nanoscale zero-valent iron, in environmental cleanup projects. While carbon nanotubes have adsorption properties in the contamination of organic and inorganic nanomaterials, zero-valent iron has been the subject of extensive research due to its high reactivity and immobility. The advantages of metal and organic nanoparticles with unique properties include metal ion adsorption and magnetic isolation. As a result, they offer rapid, simple, and efficient solutions for heavy metal and petroleum contamination. Nanotechnology was able to alleviate their issues with soil contamination and water cleanup (Alazaiza et al. 2021). Their use of metal, carbon, magnetic, and zero-valent iron nanoparticles proved environmentally benign. Carbon nanotubes may adsorb both organic and inorganic contaminants, however, because of their high reactivity and immobile impurities, zero-valent iron requires further study. The efficient magnetic separation and adsorption of metal ions make metal and magnetic nanoparticles useful.

Nanomaterials were used to reduce soil contamination. Through the processes of oxidation/reduction and precipitation, polluting material was rendered immobile with the aid of nanoparticles. Certain core-shell configurations of nanoparticles will influence various immobilization procedures. For effective field applications, zero-valent iron nanoparticles have the ability to remove soil contaminants. In remediation experiments, other nanomaterials, notably other iron oxides and carbon-based compounds, have also shown promising outcomes, especially for soils contaminated with metals and metalloids (Baragaño et al. 2021). Recent research has explored merging nano-remediation with more traditional remediation methods, including soil cleansing and phytoremediation, even though its primary objective has been pollutant immobilization. Because it can produce new compounds, nano-remediation is a revolutionary approach to remediation technology.

The accumulation of metal and metallic oxide engineered nanoparticles in agricultural soils poses a significant threat to ecosystems and soil health; literature shows that engineered nanoparticles in soils can increase hydraulic conductivity and porosity while decreasing the distance between soil particles; they also change the pH, cation exchange capacity, electrical conductivity, redox potential, and amount of soil organic matter; the engineered nanoparticles or the metal cations they release may interact with nutrients like phosphorus to produce precipitates or complexes that reduce the bioavailability of the soil solution (Suazo-Hernández et al. 2023). These impacts depend on the kind of engineered nanoparticles, concentration, exposure time, dosage, and soil characteristics. Metal and metallic oxide-engineered nanoparticle deposition in soils needs to be continuously monitored and controlled to reduce their adverse environmental consequences.

3.5 Combining Advanced Technologies with Nanotechnology

This chapter looks at the potential future, challenges, applications, and economic impacts of nanotechnology and advanced materials. The interdisciplinary approach looks at how these technologies are incorporated into a number of industries, such as electronics, manufacturing, healthcare, and energy. Successful applications that show both financial gains and technological breakthroughs are highlighted (Gidiagba et al. 2023). Discussed are the ramifications of continuing research and development in advanced materials and nanotechnology, highlighting their significance for material science and the overall economy. The revolutionary potential of these technologies and their part in creating a creative and sustainable future are highlighted by this investigation.

The application of artificial intelligence technologies in nanotechnology research has grown within the last ten years, including the interpretation of scanning probe microscopy, biological nanosystems, nanoscale material property classification, theoretical simulations in nanoscience, and nanodevice design. There is also discussion of current developments and future directions in the creation of nanocomputing hardware to improve AI applications. Using novel computer architectures, data representations, hybrid technologies that combine biological beings and nanodevices, bioengineering, neurology, and related fields, the confluence of AI and nanotechnology can propel advances in the information sciences (Sacha and Varona 2013).

Global food security is threatened by population increase, climate change, conflicting land demands for biofuels, and deteriorating soil quality. Innovative methods and the integration of information from disciplines like materials science and informatics are needed to address these problems. Opportunities for sustainable food production are presented by the combination of artificial intelligence, nanotechnology, and precision agriculture. Optimizing nutrient targeting, absorption, delivery, and capture while evaluating long-term effects on soil microbial populations is possible by combining nano informatics with nutrient cycling and agricultural production models. Nanoscale agrochemicals techniques are created for the best functioning and safety measures (Zhang et al. 2021).

Because of the distinct physicochemical interactions that nanoscale materials have with their environment, toxicity assessment is crucial for the development of safer nanomedicines as well as for regulatory compliance. How artificial intelligence and machine learning aid in our understanding of nano-bio interactions from the perspectives of human health and the environment. Precision agriculture is aided by machine learning and artificial intelligence algorithms. Through machine learning-organized nanomedicine and nanotoxicology research, the dependability of data is growing, enabling rigorous quantitative comparison and shortening meta-analysis and modeling (Singh et al. 2020). These developments present both opportunities and difficulties, particularly with regard to machine learning applications. Clinical conversion of nanotherapeutics also made use of machine learning. The goal is to expand the use of machine learning and artificial intelligence in human health and agriculture. Nanomedicine and nanotoxicology were enhanced by artificial intelligence. Precision agriculture offers a variety of advantages by combining artificial intelligence and nanotechnology in the material sciences, medicine, and energy sectors. Artificial intelligence has accelerated the development, organization, and dissemination of nanotechnology, including nanomaterials and nanosystems. Artificial intelligence, for instance, was employed to administer pesticides as efficiently as possible. By improving the properties of materials, it creates nanosensors that are utilized in artificial intelligence and biological observation, as well as energy efficiency. However, there remain issues when applying the theoretical data set to the actual world. When using artificial intelligence in the field, there are several ethical issues to consider, including societal impact, data privacy, and algorithms, among others. Beneficial and logical connections of artificial intelligence are required. Nanotechnology highlights how crucial ethical development is to the creation of regulations and investor communication. Describing the technical and moral issues with artificial intelligence. Integration nanotechnology into society requires building public trust and cooperation in the field (Hassan et al. 2023).

Modern industries have been transformed by the advancement of nanotechnology. Products developed using nanotechnology now meet key needs in both pharmaceuticals and agriculture. They also create high-tech devices utilized in precision farming. Nanoproducts should be used in order to prevent dangers. They must be used with extreme caution. Customers utilized the quick integration of nanomaterials with conventional products, where it is necessary to draw attention to the restrictions of conventional items for safety purposes. Algorithms for artificial intelligence and machine learning were introduced as modeling techniques for the preliminary examination of nanoproducts (Ahmad, Mahmood, and Muhmood 2021). Combining machine learning and nanotechnology will help us better understand how biology and nanoparticles interact. In this study, machine learning and nanotechnology are recommended for safe use and risk management in accordance with the norms and regulations, eliminating the possibility of merging them. Nowadays, agriculture has been transformed by nanotechnology, which increases output while lessening its impact on the environment. For example, precision agriculture can use technologies like GIS, GPS, and nanotechnology to improve crop health, nutrient use efficiency, and environmental pollution (Fig. 3.3).

With the use of fertilizer and agrochemicals, nanoparticles' unique properties can help plants and soil that are lacking in nutrients. The ongoing use of nanosensors is to monitor crop and soil conditions. This agricultural innovation made it simple to identify soil deterioration and the best resistance that conventional farming produces (Garg, Rumjit, and Roy 2024). Despite this difficulty, precision farming, which has transformed agriculture and improved the quality and amount of food produced, benefits greatly from nanotechnology. These features of nanotechnology have modernized agriculture, and their cost-effective application techniques have guaranteed its widespread adoption. Although the decision to integrate nanotechnology with AI and machine learning was made using a useful tool like the Bayesian probability network, agricultural practices found this challenging because it required a large amount of data that allowed for prediction (Tari 1996). The Bayesan network was used in this study to analyze winter wheat. Using experimental data and a few assumptions to fill in the data gaps, this model predicts wheat yield by spraying using ADAS.

Krouk et al. (2013) stated that simulating and predicting how systems will respond to new situations or genetic disruptions is the aim of systems biology. This study looks at the analytical and experimental methods for identifying causal relationships in gene regulatory networks. The healthcare sector is one of the largest and fastest-growing in the world because of continuous innovation. Prior to advances in machine learning, the industry primarily relied on labor, which was inefficient and time-consuming. Artificial intelligence, a branch of computer science that aims to create machines that behave like people, is increasingly being used in the medical field (Kumar, Gadag, and Nayak 2020). This article discusses the current applications of artificial intelligence in drug research, disease diagnosis, clinical trials, remote patient monitoring, and nanotechnology. Future prospects of artificial intelligence and public opinions are also taken into consideration. The assessment's last section examines the possible advantages and disadvantages of artificial intelligence in healthcare (Fig. 3.3).

3.6 Conclusions, Future Trends, and Policy Perspectives

There are numerous uses for nanotechnology in agricultural productivity. To promote this field and allow for its widespread commercial use, a number of obstacles must be overcome. Among these difficulties are:

(a) Little knowledge of the mechanisms underlying interactions between plants, soil, and nanomaterials.
(b) Insufficient knowledge of how nanoparticles are absorbed and transported by plant organelles and vascular systems.
(c) There are not enough field-scale studies to assess nanomaterials' effectiveness in actual agricultural settings.
(d) Inadequate knowledge of the hazards to human health and environmental safety posed by the use of nanomaterials.
(e) There isn't enough research looking at how implementing nanotechnology in agriculture balances its financial advantages.

There is no denying nanotechnology's uses and advantages in agriculture. Numerous nano-systems and nano products are being created and marketed as a result of the advancements in this field of study, and they have the potential to significantly improve the agricultural value chain. Nonetheless, there is still not enough information available on risk assessment to enable well-informed choices on its application. Uncertainties have resulted from a lack of research, especially about consumer health and environmental safety. Important questions are still unresolved, including whether a regulatory framework exists to guarantee the safe use of nanoparticles in crop production, how these particles may affect beneficial soil microorganisms, and how much of these materials can build up in plant tissues. The applicability of risk assessment methodology for both terrestrial and aquatic animals, the sufficiency of current toxicity assessment methods, and the availability of trustworthy techniques to detect nanomaterials in plants and food products all need more research. Human consumption is growing daily, but there is a dosage requirement that presents a real problem.

Comparing nanotechnology to other scientific fields, it is still in its infancy and has not gained widespread acceptance. This is largely due to health-related issues and a lack of regulations needed for broader adoption. Furthermore, the conflicting guidelines in the research led to additional problems with the use of nanoparticles in precision farming. These problems can be resolved by creating fair regulations to lessen health problems and create a seamless framework for using nanotechnology in agriculture.

Similar to the need to protect human health, safe soil microbes that improve soil fertility and aid in nutrient cycling are also necessary. Furthermore, in order to overcome these gaps and obtain benefits and lower environmental dangers when using nanotechnology in agriculture, a thorough investigation of the aggregation and modification of nanoparticles with plants was required. Strict laws and guidelines were desperately needed to mitigate the risks posed by nanotechnology. Establishing a well-equipped institution with scientists who specialize in nanotechnology and can assist with the biosafety calculations of nanomaterials is one way to lessen the use of nanotechnology in precision agriculture.

I. To advance scientific understanding, these institutional personnel collaborate with teams of chemical and biological researchers.
II. Certain criteria that ensure the use of nano-based products utilized by food authorities have been carefully examined and are consistent with the Food and Agriculture Organizations or World Health Organizations standards.
III. Nanomaterials' toxicity to marine life was identified, investigated, and made public.
Before using materials based on nanotechnology, safety precautions should be taken. For these safety evaluations, the sample size was crucial.
IV. Increasing the use of nanotechnology in agriculture could help farmers improve their economic structure and make farming more cost-effective. This could be made feasible by the private and public sectors contributing financial and scientific resources.
V. A competent and trained individual will be needed to address and resolve nanotechnology field issues.

References

Abd-Elgawad MM (2024) Revolutionizing nematode management–nanomaterials as a promising Aapproach for managing economically important plant-parasitic nematodes: current knowledge and future challenges. Nanotechnol Plant Dis Manag:23–44

Acharya P, Jayaprakasha GK, Crosby KM, Jifon JL, Patil BS (2020) Nanoparticle-mediated seed priming improves germination, growth, yield, and quality of watermelons (Citrullus lanatus) at multi-locations in Texas. Sci Rep 10:5037

Adeel M, Zain M, Shakoor N, Ahmad MA, Azeem I, Aziz MA, Tulcan RXS, Rathore A, Tahir M, Horton R, Xu M, Yukui R (2023) Global navigation of lithium in water bodies and emerging human health crisis. npj Clean Water 6:33

Ahmad F, Mahmood A, Muhmood T (2021) Machine learning-integrated omics for the risk and safety assessment of nanomaterials. Biomater Sci 9:1598–1608

Ahmed F, Dwivedi S, Shaalan NM, Kumar S, Arshi N, Alshoaibi A, Husain FM (2020) Development of selenium nanoparticle based agriculture sensor for heavy metal toxicity detection. Agriculture 10:610

Akintuyi OB (2024) Vertical farming in urban environments: a review of architectural integration and food security. Open Access Res J Biol Pharm 10:114–126

Alazaiza MY, Albahnasawi A, Ali GA, Bashir MJ, Copty NK, Amr SSA, Abushammala MF, Al Maskari T (2021) Recent advances of nanoremediation technologies for soil and groundwater remediation: a review. Water 13:2186

Alves S, Gonçalves M, Monteiro H, Moura B, Godina R, Almeida J (2022) Life cycle assessment of nanotechnology: carbon footprint and energy analysis. In: International Conference on Energy and Environment Research. Springer, pp 233–247

Amanatidis S, Ntziachristos L, Giechaskiel B, Bergmann A, Samaras Z (2014) Impact of selective catalytic reduction on exhaust particle formation over excess ammonia events. Environ Sci Technol 48:11527–11534

Amir A, Bibi S, Nawaz A, Ibrahim A, Ashiq MN, Saeed MQ, Hamza M, Rasheed HMZ, Islam B (2021) In vitro antibacterial response of ZnO-MgO nanocomposites at various compositions. Int J Appl Ceram Technol 18:1417–1429

Amjad S, Serajuddin M (2022) Role of nanosensors and bionanosensors in crop abiotic stress. In: Augmenting crop productivity in stress environment. Springer

Antonacci A, Arduini F, Moscone D, Palleschi G, Scognamiglio V (2018) Nanostructured (Bio) sensors for smart agriculture. TrAC Trends Anal Chem 98:95–103

Asghar R, Kazerooni EA. Current trends in the use of Controlled-Release Fertilizers for sustainable agriculture

Ashwini M, Gajera H, Hirpara DG, Savaliya DD, Kandoliya U (2024) Comparative impact of seed priming with zinc oxide nanoparticles and zinc sulphate on biocompatibility, zinc uptake, germination, seedling vitality, and antioxidant modulation in groundnut. J Nanopart Res 26:235

Askary M, Amirjani MR, Saberi T (2017) Comparison of the effects of nano-iron fertilizer with iron-chelate on growth parameters and some biochemical properties of *Catharanthus roseus*. J Plant Nutr 40:974–982

Baldwin EA, Hagenmaier R, Bai J (2011) Edible coatings and films to improve food quality. CRC Press

Banerjee J, Kole C (2016) Plant nanotechnology: an overview on concepts, strategies, and tools. In: Plant nanotechnology: principles and practices, pp 1–14

Baragaño D, Forján R, Sierra C, Gallego JLR (2021) 'Nanomaterials for soil remediation: Pollutant immobilization and opportunities for hybrid technologies. In: Sorbents materials for controlling environmental pollution. Elsevier

Bindraban PS, Dimkpa C, Nagarajan L, Roy A, Rabbinge R (2015) Revisiting fertilisers and fertilisation strategies for improved nutrient uptake by plants. Biol Fertil Soils 51:897–911

Braiek M, Rokbani KB, Chrouda A, Mrabet B, Bakhrouf A, Maaref A, Jaffrezic-Renault NJB (2012) An electrochemical immunosensor for detection of Staphylococcus aureus bacteria based on immobilization of antibodies on self-assembled monolayers-functionalized gold electrode. Biosensors 2:417–426

Campbell BM, Thornton P, Zougmoré R, Van Asten P, Lipper L (2014) Sustainable intensification: what is its role in climate smart agriculture? Curr Opin Environ Sustain 8:39–43

Chandra MS, Karthik R, Kumar MS, Lokesh C, Nthebere K, Kumar NV, Raising LP (2024) Empowering agriculture: nanobiotechnological crop fortification in the fight against malnutrition. In: Sustainable agriculture: nanotechnology, biotechnology, management and food security, p 249

Chauhan N, Urooj S (2019) The effect of Azadirachta Indica extract on the soil nutrients and the NPK value determination by electrochemical sensor. Int. J Biosens Bioelectron 5

Chen YW, Lee HV, Juan JC, Phang S-M (2016) Production of new cellulose nanomaterial from red algae marine biomass Gelidium elegans. Carbohydr Polym 151:1210–1219

Chhipa H (2017) Nanofertilizers and nanopesticides for agriculture. Environ Chem Lett 15:15–22

Dar FA, Qazi G, Pirzadah TB (2020) Nano-biosensors: NextGen diagnostic tools in agriculture. In: Nanobiotechnology in agriculture: an approach towards sustainability, pp 129–144

Delfani M, Baradarn Firouzabadi M, Farrokhi N, Makarian H (2014) Some physiological responses of black-eyed pea to iron and magnesium nanofertilizers. Commun Soil Sci Plant Anal 45:530–540

Dissanayake W, de Silva M, Munaweera I, Karunaratne V, Kottegoda N (2024) Nanotechnology-based urea delivery systems as climate-smart fertilizers: current status and beyond. In: Nanofertilizer Delivery, Effects and Application Methods, pp 71–95

Duhan JS, Kumar R, Kumar N, Kaur P, Nehra K, Duhan S (2017) Nanotechnology: the new perspective in precision agriculture. Biotechnol Reports 15:11–23

DUTTA P) Using green engineered nanoparticles for plant health management

Elizabath A, Babychan M, Mathew AM, Syriac GM (2019) Application of nanotechnology in agriculture. Indian J Pure Appl Biosci 7:131–139

Emadian SF (1988) Physiological responses of loblolly pine (Pinus taeda L.) to silicon and water stress. Texas A&M University

Falina S, Syamsul M, Rhaffor NA, Sal Hamid S, Mohamed Zain KA, Abd Manaf A, Kawarada H (2021) Ten years progress of electrical detection of heavy metal ions (hmis) using various field-effect transistor (fet) nanosensors: a review. Biosensors 11:478

Fatima F, Pathak N, Verma SR, Bajpai P (2018) Toxicity and immunomodulatory efficacy of biosynthesized silver myconanosomes on pathogenic microbes and macrophage cells. Artif Cells Nanomed Biotechnol Reports 46:1637–1645

Feijoo S, González-García S, Moldes-Diz Y, Vázquez-Vázquez C, Feijoo G, Moreira MT (2018) 'The environmental impact of magnetic nanoparticles under the perspective of carbon footprint. In: Environmental carbon footprints. Elsevier

Gammoudi I, Raimbault V, Tarbague H, Moroté F, Grauby-Heywang C, Othmane A, Kalfat R, Moynet D, Rebière D, Dejous C (2014) Enhanced bio-inspired microsensor based on microfluidic/bacteria/love wave hybrid structure for continuous control of heavy metals toxicity in liquid medium. Sensors Actuators B Chem 198:278–284

Garg S, Rumjit NP, Roy S (2024) Smart agriculture and nanotechnology: technology, challenges, and new perspective. Adv Agrochem 3:115–125

Gidiagba JO, Daraojimba C, Ofonagoro KA, Eyo-Udo NL, Egbokhaebho BA, Ogunjobi OA, Banso AA (2023) Economic impacts and innovations in materials science: a holistic exploration of nanotechnology and advanced materials. Eng Sci Technol J 4:84–100

Giraldo JP, Kruss S (2023) Nanosensors for monitoring plant health. Nat Nanotechnol:1–2

Goyal D, Yadav A, Prasad M, Singh TB, Shrivastav P, Ali A, Dantu PK, Mishra S (2020) Effect of heavy metals on plant growth: an overview. In: Contaminants in agriculture: sources, impacts management, pp 79–101

Goyal V, Rani D, Ritika MS, Deng C, Wang Y (2023) Unlocking the potential of nano-enabled precision agriculture for efficient and sustainable farming. Plants 12:3744

Han Z, Sun D, Wang H, Li R, Bao Z, Qi F (2018) Effects of ambient temperature and aeration frequency on emissions of ammonia and greenhouse gases from a sewage sludge aerobic composting plant. Bioresour Technol 270:457–466

Hassan S, Almaliki M, Hussein ZA, Albehadili HM, Banoon SR, Al-Abboodi A, Al-Saady M (2023) Development of nanotechnology by artificial intelligence: a comprehensive review. J Nanostruct 13:915–932

He X, Deng H, Hwang H-m (2019) The current application of nanotechnology in food and agriculture. J Food Drug Anal 27:1–21

Hu W, Wan L, Jian Y, Ren C, Jin K, Su X, Bai X, Haick H, Yao M, Wu W (2019) Electronic noses: from advanced materials to sensors aided with data processing. Adv Mater Technol 4:1800488

Hu X, Lu L, Guo Z, Zhu Z (2020) Volatile compounds, affecting factors and evaluation methods for rice aroma: a review. Trends Food Sci Technol 97:136–146

Huang M, Wang Z, Luo L, Wang S, Hui X, He G, Cao H, Ma X, Huang T, Zhao Y (2017) Soil testing at harvest to enhance productivity and reduce nitrate residues in dryland wheat production. Field Crop Res 212:153–164

Huang B, Chen F, Shen Y, Qian K, Wang Y, Sun C, Zhao X, Cui B, Gao F, Zeng Z (2018) Advances in targeted pesticides with environmentally responsive controlled release by nanotechnology. Nano 8:102

Ibrahim RK, Hayyan M, AlSaadi MA, Hayyan A, Ibrahim S (2016) Environmental application of nanotechnology: air, soil, and water. Environ Sci Pollut Res 23:13754–13788

Igiebor FA, Ohanmu EO, Omoregie GO, Jennifer OO, Musa S, Ejale D, Okojie N, Afe R, Ivbobie S, Ikhajiagbe B (2024) Nanobioherbicides and plant growth hormone synthesis and stress-mediated hormones. Handb Agric Biotechnol 2:125–149

Jahanbakhshi S, Rezaei M, Memarizadeh N, Hosseini M (2024) Design of biocompatible nanoformulation of dimethoate for pest control in the field condition. Int J Environ Sci Technol:1–18

Junaid M, Gokce A (2024) Global agricultural losses and their causes. Bull Biol Allied Sci Res 2024:66–66

Kaushal M, Wani SP (2017) Nanosensors: frontiers in precision agriculture. In: Nanotechnology: an agricultural paradigm, pp 279–291

Krouk G, Lingeman J, Colon AM, Coruzzi G, Shasha D (2013) Gene regulatory networks in plants: learning causality from time and perturbation. Genome Biol 14:1–7

Kumar V, Yadav SK (2009) Plant-mediated synthesis of silver and gold nanoparticles and their applications. J Chem Technol Biotechnol 84:151–157

Kumar A, Gadag S, Nayak UY (2020) The beginning of a new era: artificial intelligence in healthcare. Adv Pharma Bull 11:414

Kumari M, Gupta V, Kumar N, Arun RK (2024) Microfluidics-based nanobiosensors for healthcare monitoring. Mol Biotechnol 66:378–401

Li C, Li Y, Li Y, Fu G (2018) Cultivation techniques and nutrient management strategies to improve productivity of rain-fed maize in semi-arid regions. Agric Water Manag 210:149–157

Long F, Zhu A, Shi H, Wang H, Liu J (2013) Rapid on-site/in-situ detection of heavy metal ions in environmental water using a structure-switching DNA optical biosensor. Sci Rep 3:1–7

Mahmood I, Ansari RA, Rizvi R (2024) Nanotechnology and plant disease management. CRC Press

Mahmoud AED, Fawzy M (2021) Nanosensors and nanobiosensors for monitoring the environmental pollutants. In: Waste recycling technologies for nanomaterials manufacturing, pp 229–246

Mali SC, Raj S, Trivedi R (2020) Nanotechnology a novel approach to enhance crop productivity. Biochem Biophys Reports 24:100821

Manjaiah KM, Mukhopadhyay R, Paul R, Datta SC, Kumararaja P, Sarkar B (2019) Clay minerals and zeolites for environmentally sustainable agriculture. In: Modified clay and zeolite nanocomposite materials. Elsevier

Mariano M, El Kissi N, Dufresne A (2014) Cellulose nanocrystals and related nanocomposites: review of some properties and challenges. J Polym Sci B Polym Phys 52:791–806

Marimuthu S, Swathika N, Srivignesh S, Lakshmanan A (2023) Scope of nanotechnology in allelopathic weed management. Allelopath J 58:89

Moghaddasi S, Fotovat A, Khoshgoftarmanesh AH, Karimzadeh F, Khazaei HR, Khorassani R (2017) Bioavailability of coated and uncoated ZnO nanoparticles to cucumber in soil with or without organic matter. Ecotoxicol Environ Saf 144:543–551

Mondal R, Dam P, Chakraborty J, Paret ML, Katı A, Altuntas S, Sarkar R, Ghorai S, Gangopadhyay D, Mandal AK (2022) Potential of nanobiosensor in sustainable agriculture: the state-of-art Heliyon:e12207

Mukhopadhyay SS (2014) Nanotechnology in agriculture: prospects and constraints. Nanotechnol Sci Appl 7:63–71

Nwokediegwu ZQS, Adefemi A, Ayorinde OB, Ilojianya VI, Etukudoh EA (2024) Review of water policy and management: comparing the USA and Africa. Eng Sci Technol J 5:402–411

Prasad R, Bhattacharyya A, Nguyen QD (2017) Nanotechnology in sustainable agriculture: recent developments, challenges, and perspectives. Front Microbiol 8:1014

Pudake RN, Jain U, Kole C (2021) Biosensors in agriculture: recent trends and future perspectives. Springer

Ramesh M, Palanisamy K, Kumar N (2014) Effects of bulk & nano-titanium dioxide and zinc oxide on physio-morphological changes in Triticum Aestivum Linn. J Global Biosci 3:415–422

Raza MAS, Amin J, Valipour M, Iqbal R, Aslam MU, Zulfiqar B, Muhammad F, Ibrahim MA, Al-Ghamdi AA, Elshikh MS (2024) Cu-nanoparticles enhance the sustainable growth and yield of drought-subjected wheat through physiological progress. Sci Rep 14:14254

Rizwan M, Ali S, ur Rehman MZ, Malik S, Adrees M, Qayyum MF, Alamri SA, Alyemeni MN, Ahmad P (2019) Effect of foliar applications of silicon and titanium dioxide nanoparticles on growth, oxidative stress, and cadmium accumulation by rice (*Oryza sativa*). Acta Physiol Plant 41:35

Rodrigues AS, Batista JG, Rodrigues MÁ, Thipe VC, Minarini LA, Lopes PS, Lugão AB (2024) Advances in silver nanoparticles: a comprehensive review on their potential as antimicrobial agents and their mechanisms of action elucidated by proteomics. Front Microbiol 15:1440065

Roy E, Patra S, Madhuri R, Sharma PKJRA (2014) Simultaneous determination of heavy metals in biological samples by a multiple-template imprinting technique: an electrochemical study. RSC Adv 4:56690–56700

Ryu H, Thompson D, Huang Y, Li B, Lei YJS (2020) Electrochemical sensors for nitrogen species: a review. Sens Actuators Reports 2:100022

Sacha GM, Varona P (2013) Artificial intelligence in nanotechnology. Nanotechnology 24:452002

Šarauskis E, Kazlauskas M, Naujokienė V, Bručienė I, Steponavičius D, Romaneckas K, Jasinskas A (2022) Variable rate seeding in precision agriculture: recent advances and future perspectives. Agriculture 12:305

Sarika AA, Meena R, Mina U, Shukla A, Sharma A (2023) Adoption of the green energy technology for the mitigation of greenhouse gas emission: Embracing the goals of the Paris agreement. In: Nanomaterials and nanoliquids: applications in energy and environment. Springer

Saritha GNG, Anju T, Kumar A (2022) Nanotechnology-big impact: how nanotechnology is changing the future of agriculture? J Agric Food Res 10:100457

Sathish T, Ahalya N, Thirunavukkarasu M, Senthil T, Hussain Z, Siddiqui MIH, Panchal H, Sadasivuni KK (2024) A comprehensive review on the novel approaches using nanomaterials for the remediation of soil and water pollution. Alex Eng J 86:373–385

Schulte PA, Salamanca-Buentello F (2007) Ethical and scientific issues of nanotechnology in the workplace. Ciênc Saúde Colet 12:1319–1332

Scrinis G, Lyons K (2007) The emerging nano-corporate paradigm: nanotechnology and the transformation of nature, food and Agri-food systems. Int J Sociol Agric Food 15:22–44

Sekhon BSJN, science (2014) Nanotechnology in agri-food production: an overview. Nanotechnol Sci Appl: 31–53, 7

Shabaninejad Z, Yousefi F, Movahedpour A, Ghasemi Y, Dokanehiifard S, Rezaei S, Aryan R, Savardashtaki A, Mirzaei H (2019) Electrochemical-based biosensors for microRNA detection: nanotechnology comes into view. Anal Biochem 581:113349

Shafi U, Mumtaz R, García-Nieto J, Hassan SA, Zaidi SAR, Iqbal N (2019) Precision agriculture techniques and practices: from considerations to applications. Sensors 19:3796

Shakoor N, Adeel M, Zain M, Zhang P, Ahmad MA, Farooq T, Zhou P, Azeem I, Rizwan M, Guo K, Jilani G, Ahmar S, Maqbool S, Ming X, Rui Y (2022) Exposure of cherry radish (*Raphanus sativus* L. var. *Radculus Pers*) to iron-based nanoparticles enhances its nutritional quality by trigging the essential elements. NanoImpact 25:100388

Shakoor N, Adeel M, Ahmad MA, Hussain M, Azeem I, Zain M, Zhou P, Li Y, Xu M, Rui Y (2023a) Environment relevant concentrations of lithium influence soybean development via metabolic reprogramming. J Hazard Mater 441:129898

Shakoor N, Hussain M, Adeel M, Azeem I, Ahmad MA, Zain M, Zhang P, Li Y, Quanlong W, Horton R, Rui Y (2023b) Lithium-induced alterations in soybean nodulation and nitrogen fixation through multifunctional mechanisms. Sci Total Environ 904:166438

Sharma P, Pandey V, Sharma MMM, Patra A, Singh B, Mehta S, Husen A (2021) A review on biosensors and nanosensors application in agroecosystems. Nanoscale Res Lett 16:1–24

Shojaei TR, Salleh MAM, Tabatabaei M, Mobli H, Aghbashlo M, Rashid SA, Tan T (2019) Applications of nanotechnology and carbon nanoparticles in agriculture. In: Synthesis, technology and applications of carbon nanomaterials. Elsevier

Singh AV, Rosenkranz D, Ansari MHD, Singh R, Kanase A, Singh SP, Johnston B, Tentschert J, Laux P, Luch A (2020) Artificial intelligence and machine learning empower advanced biomedical material design to toxicity prediction. Adv Intell Syst 2:2000084

Singh N, Kumar B, Usman UL, Susan MABH (2024a) Nano revolution: exploring the frontiers of nanomaterials in science, technology, and society. Nano-Struct Nano-Objects 39:101299

Singh S, Singh R, Singh K, Katoch K, Zaeen AA, Birhan DA, Singh A, Sandhu HS, Singh H, Sahrma LK (2024b) Smart fertilizer technologies: An environmental impact assessment for sustainable agriculture. Smart Agric Technol:100504

Siontorou CG, Georgopoulos KN (2016) A biosensor platform for soil management: the case of nitrites. J Clean Prod 111:133–142

Solomon NO, Simpa P, Adenekan OA, Obasi SC (2024) Sustainable nanomaterials' role in green supply chains and environmental sustainability. Eng Sci Technol J 5:1678–1694

Srivastava AK, Dev A, Karmakar S (2018) Nanosensors and nanobiosensors in food and agriculture. Environ Chem Lett 16:161–182

Su Y, Zhou X, Meng H, Xia T, Liu H, Rolshausen P, Roper C, McLean JE, Zhang Y, Keller AA (2022) Cost–benefit analysis of nanofertilizers and nanopesticides emphasizes the need to improve the efficiency of nanoformulations for widescale adoption. Nat food 3:1020–1030

Su N, Ronga X, Xie G, Chang T, Zhang Y, Peng J, Luo G (2024) Effectiveness of a 10-year continuous reduction of controlled-release nitrogen fertilizer on production, nitrogen loss and utilization of double-cropping rice. Sci Total Environ 912:168857

Suresh S, Karthikeyan S, Jayamoorthy K (2016) Effect of bulk and nano-Fe_2O_3 particles on peanut plant leaves studied by Fourier transform infrared spectral studies. J Adv Res 7:739–747

Surya SG, Yuvaraja S, Varrla E, Baghini MS, Palaparthy VS, Salama KN (2020) An in-field integrated capacitive sensor for rapid detection and quantification of soil moisture. Sensors Actuators B Chem 321:128542

Tan J, Xu J (2020) Applications of electronic nose (e-nose) and electronic tongue (e-tongue) in food quality-related properties determination: a review. Artif Intell Agric 4:104–115

Tari F (1996) A Bayesian network for predicting yield response of winter wheat to fungicide programmes. Comput Electron Agric 15:111–121

Thakur S, Thakur S, Kumar R (2018) Bio-nanotechnology and its role in agriculture and food industry. J Mol Genet Med 12:1747–0862

Thatai S, Khurana P, Prasad S, Kumar D (2014) A new way in nanosensors: gold nanorods for sensing of Fe (III) ions in aqueous media. Microchem J 113:77–82

Ti C, Xia L, Chang SX, Yan X (2019) Potential for mitigating global agricultural ammonia emission: a meta-analysis. Environ Pollut 245:141–148

Trippa D, Scalenghe R, Basso MF, Panno S, Davino S, Morone C, Giovino A, Oufensou S, Luchi N, Yousefi S (2024) Next-generation methods for early disease detection in crops. Pest Manag Sci 80:245–261

Umapathi R, Ghoreishian SM, Sonwal S, Rani GM, Huh YS (2022) Portable electrochemical sensing methodologies for on-site detection of pesticide residues in fruits and vegetables. Coord Chem Rev 453:214305

Wang X, Cheng M, Yang Q, Wei H, Xia A, Wang L, Ben Y, Zhou Q, Yang Z, Huang X (2019) A living plant cell-based biosensor for real-time monitoring invisible damage of plant cells under heavy metal stress. Sci Total Environ 697:134097

Wang C, Lv Z, Yang W, Feng X, Wang B (2023) A rational design of functional porous frameworks for electrocatalytic CO 2 reduction reaction. Chem Soc Rev 52:1382–1427

Wong LS, Wong CS (2014) A new method for heavy metals and aluminium detection using biopolymer-based optical biosensor. IEEE Sensors J 15:471–475

Wu H, Nißler R, Morris V, Herrmann N, Hu P, Jeon S-J, Kruss S, Giraldo JP (2020) Monitoring plant health with near-infrared fluorescent H_2O_2 nanosensors. Nano Lett 20:2432–2442
Yadav S (2010) Heavy metals toxicity in plants: an overview on the role of glutathione and phytochelatins in heavy metal stress tolerance of plants. S Afr J Bot 76:167–179
Yang G, Zhao FJB (2015) Electrochemical sensor for chloramphenicol based on novel multiwalled carbon nanotubes@ molecularly imprinted polymer. Biosens Bioelectron 64:416–422
Zhang H, Zhang G, Xu J, Wen Y, Lu B, Zhang J, Ding WJS (2016) Novel highly selective fluorescent sensor based on electrosynthesized poly (9-fluorenecarboxylic acid) for efficient and practical detection of iron (III) and its agricultural application. Sens Actuators B 230:123–129
Zhang P, Guo Z, Ullah S, Melagraki G, Afantitis A, Lynch I (2021) Nanotechnology and artificial intelligence to enable sustainable and precision agriculture. Nat Plants 7:864–876
Zheng Z, Zhou Y, Li X, Liu S, Tang Z (2011) Highly-sensitive organophosphorous pesticide biosensors based on nanostructured films of acetylcholinesterase and CdTe quantum dots. Biosens Bioelectron 26:3081–3085

Nanotechnology for Insect Pest Management

4

Al-kazafy Hassan Sabry and Zakaria Abdelhalim Attia

Abstract

Conventional pesticide formulations exert significant adverse effects on environmental quality, including soil degradation, groundwater contamination, and the accumulation of toxic residues in agricultural produce. Nanopesticide formulations have emerged as a promising alternative for mitigating these impacts through enhanced delivery efficiency, controlled release, and targeted action, thereby reducing the overall concentration of active ingredients required for effective pest control. By improving bioavailability and reducing off-target losses, nanopesticides can substantially minimize pesticide contamination of water resources and lower residual levels in crops, vegetables, and fruits. Furthermore, the application of nanotechnology in pesticide production has the potential to decrease manufacturing and application costs. Nanopesticides include different types, such as nanoinsecticides, nanoherbicides, nanofungicides, and nanoacaricides. Current global pesticide consumption exceeds 3.53 million metric tons annually in conventional formulations; adoption of nanopesticide technology could potentially reduce this volume by more than 80%. This chapter has its focus on the functional roles and applications of nanoinsecticides, nanoherbicides, and nanofungicides within integrated pest management systems.

Keywords

Agricultural development · Nanopesticides · Plant production · Plant protection

A.-k. H. Sabry (✉) · Z. A. Attia
Department of Pests and Plant Protection, National Research Centre, Cairo, Egypt

S. S. Purewal et al. (eds.), *Agri-Nanotechnology: Innovations for Sustainable Agriculture and Environmental Restoration*,
https://doi.org/10.1007/978-981-96-9756-4_4

4.1 Introduction

In recent years, nanotechnology has become an important tool that supports progress in many fields, including medicine, manufacturing, and agriculture. Agriculture, which plays a key role in global food security, has gained many benefits from nanotechnology. These include making nanofertilizers and nanopesticides that improve nutrient delivery and make pest control more effective. Conventional pesticide formulations necessitate high application rates to achieve adequate pest suppression, thereby contributing to environmental contamination, accumulation of persistent chemical residues, and elevated production expenditures (Shrone et al. 2010). In contrast, nanopesticide formulations facilitate a reduction in the quantity of active ingredients required by up to threefold through improved physicochemical stability, controlled-release kinetics, and site-specific delivery mechanisms, while simultaneously enhancing pesticidal efficacy (Huang et al. 2015). The integration of nanotechnology into agrochemical development is accelerating globally. In 2021, the international agrochemical market registered 230 nanotechnology-based products spanning 37 distinct categories, commercialized by 75 corporations across 26 nations. Nanotechnology-based products find diverse applications across animal husbandry, fertilizers, pesticides, and plant breeding (Zobir et al. 2021; Wang et al. 2022). The present chapter explains the contribution of nanopesticide technologies to pest control, with particular focus on their incorporation into sustainable Integrated Pest Management (IPM) strategies.

4.2 Nanopesticide Market

The global nanopesticide market includes various types, such as nanoinsecticides, nanoherbicides, nanofungicides, miticides, molluscicides, nematicides, rodenticides, and virucides.

According to the Food and Agriculture Organization (FAO), the valuation of this market was approximately USD 0.6 billion in 2021, with projections indicating expansion to USD 1.9 billion by 2030 (Fig. 4.1). Market segmentation data reveal that nanoinsecticides constitute the predominant share at 44%, followed by nanofungicides (21%) and nanoherbicides (12%). The residual 23% comprises other nanopesticide categories, including miticides, molluscicides, nematicides, rodenticides, and virucides (Fig. 4.2).

4.3 Nanoinsecticides and Pest Management

Nanoinsecticide formulations should exhibit four key traits: rapid absorption, rapid distribution, a large surface area, and rapid excretion. These characteristics enhance their effectiveness compared to conventional formulations. Nanoinsecticides are classified into two main categories: inorganic nanoinsecticides and organic nanoinsecticides.

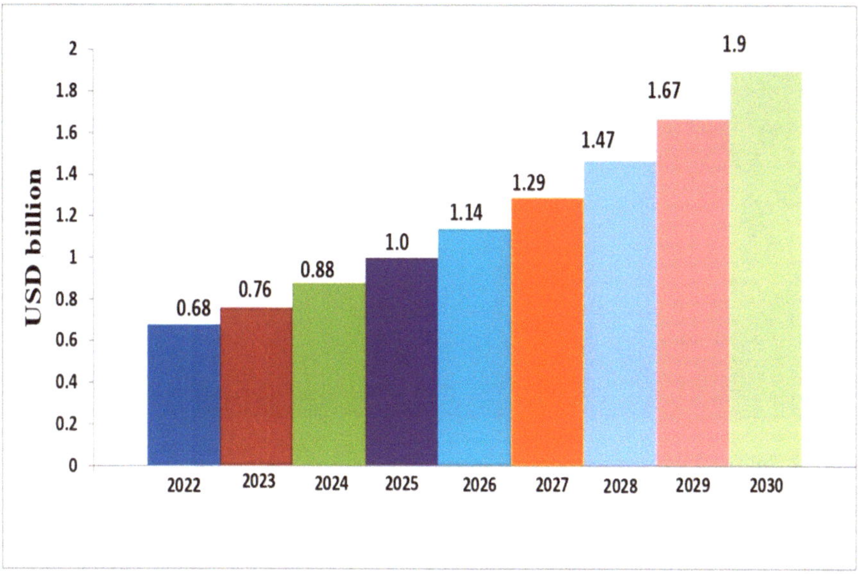

Fig. 4.1 Global nanopesticides market size

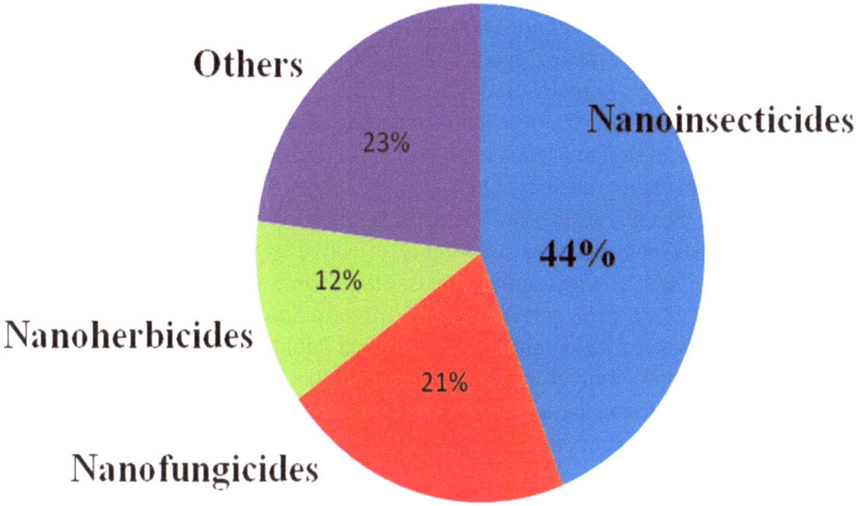

Fig. 4.2 Nanopesticides market by types

Fig. 4.3 Silica dioxide powder

4.3.1 Inorganic Nanoinsecticides

4.3.1.1 Role of Silica Nanoparticles in Insect Pest Management

Nanosilica functions not only as an effective insecticide against specific insect pests but also as a synergistic agent for various organic insecticides, enhancing plant resistance to insect attacks (Saw et al. 2023). In insect control, silica is typically utilized in three forms: silicon dioxide powder (SiO_2) (Fig. 4.3), mineral silicates, and alumino-silicates (Saw et al. 2023). The advantages of nanosilica in pest management include high efficacy against target insect species, low toxicity to non-target organisms, high porosity, and a large surface area (500–700 m^2/g) (Sharma et al. 2022). Common sources of silica include agricultural by-products such as rice husk, corn cobs, and sugarcane bagasse (Sharma et al. 2022). The insecticidal action of nanosilica is primarily attributed to two mechanisms: blocking of insect spiracles and abrasion of the cuticle surface (Shoaib et al. 2018). Thabet et al. (2021) demonstrated that silica nanoparticles are highly effective against several agricultural pests, including the cowpea aphid *Aphis craccivora*, the American serpentine leafminer *Liriomyza trifolii*, and the cotton leafworm *Spodoptera littoralis*. In addition, nanosilica has shown potential in controlling public health pests such as mosquitoes. For example, Baz et al. (2022) reported that silica nanoparticles caused 100% mortality in first instar larvae of Culex pipiens and 81.7% mortality in fourth instar larvae. El-Naggar et al. (2020) further observed that silica nanoparticles are effective not only against stored product pests such as *Sitophilus oryzae*, *Rhyzopertha dominica*, *Tribolium castaneum*, and *Oryzaephilus surinamensis* but also function as efficient fertilizers, thereby providing dual benefits in agricultural systems.

4.3.1.2 Role of Silver Nanoparticles in Insect Pest Management

Silver nanoparticles (AgNPs) are recognized as among the most potent nanomaterials with broad applications in medicine, pharmaceuticals, and agriculture (Fig. 4.4). Their

Fig. 4.4 Silver nanoparticles

insecticidal action is primarily attributed to the inhibition of protein synthesis in treated organisms. Recently, green synthesis methods have enabled the production of AgNPs from natural plant sources. Rehman et al. (2021) synthesized AgNPs from *Camelina sativa* leaves and evaluated their efficacy against *Sitophilus granarius* and *Oryzaephilus surinamensis*. Zavala-Zapata et al. (2024) produced green AgNPs and applied them against Diaphorina citri under laboratory and greenhouse conditions, achieving 100% and 80% efficacy, respectively. Similarly, El-Ashmouny et al. (2022) reported high effectiveness of AgNPs against the cotton leafworm (*Spodoptera littoralis*), while Bhadani et al. (2022) demonstrated their efficacy against whitefly. Santos et al. (2021) synthesized AgNPs using entomopathogenic fungi, and Santhoshkumar et al. (2024) synthesized them from *Cissus quadrangularis* stem, subsequently testing them against the cowpea beetle (*Callosobruchus maculatus*). In this case, AgNPs were found to be threefold more effective than the crude plant extract, with LC_{50} values of 1.87 mg/L compared to 5.22 mg/L for the extract. AgNPs have also been employed against public health insect vectors such as mosquitoes. Wang et al. (2024a, b) evaluated their use in mosquito control, while Madhiyazhagan et al. (2015) demonstrated that AgNPs were highly effective against larvae and pupae of *Aedes aegypti*, *Anopheles stephensi*, and *Culex quinquefasciatus*. Furthermore, they exhibited strong ovicidal activity, with complete inhibition of egg hatchability (0% hatch rate) at a concentration of 30 ppm for all three mosquito species.

Fig. 4.5 Titanium dioxide nanoparticles

4.3.1.3 Role of Titanium Dioxide Nanoparticles in Insect Pest Management

Titanium dioxide (TiO_2) is a highly versatile material with broad applications in food, medicine, and agriculture. In the food industry, TiO_2 is commonly employed as a whitening agent to enhance the color and brightness of products (Fig. 4.5). Narayanan et al. (2021) synthesized TiO_2 nanoparticles from Pouteria campechiana leaves and evaluated their insecticidal activity against the larvae and pupae of Aedes aegypti, demonstrating significant efficacy. Thus, titanium dioxide is effective not only against agricultural insect pests but also against public health vectors. Gutiérrez-Ramírez et al. (2024) assessed the activity of TiO_2 nanoparticles against the American cockroach (*Periplaneta americana*), while Gutiérrez-Ramírez et al. (2021) reported 99% mortality of second-instar nymphs of *Bactericera cockerelli* (Hemiptera: Triozidae) following TiO_2 nanoparticle exposure. Ishaq et al. (2020) demonstrated the effectiveness of TiO_2 nanoparticles against two stored grain pests, *Tribolium castaneum* and *Trogoderma granarium*. Nasser et al. (2024) found that TiO_2 nanoparticles caused 100% mortality in the subterranean termite *Odontotermes formosanus* within 48 hours of treatment. Furthermore, Chinnaperumal et al. (2018) evaluated the larvicidal, pupicidal, and antifeedant properties of TiO_2 nanoparticles against *Helicoverpa armigera*. The results indicated 100% mortality for both first- and second-instar larvae, and 92.34% mortality for third-instar larvae at 100 ppm. TiO_2 nanoparticles were found to be safe for non-target organisms such as the earthworm *Eudrilus eugeniae*.

4.3.1.4 Iron Oxide Nanoparticles and Insect Control

Iron oxide, an inorganic compound, exhibits significant insecticidal activity against various insect pests. This nanoparticle can be synthesized from natural resources, including plant leaves. Ramkumar et al. (2021) synthesized iron oxide nanoparticles from the green leaves of *Trigonella foenum-graecum*. In their method, 5 g of dried leaf powder was mixed with 100 mL of distilled water and boiled at 60 °C. The resulting mixture was filtered and stored at 4 °C. Subsequently, 10 mL of the filtrate was combined with 1.5 mL of 0.1 M ferric chloride solution and stirred for 5 h. After 24 h of storage, the solution changed from transparent yellow to intense black, indicating the formation of iron oxide nanoparticles (Fig. 4.6).

The resulting solution was centrifuged at 5000 rpm for 20 min, and the obtained iron oxide nanoparticles were evaluated against the tomato leaf miner (Tuta absoluta). Mortality rates ranged from 50% to 72% after four days of treatment. Similarly, Zargham et al. (2023) synthesized iron oxide nanoparticles from the aqueous leaf extract of *Grevillea robusta* and tested them against the second- and fourth-instar larvae of *Anopheles stephensi* and *Aedes aegypti*. For *A. aegypti*, mortality ranged from 64% to 96% in second-instar larvae and 65% to 98% in fourth-instar larvae. Corresponding values for *A. stephensi* were 56%–84% for second instars and 56%–88% for fourth instars. These findings demonstrate that iron oxide nanoparticles are highly effective against both *A. stephensi* and *A. aegypti*. In addition to their direct insecticidal activity, iron oxide nanoparticles can act as synergistic agents when combined with synthetic insecticides. Muthusamy et al. (2023) reported that combining iron oxide nanoparticles with fenvalerate significantly enhanced toxicity against the cotton bollworm (*Helicoverpa armigera*) and the tobacco cutworm (*Spodoptera litura*), achieving mortality rates of 91.41% and 92.83%, respectively.

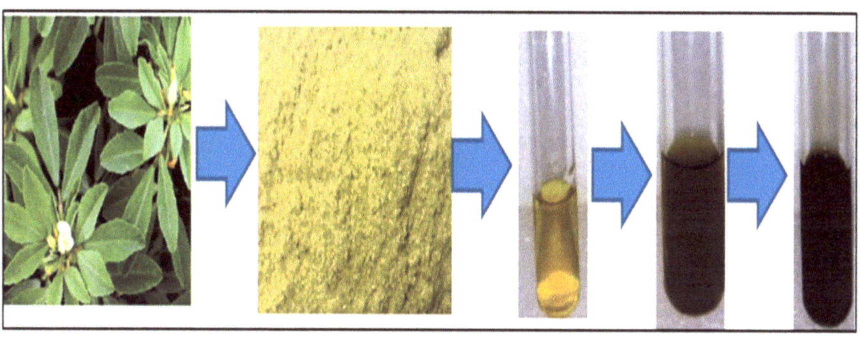

Fig. 4.6 Synthesis of iron oxide nanoparticles from the green leaves of *T. foenum-graecum*

4.3.1.5 Copper Oxide Nanoparticles and Insect Control

Copper oxide (CuO) is an inorganic compound with broad-spectrum pesticidal properties, effective against both insect and fungal pests. Its primary advantages include low production cost and greater accessibility compared to many non-metal-based alternatives. Rahman et al. (2022) evaluated CuO nanoparticles against the third-, fourth-, and fifth-instar larvae of *Spodoptera frugiperda.* The nanoparticles exhibited both high toxicity and strong antifeedant activity, with mortality rates of 97%, 94%, and 81%, and corresponding antifeedant activities of 98.25%, 98.01%, and 98.42%, respectively. El-Saadony et al. (2020) synthesized CuO nanoparticles from *Pseudomonas fluorescens* isolated from soil and tested them against the red flour beetle (*Tribolium castaneum*), reporting an LC_{50} value of 37 ppm. Malandrakis et al. (2024) combined CuO nanoparticles with deltamethrin for the control of the olive fruit fly (*Bactrocera oleae*), resulting in increased susceptibility compared to deltamethrin alone. The combination also shortened the insect's life cycle, reducing larval duration, pupal period, and adult longevity. Badawy et al. (2021) demonstrated high efficacy of CuO nanoparticles against *Sitophilus granarius* and *Rhyzopertha dominica*, with mortality rates of 94.4% and 90%, respectively, and further noted their potential as an agricultural fertilizer.

Nenaah et al. (2025) assessed CuO nanoparticles against the larvae and adult females of *Aedes aegypti*. At 40 μg/mL, the nanoparticles caused 100% mortality in first-, second-, and third-instar larvae, as well as in adult stages, while exhibiting low toxicity toward non-target species such as *Eisenia fetida* and *Gambusia affinis*. Kamaraj et al. (2022) tested CuO nanoparticles against larvae of three cattle tick species—*Hippobosca maculata*, *Haemaphysalis bispinosa*, and *Rhipicephalus microplus*—reporting LC_{50} values of 11.13, 9.50, and 4.30, respectively. Abou El-Ela et al. (2022) investigated the effects of CuO nanoparticles on the brown planthopper (*Nilaparvata lugens*) in rice fields, finding increased mortality within 24 hours and a reduction in LC_{50}, likely due to inhibition of detoxification enzymes such as acetylcholinesterase. Furthermore, Wang et al. (2024a, b) synthesized CuO nanoparticles from tomato leaves and applied them against *Helicoverpa armigera*, confirming their strong insecticidal potential.

4.3.2 Organic Nanoinsecticides

Organic insecticides play a critical role in the suppression of insect pests. The application of nanoinsecticides significantly reduces the quantity of active ingredients released into the environment, thereby minimizing environmental contamination. All conventional insecticide classes can be formulated at the nanoscale, and this approach is typically achieved through two main strategies:

- Nanocarrier-based delivery systems, in which the active ingredient is encapsulated or adsorbed onto a carrier material (e.g., chitosan, starch, cellulose) to improve stability, controlled release, and target specificity (Fig. 4.7).

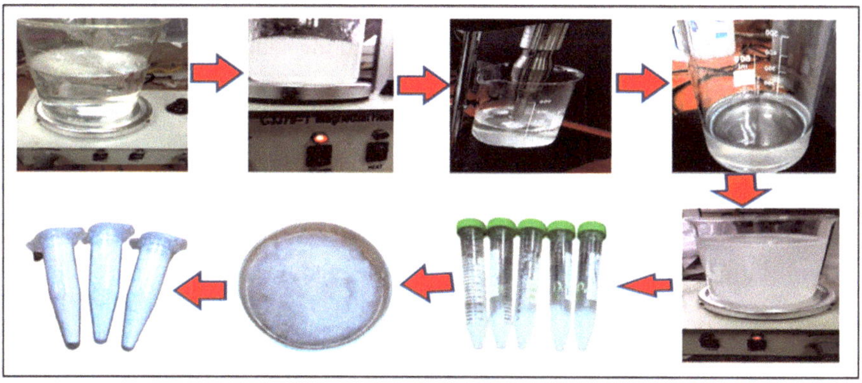

Fig. 4.7 Preparation of insecticides loaded on chitosan nanoparticles

- Particle size reduction techniques, in which microparticles of insecticides are directly converted into nanoparticles using specialized equipment, such as a sonicator.

4.3.2.1 Nano-Organophosphorus Insecticides

Organophosphorus insecticides are among the most extensively used chemical agents for insect pest control worldwide (Pu et al. 2023). However, a major limitation of this group is the persistence of pesticide residues in the environment. Nanoformulation approaches offer a potential solution by enabling controlled release and reduced environmental impact. Maan et al. (2024) utilized chitosan as a nanocarrier to develop a nanoformulation of chlorpyrifos. Compared with the conventional formulation, the nanochlorpyrifos exhibited a significantly slower release rate of the active ingredient and persisted in the soil for over 15 days. Furthermore, it demonstrated reduced toxicity toward beneficial soil microbiota.

Vallim et al. (2022) assessed the toxicity of nano-dimethoate on non-target organisms, reporting that the nanoformulation exhibited lower toxicity toward zebrafish larvae, highlighting the potential of nanotechnology to improve the environmental safety profile of pesticides. Farhan et al. (2024a, b) found that triazofos nanoparticles were more effective against *Anopheles gambiae* than the conventional formulation. Ahmed et al. (2019) compared the effects of profenofos alone and in combination with silver nanoparticles, observing a substantial increase in efficacy against second-instar larvae of Spodoptera littoralis. The LC_{50} value for profenofos alone was 79.5 ppm, whereas the combination with silver nanoparticles reduced the LC_{50} to 0.94 ppm, representing an 85-fold increase in potency.

4.3.2.2 Nano-Carbamates Insecticides

Carbamates are regarded as the second most effective class of organic insecticides after pyrethroids. Farhan et al. (2024a, b) compared the efficacy of nanoformulations and conventional formulations of several carbamate insecticides, including carbosulfan, propoxur, and methomyl. Their findings indicated that the nanoparticle formulation of carbosulfan exhibited the highest efficacy against *Anopheles gambiae* while demonstrating reduced toxicity toward the adults of Coccinella septempunctata, a beneficial non-target species.

4.3.2.3 Nano-Pyrethroids Insecticides

Pyrethroid insecticides are predominantly employed for the control of public health pests. The use of nanoformulations can reduce the required concentrations for effective pest suppression. Taktak et al. (2021) evaluated nanoformulations of alpha-cypermethrin, deltamethrin, lambda-cyhalothrin, and permethrin against larvae of Culex pipiens. All nanoformulations demonstrated 1.5- to 2.0-fold higher efficacy compared to their conventional counterparts, while exhibiting lower toxicity to the freshwater alga *Raphidocelis subcapitata*. Similarly, Oliveira et al. (2019) reported that a nanoformulation of pyrethrum was less toxic to honeybees (Apis mellifera) than the traditional formulation.

Shawer et al. (2022) prepared various nanoformulations of lambda cyhalothrin, including nanocapsules and nanoparticles, and found that the nanoparticle formulation achieved 86% mortality in *Drosophila suzukii*, compared with 74% for the nanocapsule formulation. Feng et al. (2021) demonstrated that a nanoformulation of lambda cyhalothrin was more effective against *Spodoptera frugiperda* larvae than the conventional formulation. In addition to improved efficacy, some pyrethroid nanoformulations exhibit reduced toxicity; for example, Khalifa et al. (2022) reported that a nanoformulation of deltamethrin induced less neurobehavioral toxicity in male albino rats compared with the standard formulation.

Galloux et al. (2024) found that a nanoformulation of deltamethrin was more stable and more effective against *Acyrthosiphon pisum* than the normal formulation. Ziaee et al. (2020) used silica nanoparticles (SNPs) as carriers for deltamethrin, applying the formulation against small and large larvae of Trogoderma granarium. Mortality rates were 70.5% and 55.5% for small and large larvae, respectively.

4.3.2.4 Nano-Neonicotinoids Insecticides

Neonicotinoids represent a key class of insecticides in integrated pest management programs. However, concerns over their toxicity to non-target organisms have driven the development of nanoformulations, which can reduce both the required concentrations and the overall production costs. Imidacloprid is one of the most widely used neonicotinoid insecticides. Sabry et al. (2021) evaluated a nanoformulation of imidacloprid against the Egyptian cotton leafworm (*Spodoptera littoralis*) and compared its efficacy with that of the conventional formulation (Fig. 4.7). The LC_{50} value for the nanoformulation was 15.9 ppm, compared with 66.5 ppm for the conventional formulation, indicating a fourfold increase in potency. Moradi et al. (2019) reported that nano-imidacloprid exhibited greater efficacy against Myzus persicae compared with the conventional formulation. Ebadollahi et al. (2022) found that nanoencapsulated acetamiprid was highly effective against third-instar larvae of the elm leaf beetle (*Xanthogaleruca luteola*). Wang et al. (2021) further demonstrated that a nanoformulation of acetamiprid outperformed the conventional formulation in terms of insecticidal activity, plant residue profile, and absorption in tea seedlings. Thiamethoxam has also been formulated at the nanoscale for pest control. Elabasy et al. (2020) loaded thiamethoxam onto nanocellulose and applied it against second-instar nymphs of *Phenacoccus solenopsis*. Li et al. (2021) investigated the effect of pH on thiamethoxam nanoparticle performance, finding that the formulation was highly pH-sensitive, with release rates increasing at pH values above 10. The efficacy of thiamethoxam nanoparticles against peach aphid was 1.44-fold greater than that of the conventional formulation.

4.3.2.5 Nano-Botanical Extracts

Botanical extracts are natural products derived from various parts of plants, including leaves, stems, bark, roots, flowers, and fruits. Many of these extracts possess insecticidal properties and are employed as alternative agents for controlling insect pests. Common examples include neem extract and chili pepper extract. Madasamy et al. (2023) evaluated four botanical extracts—*Justicia adhatoda* (a), *Ipomoea carnea* (b), *Pongamia glabra* (c), and *Annona squamosa* (d) (Fig. 4.8)—both individually and in combination with silver nanoparticles, against *Phenacoccus solenopsis*.

The results indicated that all tested botanical extracts significantly reduced *Phenacoccus solenopsis* populations, both when applied alone and when combined with silver nanoparticles. Qari et al. (2020) evaluated the efficacy of silica nanoparticles in combination with four plant extracts—*Cinnamomum camphora*, *Matricaria chamomilla*, *Mentha arvensis*, and *Trigonella foenum-graecum*—against *Thrips tabaci*. Monteiro et al. (2021) investigated the effects of zein nanoparticles loaded with the botanical compounds limonene and carvacrol on *Spodoptera frugiperda*. In their study, larvae of *S. frugiperda* were significantly affected by all tested botanical extracts and nanoparticle formulations.

Fig. 4.8 Nanoparticles of imidacloprid

4.3.2.6 Nanodiamide Insecticides

Diamides are a relatively new class of insecticides with activity against a wide range of target pests. Nanoformulation of diamides offers the potential to reduce production costs while enhancing performance. Ding et al. (2024) developed a nanoformulation of chlorantraniliprole, a diamide insecticide, that exhibited lower production costs and improved dispersibility compared with the conventional formulation. When applied against *Cnaphalocrocis medinalis*, the nanoformulation required only 40% of the dose used in the standard formulation, yet achieved higher efficacy. Tang et al. (2024) prepared a nanosuspension of chlorantraniliprole and tested it against first-instar larvae of the rice leaf roller (*C. medinalis*), reporting superior performance over the conventional formulation.

Fan et al. (2023) utilized carboxymethyl chitosan (CMCS) as a polymer matrix for encapsulating chlorantraniliprole. This nanoformulation was highly effective against the fall armyworm (Spodoptera frugiperda) while exhibiting reduced toxicity toward zebrafish (Danio rerio), a non-target aquatic organism.

4.3.2.7 Nanoavermectin Insecticides

Avermectin insecticides are widely employed for the control of numerous insect pests, with emamectin benzoate being one of the most commonly used compounds in this group. However, its stability under normal environmental conditions is relatively low. To address this limitation, Cui et al. (2021) developed emamectin benzoate nanoparticles using an antisolvent precipitation method combined with orthogonal experimental design. The resulting nanoformulation exhibited improved stability and solubility compared with the conventional formulation.

Yang et al. (2017) reported that emamectin benzoate nanoparticles demonstrated greater retention on rice, cabbage, and cucumber leaves than the standard formulation. Furthermore, the nanoformulation exhibited enhanced insecticidal activity against the diamondback moth (*Plutella xylostella*) and the green peach aphid (*Myzus persicae*) compared with the conventional product.

4.3.2.8 Nanooxadiazine Insecticides

Indoxacarb is the sole representative of its insecticide class and acts as a sodium–potassium channel blocker. Nanoformulations of indoxacarb have been developed to enhance its efficacy against target insect pests. Sabry et al. (2021) prepared a nanoformulation of indoxacarb and evaluated its activity against second-instar larvae of the Egyptian cotton leafworm (*Spodoptera littoralis*) in comparison with the conventional formulation. The nanoformulation was 12 times more effective than the standard formulation, with particle sizes ranging from 200 to 400 nm (Fig. 4.9). Bilal et al. (2020) investigated indoxacarb loaded onto silica nanoparticles for controlling the diamondback moth (Plutella xylostella), finding it to be more effective than indoxacarb alone. Similarly, Chen et al. (2024) assessed nano-indoxacarb against two destructive pests, *P. xylostella* and *Pieris rapae*, and reported that the LC_{50} values were reduced by factors of 6.784 and 1.931, respectively, compared with the conventional formulation.

Fig. 4.9 (**a**) *Justicia adhatoda* (**b**) *Ipomea carnea* (**c**) *Pongamia glabra* (**d**) *Annona squamosa*

4.3.2.9 Nanophenyl Pyrazole Insecticides

Fipronil is the most widely used insecticide in its class and is highly effective against a range of insect pests, particularly public health vectors such as mosquitoes, houseflies, and cockroaches. Its mode of action involves blocking chloride channels in the insect nervous system. Tian et al. (2022) applied nanotechnology to reduce the toxicity of fipronil toward non-target organisms while maintaining or enhancing its efficacy against target pests. In their study, fipronil was combined with regular nanomicelles and evaluated against the diamondback moth (*Plutella xylostella*) as the target organism and zebrafish (*Danio rerio*) as the non-target organism. The nanoformulation reduced toxicity to zebrafish by a factor of four compared with the conventional formulation, while maintaining equivalent toxicity against *P. xylostella*.

Peters et al. (2019) encapsulated fipronil within silica nanocapsules for termite control, achieving a controlled release profile that reduced environmental contamination. Kumar et al. (2019) evaluated a nanoformulation of fipronil against the brown planthopper (Nilaparvata lugens) in rice fields, reporting a 93.47% reduction in pest population compared with 80.47% for the conventional formulation.

4.3.2.10 Nanopyrethroid Ethyl Insecticides

The pyrethroid ethyl insecticide group, also known as synthetic non-ester pyrethroids, acts by disrupting nerve impulse transmission in insect axons. Etofenprox is one of the most widely used insecticides in this group and functions as a sodium–potassium channel blocker. Sabry et al. (2023) evaluated a nanoparticle formulation of etofenprox against adults of the silverleaf whitefly (*Bemisia tabaci*) in eggplant fields, comparing its performance with that of the conventional formulation (Figs. 4.10 and 4.11). The nanoformulation was found to be four times more effective, with LC_{50} values of 0.9 ppm and 3.5 ppm for the nanoformulation and conventional formulation, respectively. Residue analysis in eggplant fruits revealed lower residue levels for the nanoformulation compared with the conventional formulation.

Etofenprox has also been developed as a controlled-release formulation (CRF) to regulate the release of the active ingredient (Hwang et al. 2011). In this form, etofenprox nanocapsules were tested against *Spodoptera litura*, showing significantly higher efficacy against larval stages compared with the conventional formulation.

Fig. 4.10 Nanoparticles of indoxacarb

Fig. 4.11 Zeta particle size of etofenprox nanoparticles

4.3.2.11 Nanopyrroles Insecticides

This class of insecticides exerts its action by disrupting respiratory pathways and proton gradients in mitochondria, thereby uncoupling oxidative phosphorylation (Agumba et al. 2019). Chlorfenapyr is the most widely used insecticide in this category. Farhan et al. (2024a, b) evaluated a nanoformulation of chlorfenapyr against the mosquitoes *Anopheles funestus*, *Anopheles coluzzii*, and *Culex quinquefasciatus*, achieving mortality rates of 100%, 98%, and 98%, respectively. The nanoformulation demonstrated greater efficacy than the conventional formulation. Zheng et al. (2023) compared the physical properties of nanochlorfenapyr—such as thermal stability, adhesion, permeability, and systemic activity—with those of the standard formulation in maize fields. The nanoformulation exhibited superior performance in all parameters, with a deposition rate of 12.28 mg/kg on maize leaves. When tested against *Spodoptera frugiperda*, nanochlorfenapyr produced mortality rates 7.5–17.5 times higher than those of the conventional formulation. Sabry and Ali (2022) reported that nanochlorfenapyr, with a particle size of 99.8 nm (Figs. 4.12, 4.13, and 4.14), was highly effective against the conical snail (*Cochlicella acuta*) and the chocolate-banded snail (*Massylaea vermiculata*) compared with the standard product. Similarly, Yan et al. (2022) demonstrated that nanoparticle-based chlorfenapyr improved control efficacy against the fruit fly *Drosophila melanogaster*.

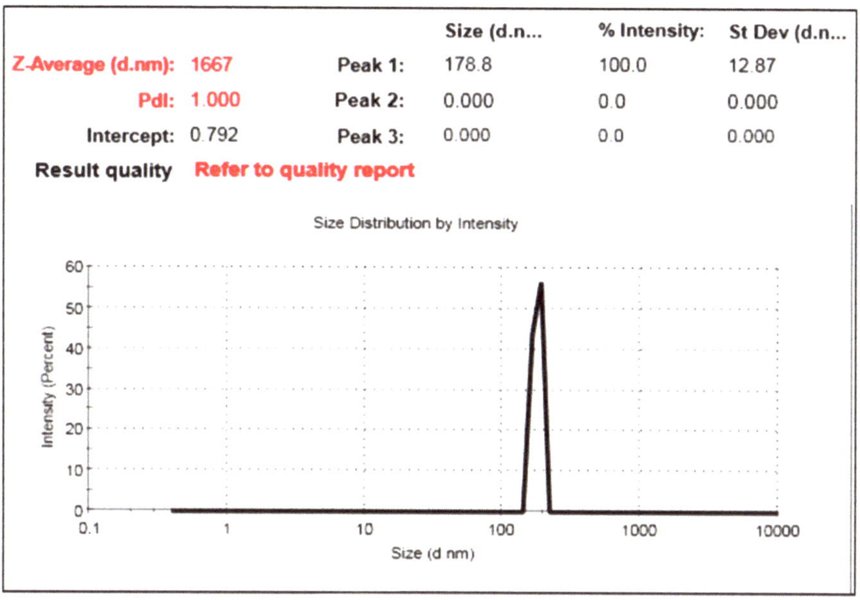

Fig. 4.12 Nanoparticles of etofenprox

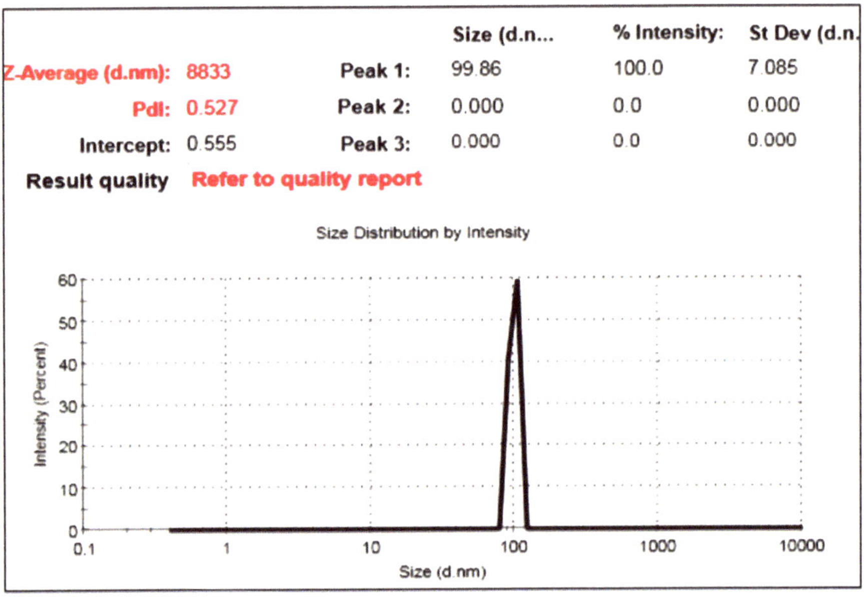

Fig. 4.13 Zeta particle size of chlorfenapyr

Fig. 4.14 Nanoparticles of chlorfenapyr

References

Abou El-Ela AS, Ntiri ES, Munawar A, Shi XX, Zhang C, Pilianto J, Zhang Y, Chen M, Zhou W, Zhu ZR (2022) Silver and copper-oxide nanoparticles prepared with GA3 induced defense in rice plants and caused mortalities to the brown planthopper, Nilaparvata lugens (Stål). Nano Impact 28:100428. https://doi.org/10.1016/j.impact.2022.100428

Agumba S, Gimnig JE, Ogonda L, Ombok M, Kosgei J, Munga S, Guyah B, Omondi S, Ochomo E (2019) Diagnostic dose determination and efficacy of chlorfenapyr and clothianidin insecticides against *Anopheles malaria* vector populations of western Kenya. Malar J 18:243. https://doi.org/10.1186/s12936-019-2858-z

Ahmed KS, Mikhail WZA, Sobhy HM, Radwan EMM, Salaheldin TA (2019) Impact of nanosilver-profenofos on cotton leafworm, Spodoptera littoralis (Boisd.) larvae. Bull Natl Res Cent 43:46. https://doi.org/10.1186/s42269-019-0076-z

Badawy AA, Abdelfattah NAH, Salem SS, Awad MF, Fouda A (2021) Efficacy assessment of biosynthesized copper oxide nanoparticles (CuO-NPs) on stored grain insects and their impacts on morphological and physiological traits of wheat (Triticum aestivum L.). Biology 10(3):233. https://doi.org/10.3390/biology10030233

Baz MM, El-Barkey NM, Kamel AS, El-Khawaga AH, Nassar MY (2022) Efficacy of porous silica nanostructure as an insecticide against filarial vector Culex pipiens (Diptera: Culicidae). Int J Tropic Insect Sci 42:2113–2125. https://doi.org/10.1007/s42690-022-00732-7

Bhadani RV, Gajera HP, Hirpara DG, Savaliya DD, Anuj SA (2022) Biosynthesis and characterization of extracellular metabolites-based nanoparticles to control the whitefly. Arch Microbiol 204(6):311. https://doi.org/10.1007/s00203-022-02917-7

Bilal M, Xu C, Cao L, Zhao P, Cao C, Li F, Huang Q (2020) Indoxacarb-loaded fluorescent mesoporous silica nanoparticles for effective control of Plutella xylostella L. with decreased detoxification enzymes activities. Pest Manag Sci 76(11):3749–3758. https://doi.org/10.1002/ps.5924

Chen M, Zhang J, Wang H, Li L, Yin M, Shen J, Yan S, Liu B (2024) Preparation of nanoscale indoxacarb by using star polymer for efficiency pest management. Agric 14(7):1006. https://doi.org/10.3390/agriculture14071006

Chinnaperumal K, Govindasamy B, Paramasivam D, Dilipkumar A, Dhayalan A, Vadivel A, Sengodan K, Pachiappan P (2018) Bio-pesticidal effects of Trichoderma viride formulated titanium dioxide nanoparticle and their physiological and biochemical changes on Helicoverpa armigera (hub.). Pestic Biochem Physiol 149:26–36. https://doi.org/10.1016/j.pestbp.2018.05.005

Cui B, Li J, Lai Z, Gao F, Zeng Z, Zhao X, Liu G, Cui H (2021) Emamectin benzoate-loaded zein nanoparticles produced by antisolvent precipitation method. Polym Test 94:107020. https://doi.org/10.1016/j.polymertesting.2020.107020

Ding X, Guo L, Du Q, Wang T, Zeng Z, Wang Y, Cui H, Gao F, Cui B (2024) Preparation and comprehensive evaluation of the efficacy and safety of chlorantraniliprole nanosuspension. Toxics 12(1):78. https://doi.org/10.3390/toxics12010078

Ebadollahi A, Valizadeh B, Panahandeh S, Mirhosseini H, Zolfaghari M, Changbunjong T (2022) Nanoencapsulation of acetamiprid by sodium alginate and polyethylene glycol enhanced its insecticidal efficiency. Nano 12(17):2971. https://doi.org/10.3390/nano12172971

Elabasy A, Shoaib A, Waqas M, Shi Z, Jiang M (2020) Cellulose nanocrystals loaded with thiamethoxam: fabrication, characterization, and evaluation of insecticidal activity against Phenacoccus solenopsis Tinsley (Hemiptera: Pseudococcidae). Nano 10(4):788. https://doi.org/10.3390/nano10040788

El-Ashmouny RS, Rady MH, Merdan BA, El-Sheikh TAA, Hassan RE, El Gohary EGE (2022) Larvicidal and pathological effects of green synthesized silver nanoparticles from Artemisia herba alba against Spodoptera littoralis through feeding and contact application. Egypt J Basic Appl Sci 9(1):239–253. https://doi.org/10.1080/2314808X.2022.2063012

El-Naggar ME, Abdelsalam NR, Fouda MMG, Mackled MI, Al-Jaddadi MAM, Ali HM, Siddiqui MH, Kandil EE (2020) Soil application of nano silica on maize yield and its insecticidal activ-

ity against some stored insects after the post-harvest. Nano 10(4):739. https://doi.org/10.3390/nano10040739
El-Saadony MT, Abd El-Hack ME, Taha AE, Fouda MMG, Ajarem JS, Maodaa SN, Allam AA, Elshaer N (2020) Ecofriendly synthesis and insecticidal application of copper nanoparticles against the storage pest Tribolium castaneum. Nano 10(3):587. https://doi.org/10.3390/nano10030587
Fan T, Meng Z, Chen X, Liang Y, Zhao M, Wu Q, Cui J, Xu W, Wang J (2023) Fabrication of stimuli-responsive nanoparticles for high-efficiency chlorantraniliprole delivery and smart control of Spodoptera frugiperda. Ind Crop Prod 205:117427. https://doi.org/10.1016/j.indcrop.2023.117427
Farhan M, Akhtar S, Ahmad I, Maryam, Hussain H, Yasin M, Farooqi MA, Zhang S (2024a) Assessing the potential of nano-formulated chlorfenapyr and clothianidin insecticides-treated sugar baits against Anopheles funestus, Anopheles coluzzii and Culex quinquefasciatus mosquitoes. Acta Trop 256:107269. https://doi.org/10.1016/j.actatropica.2024.107269
Farhan M, Zhao C, Akhtar S, Ahmad I, Jilong P, Zhang S (2024b) Assessment of nano-formulated conventional insecticide-treated sugar baits on mosquito control and the effect on non-target aphidophagous Coccinella septempunctata. Insects 15(1):70. https://doi.org/10.3390/insects15010070
Feng B, Zhi H, Chen H, Cui B, Zhao X, Sun C, Wang Y, Cui H, Zeng Z (2021) Development of chlorantraniliprole and lambda cyhalothrin double-loaded nano-microcapsules for synergistical pest control. Nano 11(10):2730. https://doi.org/10.3390/nano11102730
Galloux M, Bastiat G, Lefrancois C, Apaire-Marchais V, Deshayes C (2024) Nanoencapsulated deltamethrin combined with indoxacarb: an effective synergistic association against aphids. J Pestic Sci 49(3):168–178. https://doi.org/10.1584/jpestics.D24-003
Gutiérrez-Ramírez JA, Betancourt-Galindo R, Aguirre-Uribe LA, Cerna-Chávez E, Sandoval-Rangel A, Ángel EC, Chacón-Hernández JC, García-López JI, Hernández-Juárez A (2021) Insecticidal effect of zinc oxide and titanium dioxide nanoparticles against Bactericera cockerelli Sulc. (Hemiptera: Triozidae) on tomato Solanum lycopersicum Agronomy 11(8):1460. https://doi.org/10.3390/agronomy11081460
Gutiérrez-Ramírez JA, Betancourt-Galindo R, Aguirre-Uribe LA, Cerna-Chávez E, Sandoval-Rangel A, Castro-del Ángel E, Chacón-Hernández JC, García-López JI, Hernández-Juárez A (2024) Studying the insecticidal effect of titanium oxide nanoparticles on the Periplaneta americana. Entomol Appl Sci Letters 11(1):40–47. https://doi.org/10.51847/OUUOpwl7Er
Huang S, Wang L, Liu L, Hou Y, Li L (2015) Nanotechnology in agriculture, livestock, and aquaculture in China. A review. Agron Sustain Dev 35:369–400. https://doi.org/10.1007/s13593-014-0274-x
Hwang IC, Kim TH, Bang SH, Kim KS, Kwon HR, Seo MJ, Youn YN, Park HJ, Yasunaga-Aoki C, Yym (2011) Insecticidal effect of controlled release formulations of etofenprox based on nano-bio technique. J Fac Agric Kyushu Univ 56(1):33–40
Ishaq M, Ismail K, Ahmad N, Ayub AR, Iqbal A, Ashraf F, Shafiq S, Adnan M, Mukhtar MA, Haider G, Umar M, Khadim M (2020) Synthesis and characterization of titanium dioxide nanoparticles against two insect pests. Int J Sci Res 10(5):475–493. https://doi.org/10.29322/IJSRP.10.05.2020.p10155
Kamaraj C, Gandhi PR, Kumar RCS, Balasubramani G, Malafaia G (2022) Biosynthesis and extrinsic toxicity of copper oxide nanoparticles against cattle parasites: an eco-friendly approach. Environ Res 214(Part 4):114009. https://doi.org/10.1016/j.envres.2022.114009
Khalifa AG, Moselhy WA, Mohammed HM, Khalil F, Shaban M, El-Nahass ES, Al-Muzafar HM, Adel Amin K, Abdou KA (2022) Deltamethrin and its nanoformulations induce behavioral alteration and toxicity in rat brain through oxidative stress and JAK2/STAT3 signaling pathway. Toxics 10(6):303. https://doi.org/10.3390/toxics10060303
Kumar N, Kumar R, Shakil NA, Sarkar DJ, Chander S (2019) Evaluation of fipronil nanoformulations for effective management of brown plant hopper (*Nilaparvata lugens*) in rice. Int J Pest Manag 65:86–93. https://doi.org/10.1080/09670874.2018.1468046

Li W, Wang Q, Zhang F, Shang H, Bai S, Sun J (2021) pH-sensitive thiamethoxam nanoparticles based on bimodal mesoporous silica for improving insecticidal efficiency. R Soc Open Sci 8:201967. https://doi.org/10.1098/rsos.201967
Maan S, Jatrana A, Kumar V, Sindhu M, Mondal S (2024) Controlled release of chlorpyrifos through crosslinked chitosan–guargum biopolymer based nano-formulation. Sustain Chem Pharm 37:101378. https://doi.org/10.1016/j.scp.2023.101378
Madasamy M, Sahayaraj K, Sayed SM, Al-Shuraym LA, Selvaraj P, El-Arnaouty SA, Madasamy K (2023) Insecticidal mechanism of botanical crude extracts and their silver nanoliquids on Phenacoccus solenopsis. Toxics 11(4):305. https://doi.org/10.3390/toxics11040305
Madhiyazhagan P, Murugan K, Kumar AN, Nataraj T, Dinesh D, Panneerselvam C, Subramaniam J, Mahesh Kumar P, Suresh U, Roni M, Nicoletti M, Alarfaj AA, Higuchi A, Munusamy MA, Benelli G (2015) Sargassum muticum-synthesized silver nanoparticles: an effective control tool against mosquito vectors and bacterial pathogens. Parasitol Res 114(11):4305–4317. https://doi.org/10.1007/s00436-015-4671-0
Malandrakis AA, Varikou K, Kavroulakis N, Nikolakakis A, Dervisi I, Reppa CI, Papadakis S, Holeva MC, Chrysikopoulos CV (2024) Copper nanoparticles interfere with insecticide sensitivity, fecundity and endosymbiont abundance in olive fruit fly Bactrocera oleae (Diptera: Tephritidae). Pest Manag Sci 80:3640–3649. https://doi.org/10.1002/ps.8068
Monteiro RA, Camara MC, de Oliveira JL, Campos EVR, Carvalho LB, Proença PLF, Casagrande MG, Lima R, do Nascimento J, Gonçalves KC, Polanczyk RA, Fraceto LF (2021) Zein based-nanoparticles loaded botanical pesticides in pest control: an enzyme stimuli-responsive approach aiming sustainable agriculture. J Hazard Mater 417:126004. https://doi.org/10.1016/j.jhazmat.2021.126004
Moradi FG, Hejazi MJ, Hamishehkar H, Enayati AA (2019) Co-encapsulation of imidacloprid and lambda-cyhalothrin using biocompatible nanocarriers: characterization and application. Ecotoxicol Environ Saf 175:155–163. https://doi.org/10.1016/j.ecoenv.2019.02.092
Muthusamy R, Ramkumar G, Kumarasamy S, Chi NTL, Obaid SA, Alfarraj S, Karuppusamy I (2023) Synergism and toxicity of iron nanoparticles derived from Trigonella foenum-graecum against pyrethroid treatment in Spodoptera litura and Helicoverpa armigera (Lepidoptera: Noctuidae). Environ Res 231(Part 1):116079. https://doi.org/10.1016/j.envres.2023.116079
Narayanan M, Devi PG, Natarajan D, Kandasamy S, Devarayan K, Alsehli M, Elfasakhany A, Pugazhendhi A (2021) Green synthesis and characterization of titanium dioxide nanoparticles using leaf extract of Pouteria campechiana and larvicidal and pupicidal activity on Aedes aegypti. Environ Res 200:111333. https://doi.org/10.1016/j.envres.2021.111333
Nasser R, Ibrahim E, Fouad H, Ahmad F, Li W, Zhou Q, Yu T, Chidwala N, Mo J (2024) Termiticidal effects and morpho-histological alterations in the subterranean termite (Odontotermes formosanus) induced by biosynthesized zinc oxide, titanium dioxide, and chitosan nanoparticles. Nano 14(11):927. https://doi.org/10.3390/nano14110927
Nenaah GE, Mahfouz ME, Almadiy AA, Albogami BZ, Alasmari SM, Gazzy AA, Fadl AEA (2025) Fabrication, characterization and mosquitocidal activity of CuO nanoparticles against Aedes aegypti L. (Diptera: Culicidae), the main vector of dengue arbovirus with environmental risk assessment. BioNanoScience 15:93. https://doi.org/10.1007/s12668-024-01634-3
Oliveira CR, Domingues CEC, de Melo NFS, Roat TC, Malaspina O, Costa MJ, Silva-Zacarin ECM, Fraceto LF (2019) Nanopesticide based on botanical insecticide pyrethrum and its potential effects on honeybees. Chemosphere 236:124282. https://doi.org/10.1016/j.chemosphere.2019.07.013
Peters BC, Wibowo D, Yang GZ, Hui Y, Middelberg APJ, Zhao CX (2019) Evaluation of baiting fipronil-loaded silica nanocapsules against termite colonies in fields. Heliyon 5(8):e02277. https://doi.org/10.1016/j.heliyon.2019.e02277
Pu H, Fang T, Wu Z, Sun DW (2023) Advancements in recyclable photocatalytic semiconductor substrates for SERS detection in food safety applications. Trends Food Sci Technol 138:697–707. https://doi.org/10.1016/j.tifs.2023.07.005
Qari SH, Khalil AH, Abdelfattah NAH, Ayman A, Shehawy (2020) Effect of different plant extracts and nanoparticles on Thrips tabaci (Lind.) (Thysanoptera: Thripidae) under field con-

ditions and their allelopathic potential on the onion, Allium cepa L. using bioassays and RAPD analysis. Egypt J Biol Pest Control 30:13. https://doi.org/10.1186/s41938-020-00216-1

Rahman A, Pittarate S, Perumal V, Rajula J, Thungrabeab M, Mekchay S, Krutmuang P (2022) Larvicidal and antifeedant effects of copper nano-pesticides against Spodoptera frugiperda (J.E. Smith) and its immunological response. Insects 13(11):1030. https://doi.org/10.3390/insects13111030

Ramkumar G, Asokan R, Ramya S, Gayathr G (2021) Characterization of Trigonella foenum-graecum-derived iron nanoparticles and its potential pesticidal activity against Tuta absoluta (Lepidoptera). J Clust Sci 32:1185–1190. https://doi.org/10.1007/s10876-020-01867-8

Rehman HU, Majeed B, Farooqi MA, Rasul A, Sagheer M, Ali Q, Akhtar ZR (2021) Green synthesis of silver nitrate nanoparticles from Camelina sativa (L.) and its effect to control insect pests of stored grains. Int J Tropic Insect Sci 41:1–9. https://doi.org/10.1007/s42690-021-00495-7

Sabry AKH, Ali RF (2022) Development and using of some nanopesticide formulations against the conical snail, Cochlicella acuta, and the chocolate banded snail, Massylaea vermiculata. Bull Natl Res Cent 46:15. https://doi.org/10.1186/s42269-021-00695-8

Sabry AKH, Salem HAN, Metwally HM (2021) Development of imidacloprid and indoxacarb formulations to nanoformulations and their efficacy against Spodoptera littoralis (Boisd). Bull Natl Res Cent 45:16. https://doi.org/10.1186/s42269-020-00477-8

Sabry AH, Mohamady AH, Sleem RA, Abolmaaty SM, Helmy RM (2023) Role of etofenprox nanoformulation in suppression of the silver whitefly, Bemisia tabaci and its residue in eggplant fruits. J Plant Protect Res 63(1):30–38. https://doi.org/10.24425/jppr.2023.144501

Santhoshkumar T, Govindarajan RK, Kamaraj C, Ragavendran C, Kamal MA, Moglad EH, Zaki RM, Priyadharsan A, Baek K (2024) Green fabricated silver nanoparticles as a new eco-friendly insecticide for controlling stored cowpea bug Callosobruchus maculatus (Coleoptera: Bruchidae). Biocatal Agric Biotechnol 56:103023. https://doi.org/10.1016/j.bcab.2024.103023

Santos TS, Silva TM, Cardoso JC, Albuquerque-Júnior RLC, Zielinska A, Souto EB, Severino P, Mendonça MDC (2021) Biosynthesis of silver nanoparticles mediated by entomopathogenic fungi: antimicrobial resistance, nanopesticides, and toxicity. Antibiotics 10(7):852. https://doi.org/10.3390/antibiotics10070852

Saw G, Nagdev P, Jeer M, Murali-Baskaran RK (2023) Silica nanoparticles mediated insect pest management. Pestic Biochem Physiol 194:105524. https://doi.org/10.1016/j.pestbp.2023.105524

Sharma P, Prakash J, Kaushal R (2022) An insight into the green synthesis of SiO_2 nanostructures as a novel adsorbent for removal of toxic water pollutants. Environ Res 212:113328. https://doi.org/10.1016/j.envres.2022.113328

Sharon M, Choudhary AK, Kumar R (2010) Nanotechnology in agricultural diseases and food safety. J Phytopathol 2:83–92. https://doi.org/10.4236/anp.2020.91001

Shawer R, El-Leithy ES, Abdel-Rashid RS, Eltaweil AS, Baeshen RS, Mori N (2022) Preparation of lambda-cyhalothrin-loaded chitosan nanoparticles and their bioactivity against Drosophila suzukii. Nano 12(18):3110. https://doi.org/10.3390/nano12183110

Shoaib A, Elabasy A, Waqas M, Lin L, Cheng X, Zhang Q, Shi ZH (2018) Entomotoxic effect of silicon dioxide nanoparticles on Plutella xylostella (L.) (Lepidoptera: Plutellidae) under laboratory conditions. Toxicol Environ Chem 100(1):80–91. https://doi.org/10.1080/02772248.2017.1387786

Taktak NEM, Badawy MEI, Awad OM, Abou El-Ela NE, Abdallah SM (2021) Enhanced mosquitocidal efficacy of pyrethroid insecticides by nanometric emulsion preparation towards Culex pipiens larvae with biochemical and molecular docking studies. J Egypt Public Health Assoc 96:21. https://doi.org/10.1186/s42506-021-00082-1

Tang Q, Sun Y, Du L, Zhou M, Yang D (2024) Preparation and performance of 10 wt% of chlorantraniliprole nano-suspension concentrate (SC) using polyether amine-bridged sulfonated alkali lignin (PSAL) as a dispersant. Ind Crop Prod 214:118539. https://doi.org/10.1016/j.indcrop.2024.118539

Thabet AF, Boraei HA, Galal OA, El-Samahy MFM, Mousa KM, Zhang YZ, Tuda M, Helmy EA, Wen J, Nozaki T (2021) Silica nanoparticles as pesticide against insects of different feeding types and their non-target attraction of predators. Sci Rep 11:14484. https://doi.org/10.1038/s41598-021-93518-9

Tian Y, Zhang X, Huang Y, Tang G, Gao Y, Chen X, Zhou Z, Li Y, Li X, Wang H, Yu X, Liu Y, Yan G, Wang J, Cao Y (2022) Amphiphilic prodrug nano-micelles of fipronil coupled with natural carboxylic acids for improving physicochemical properties and reducing the toxicities to aquatic organisms. Chem Eng J 439:135717. https://doi.org/10.1016/j.cej.2022.135717

Vallim JH, Clemente Z, Castanha RF, Pereira AES, Campos EVR, Assalin MR, Maurer-Morelli CV, Fraceto LF, de Castro VLSS (2022) Chitosan nanoparticles containing the insecticide dimethoate: a new approach in the reduction of harmful ecotoxicological effects. NanoImpact 27:100408. https://doi.org/10.1016/j.impact.2022.100408

Wang X, Yan M, Zhou J, Song W, Xiao Y, Cui C, Gao W, Ke F, Zhu J, Gu Z, Hou R (2021) Delivery of acetamiprid to tea leaves enabled by porous silica nanoparticles: efficiency, distribution and metabolism of acetamiprid in tea plants. BMC Plant Biol 21:337. https://doi.org/10.1186/s12870-021-03120-4

Wang D, Saleh NB, Byro A, Zepp R, Sahle-Demessie E, Luxton TP, Ho KT, Burgess RM, Flury M, White JC (2022) Nano-enabled pesticides for sustainable agriculture and global food security. Nat Nanotechnol 17:347–360. https://doi.org/10.1038/s41565-022-01082-8

Wang Z, Fan N, Li X, Yue L, Wang X, Liao H, Xiao Z (2024a) Trophic transfer of metal oxide nanoparticles in the tomato–Helicoverpa armigera food chain: effects on phyllosphere microbiota, insect oxidative stress, and gut microbiome. ACS Nano 18(39):26631–26642. https://doi.org/10.1021/acsnano.4c05063

Wang C, Jiang Y, He K, Wāng Y (2024b) Eco-friendly synthesis of silver nanoparticles against mosquitoes: pesticidal impact and indispensable biosafety assessment. Sci Total Environ 953:176006. https://doi.org/10.1016/j.scitotenv

Yan S, Yin H, Li N, Chen Y, Ji C, Jiang Q, Du J, Yin M, Shen J, Zhang J (2022) Combination of a nanocarrier delivery system with genetic manipulation further improves pesticide efficiency: a case study with chlorfenapyr. Environ Sci Nano 9:2020–2031. https://doi.org/10.1039/D2EN00126H

Yang D, Cui B, Wang C, Zhao X, Zeng Z, Wang Y, Sun C, Liu G, Cui H (2017) Preparation and characterization of emamectin benzoate solid nanodispersion. J Nanomater 6560780:1–9. https://doi.org/10.1155/2017/6560780

Zargham F, Afzal M, Rasool K, Manzoor S, Qureshi NA (2023) Larvicidal activity of green synthesized iron oxide nanoparticles using Grevillea robusta Cunn. leaf extract against vector mosquitoes and their characterization. Exp Parasitol 252:108586. https://doi.org/10.1016/j.exppara.2023.108586

Zavala-Zapata V, Ramírez-Barrón SN, Sánchez-Borja M, Aguirre-Uribe LA, Delgado-Ortiz JC, Sánchez-Peña SR, Mayo-Hernández J, García-López JI, Vargas-Tovar JA, Hernández-Juárez A (2024) Insecticide efficacy of green synthesis silver nanoparticles on Diaphorina citri Kuwayama (Hemiptera: Liviidae). Insects 15(7):469. https://doi.org/10.3390/insects15070469

Zheng Q, Wu J, Yan W, Zhu S, Miao X, Wang R, Huang S, Cheng D, Zhang P, Zhang Z (2023) Green synthesis of a chlorfenapyr chitosan nanopesticide for maize root application: reducing environmental pollution and risks to nontarget organisms. Int J Biol Macromol 253(Pt 4):126988. https://doi.org/10.1016/j.ijbiomac.2023.126988

Ziaee M, Babamir-Satehi A (2020) Characterization of nanostructured silica as carrier for insecticides deltamethrin, pyriproxyfen, and chlorpyrifos and testing the insecticidal efficacy against Trogoderma granarium (Coleoptera: Dermestidae) larvae. J Econ Entomol 113(1):511–517. https://doi.org/10.1093/jee/toz264

Zobir SAM, Ali AF, Sulaiman MR, Ahmad KA (2021) Review on nanopesticides for plant protection synthesized using the supramolecular chemistry of layered hydroxide hosts. Biology 10:1077. https://doi.org/10.3390/biology10111077

Application of Nanotechnology for Insect Pest Management

5

Disha Mallick and Md. Mamunur Rahman

Abstract

The intensification of agricultural production is significantly hindered by insect pests, necessitating widespread use of chemical pesticides. However, this has resulted in several detrimental effects, including the development of pesticide resistance, destruction of beneficial organisms, environmental contamination, biodiversity loss, and risks to human health. In response, nanotechnology has emerged as a revolutionary tool in insect pest management, offering innovative solutions to address the limitations of conventional pest control methods. By leveraging molecular-level engineering, nanotechnology focuses on the development of nanomaterials such as nanoemulsions, nanocapsules, and nanopesticides, which enhance the efficacy, stability, and precision of traditional pesticides. Nanoemulsions, with their ability to increase solubility and enhance penetration of active ingredients, offer a promising method for improving the bioavailability of pesticides. Nanocapsules, through controlled and sustained release, allow for prolonged insecticidal action while minimizing nontarget effects. Nanopesticides, designed to deliver active components with improved specificity and reduced environmental footprint, present new possibilities for reducing pest populations while safeguarding crop yields and ecosystem health. The mode of action of these nanomaterials involves improved penetration of insect cuticles, disruption of critical physiological pathways, and the capacity to bypass established resistance mechanisms. These nanoscale formulations provide enhanced insecticidal activity, reduced environmental contamination, and more precise application methods, ultimately contributing to sustainable agricultural practices. Despite the advantages, the integration of nanotechnology into pest management systems

D. Mallick · M. M. Rahman (✉)
Department of Entomology, Gazipur Agricultural University, Gazipur, Bangladesh
e-mail: mamun@gau.edu.bd

S. S. Purewal et al. (eds.), *Agri-Nanotechnology: Innovations for Sustainable Agriculture and Environmental Restoration*,
https://doi.org/10.1007/978-981-96-9756-4_5

is not without challenges. High development costs, regulatory uncertainties, and concerns over the long-term environmental fate of nanoparticles represent significant barriers to widespread adoption. A comprehensive evaluation of the benefits and limitations of nanotechnology in pest management is crucial for ensuring its viability. Nanotechnology holds great potential to transform insect pest management by offering more effective and environmentally friendly alternatives to conventional pesticides. However, ongoing research is essential to address the ecological and regulatory challenges associated with nanomaterial use, ensuring that this innovative approach contributes to both agricultural sustainability and environmental protection.

Keywords
Nanotechnology · Nanomaterials · Nanoemulsions · Nanocapsules · Nanopesticides · Insect pest management · Sustainable agriculture

5.1 Introduction

Insect pests cause significant losses in agricultural production worldwide. Approximately 30–40% of vegetable crops and 10–30% of overall agricultural produce are damaged, leading to staggering financial impacts exceeding $470 billion worldwide (Kaur 2024; Sharma et al. 2017; Culliney 2014). A variety of insecticides are necessary for efficient pest control to increase agricultural output (Schmitz et al. 2022). The use of 2.7 million tons of pesticides in 2020, with an average of 1.8 kg/ha, represents a 50% increase over the last three decades (Wisnujati 2023). Approximately $32–41.1 billion is the estimated value of the worldwide pesticide market (Pimentel and Hart 2001). Soil, water, and wildlife are among the nontarget environments that can be negatively impacted by insecticides, which are chemicals that either kill or stop the growth of pests. Researchers fear that excessive use of pesticides damages ecosystems and deteriorates water quality, which is vital for the survival of humans and animals (Tang et al. 2021). The situation has worsened as exotic pests and pesticide resistance have emerged (Kaur 2024). Pesticide pollution affects biodiversity, soil, and water in ecosystems (Patil and Patil 2022). Because these persistent compounds are harmful to both humans and wildlife, their usage must be performed responsibly (Sharma et al. 2017). Large-scale pesticide use leads to issues such as resistance, the extinction of nontarget species, environmental contamination, and health risks. Pesticide efficacy is reduced when pests develop resistance to common pesticides (Sparks and Nauen 2015; Isman 2020). To solve the world's problems, other approaches, such as precision pest control via nanotechnology, are essential. Traditional pesticides can be replaced with safer alternatives via nanotechnology (Tirunagaru et al. 2024).

Alternative techniques, especially biological approaches, are necessary for effective weed management because conventional pest control methods, such as manual, mechanical, and chemical approaches, have drawbacks such as labor shortages, environmental impacts, herbicide resistance, and health risks (Abbas et al. 2018). These methods necessitate a thorough understanding of pests, frequently lack consistent safety evaluations, and often result in over-application, inadequate targeting, and hazardous pesticide residues, underscoring the need for enhanced application protocols (Dhananjayan et al. 2020). The dependence on synthetic pesticides results in insect resistance and pest resurgence, highlighting the need for sustainable practices that integrate biotechnological approaches and improve ecosystem resilience (Sharma 2012). Integrated pest management is crucial because chemical insecticides can adversely affect natural enemies, disrupting pest control and variable outcomes (Riyaz and Kathiravan 2019). Conventional pest control frequently simplifies intricate agricultural systems. It overlooks wider social and ecological consequences, resulting in environmental damage, health problems, and insect resistance, marginalizing subsistence farmers in developing nations (Atreya et al. 2011). The use of 2.25 treatments per crop cycle in non-IPM systems, compared to 0.13 in IPM systems, suggests that reliance on synthetic insecticides increases both costs and environmental issues (Araujo Robusti et al. 2023).

Nanotechnology improves agricultural productivity and food security by applying eco-friendly fertilizers that enhance plant growth, stress resilience, and seed germination (Al-Rawi et al. 2024). Innovations mitigate the shortcomings of conventional pesticides and fertilizers by improving nutrient efficiency, decreasing toxicity, and enabling targeted application (Bhushan et al. 2024). Nanoscale material manipulation (1–100 nm) enables precise input delivery and swift plant health monitoring while addressing regulatory and environmental issues (Santos et al. 2024). Nanotechnology-enhanced insecticides, fertilizers, and pesticides contribute to sustainability and production efficiency while reducing adverse environmental effects and addressing the rising demand for high-quality agricultural products (Singh et al. 2024a, b).

The efficiency of agriculture has been significantly enhanced through nanotechnology, which has a substantial impact on improving various sectors such as disease prevention, nutrient absorption, and tolerance to abiotic stresses. To achieve these benefits, nanotechnology involves the application of different nanosensors, nanoparticles, and nanofertilizers, all while addressing environmental concerns (Faizan et al. 2024). The primary aim of this chapter is to explore the various applications of nanotechnology, particularly nanoemulsions, nanocapsules, and nanopesticides, in the context of sustainable pest management, with careful consideration of environmental impacts. This work aims to provide a comprehensive understanding that encompasses all aspects—from mechanisms of action to practical applications—and to highlight the comparative advantages of these nanotechnological approaches over conventional pest control methods, while also addressing current challenges in pest management.

5.2 Categories of Nanomaterials for Insect Pest Management

Nanomaterials offer remarkable opportunities for developing advanced pest management strategies. By addressing and overcoming the limitations associated with conventional synthetic pesticides, they enable more precise and targeted pest control. Owing to their unique characteristics and nanoscale properties, nanomaterials demonstrate rapid and effective pest control while minimizing adverse impacts on the environment.

5.2.1 Nanoemulsions

5.2.1.1 Structure and Properties

Nanoemulsions can be characterized as isotropic, mixture of oil, water, surfactants, and cosurfactants, which are stable both thermodynamically and kinetically and range between 20 and 200 nm but often below 100 nm for better accuracy (Dhumal et al. 2022; Anton and Vandamme 2011; Shafiq et al. 2007; Mei et al. 2003). Surfactants are used to strengthen these ultrafine, diverse, biphasic systems (Jacob et al. 2024). The appropriate combination of the three main components—oil, water, and surfactants—contributes to the desired properties of nanomaterials, including their stability and functionality (Chavda et al. 2024). Two manufacturing methods, high-energy methods and low-energy methods, utilize various types of surfactants. Advantageous characteristics, such as large surface area per unit volume, strong stability, optical transparency, and adjustable rheology, have been noted and can resist common destabilization issues such as creaming and sedimentation because of their small size. Oil molecules are hydrophobic and have relatively high solubility and dispersibility under aqueous conditions (Dasgupta et al. 2019). Because they maintain their stability in the face of environmental changes, nanoemulsions are appropriate for use (Anton and Vandamme 2011).

5.2.1.2 Operational Method

Nanoemulsions offer numerous benefits across various dimensions, including optimizing crop protection, enhancing pesticide efficiency, and mitigating adverse environmental impacts. These advantages have contributed to their growing popularity in modern agricultural practices (Gupta et al. 2024). Precision in pest control and controlled release are two of the most valuable characteristics of nanoemulsions, establishing them as optimized carriers for pesticides. These properties help address concerns related to chemical hazards and reduce pesticide runoff into soil and water, thereby minimizing environmental impact (Gupta et al. 2024).

5.2.1.3 Applications in Field Trials

The application of nanoemulsions containing anise oil on soybean leaves has been found to be convenient and a promising alternative to conventional pesticide formulations, offering enhanced wetting and spreading on hydrophobic surfaces. These properties help to effectively address issues such as rebound and spatter during pesticide application (Wu et al. 2023). Food safety and food security can be supported through efficient pest management enabled by the unique properties of nanoemulsions, which exhibit enhanced antibacterial, antifungal, and insecticidal activities (Patir 2022). Nanoemulsions contribute to the principles of sustainable agriculture, as they serve as advanced plant protection agents that precisely target specific cells while minimizing collateral effects. This results in improved crop preservation, reduced chemical usage, and more efficient pesticide delivery (Singh et al. 2023). The common storage problem (fungal development and mycotoxin generation) can be solved by applying oils generated via nanoemulsions. This action improves not only the safety but also the quality of agricultural products and also maintains grain quality and lowers the possibility of infection (Padrilah et al. 2024). In spite of all the benefits of nanoemulsion, it is not completely free from detrimental effect, so the continuous research and regulation of the residual impacts on the environment and the toxic effect should be under control to ensure environmental stewardship (Bhaskar et al. 2023).

5.2.2 Nanocapsules

5.2.2.1 Structure and Function

The structure of nanocapsules is typically characterized by nanoscale dimensions ranging between 100 and 500 nm, designed to enhance the stability and controlled release properties of active ingredients through encapsulation. This controlled release of nutrients and beneficial microorganisms, as demonstrated by chitosan–alginate nanocapsules, promotes plant growth while reducing environmental impacts and dependence on conventional fertilizers (Hamed et al. 2024). Polysaccharide-derived nanoparticles ensure biocompatibility, controlled release, and reduced biochemical consumption by encapsulating natural materials such as essential oils and plant extracts, thus supporting sustainable agricultural practices (Mondéjar-López et al. 2024). These natural polymer-based nanocapsules reduce toxicity and enhance the effectiveness of fertilizers and pesticides by improving plant development and stress resistance through site-specific nutrient release, thereby minimizing crop losses (Ngangom et al. 2024; Singh et al. 2024a, b). Reduced postharvest losses; improved plant health, development, and germination rates; and sustainable resource management are all benefits of precise pesticide delivery by nanocapsules (Tamta et al. 2024). Plant protein-based nanocapsules increase the transport, solubility, stability, bioavailability of bioactive components and hence help in customized pesticide delivery, lowering the dose and environmental effects while reacting to certain stimuli for increased efficacy and sustainability (Kumar et al. 2024; Rajić et al. 2022). These acts of nanocapsules finally help to distribute agrochemicals precisely, improve their efficiency by reducing their dosage, allowing their controlled release, and lessen soil contamination from artificial chemicals (Bhaskar et al. 2023).

5.2.2.2 Sustained Release and Environmental Protection

Nanocapsules allow prolonged pesticide release by shielding active ingredients and reacting to environmental cues such as pH and temperature, minimizing leaching and environmental damage, and ensure increasing crop yields through effective control (Kumar et al. 2024).

The pH-dependent release mechanism offers additional benefits in copper (Cu) nanocapsules by enhancing environmental safety and pesticide efficacy. These nanocapsules reduce leaching, increase wash persistence, and do not pose phytotoxicity risks to crops (Qu et al. 2023). Furthermore, they provide added advantages such as increased photostability and reduced toxicity to nontarget organisms. Ma et al. (2023) also demonstrated that pH-responsive nanocapsules improve disease management and contribute to sustainable agriculture by enabling continuous pyraclostrobin release in acidic environments. Among other types of nanocapsules, ammonium persulfate (APS)-based nanocapsules have been shown to enhance hydraulic fracturing fluid degradation through regulated APS release, utilizing thermoresponsive properties for sustained delivery in reservoirs under moderate temperature conditions (Pu et al. 2024). Similarly, hollow silica nanoparticles have been reported to enable prolonged anionic surfactant release, enhance oil recovery, and contribute to environmental remediation by altering surface wettability and improving surfactant efficiency in subsurface applications (Alsmaeil et al. 2021). Compared to traditional formulations, these nanocapsule systems offer on-demand, controlled release, thereby maximizing the utilization of biocides and bio-based binders for safer and more effective applications across various sectors. These systems can be triggered by a range of stimuli, ensuring sustained efficacy through targeted distribution, prolonged release, improved absorption, reduced chemical exposure, and enhanced ecological safety, all while improving pest and pathogen management (Chen et al. 2023; Gaur et al. 2022).

5.2.2.3 Examples of Applications

Pesticide nanocapsules promote sustainable agriculture with less toxicity to nontarget organisms by enabling targeted distribution, increased biological activity, pH-responsive release, and greater retention and resistance to rain erosion (Zhou et al. 2024). Chen et al. (2023) demonstrated great efficacy against *Plutella xylostella* when loaded with avermectin and against *Rhizoctonia solani* and *Fusarium graminearum* when loaded with tebuconazole. When UV-triggered release is used, the light-responsive nanocapsules for controlled release of imidacloprid reduce the necessary concentrations and increase pest control effectiveness against *Aphis craccivora* Koch (Gao et al. 2024). By facilitating tailored pesticide application, nanocapsules improve efficacy while minimizing their environmental impact. To maximize pesticide application while reducing toxicity and leaching, pesticides can be released in response to stimuli (Kumar et al. 2024).

Lambda-cyhalothrin-loaded polymer nanocapsules improve foliar pest control by increasing dispersion, adhesion, and sustained release; this results in improved insecticidal efficacy and decreased aquatic toxicity compared with conventional formulations (Luo et al. 2021). Neem oil nanoencapsulation improves its effectiveness in controlling pests, shortens its duration of action on target pests, lowers environmental toxicity, and aids in the fight against insect and pathogen resistance (Charpe 2022). According to Dwivedi and Tomer (2020), the use of nanocapsules containing insecticides, fungicides, and nematicides enhance their effectiveness against pests such as *Helicoverpa armigera,* while also reducing pesticide residues in crops and soil through controlled release and fewer applications. Owing to improved pesticide efficiency, the dual-functionalized nanocapsule delivery strategy for rice sheath blight, which employs validamycin and thifluzamide, improves bioactivity and spread (Cui et al. 2020).

5.2.3 Nanopesticides

5.2.3.1 Design and Formulation

Nanopesticides represent advanced formulations that protect plants by employing active chemicals at the nanoscale level. The authors address food demand, climate change, and biodiversity loss by enhancing solubility, release control, and degradation inhibition (Olusola and Adetunji 2024). Nanopesticides facilitate efficient pest management with reduced toxicity to nontarget organisms, attributed to their smaller particle sizes and improved absorption and stability (Xu et al. 2024). Nanopesticides demonstrate lower costs and improved control efficacy compared to traditional pesticides, and they are acknowledged for their potential to alter agricultural practices substantially (Wu et al. 2024a, b). Nanomaterials or nanostructures are utilized to improve the efficacy of active chemicals while reducing negative impacts on the environment and human health owing to their large surface area, increased solubility, and controlled release properties (Belagalla et al. 2024). Nanopesticides improve antifeedant characteristics and reduce toxicity to nontarget organisms, promoting sustainable pest management (Adetuyi et al. 2024). According to Dong et al. (2024), nanopesticides enhance agricultural productivity and address environmental challenges through improved adhesion, solubility, stability, and targeted delivery.

Tong et al. (2024) characterize nanopesticides as insecticide formulations at the nanoscale to improve leaf deposition and efficacy. This is accomplished through uniform distribution, increased resistance to degradation, and enhanced absorption by plants and fungi, all contributing to effective pest management. Nanopolymers, nanoemulsions, and nanospheres represent advancements in agriculture that improve the stability and effectiveness of pest control while minimizing the application of active substances and their adverse environmental effects (Nanda et al. 2024). Nanopesticides are functional molecules characterized by a nanoscale size of 1 to 200 nm, acting as agrochemical carriers. These innovations enhance stability, mobility, solubility, and biodegradability, resulting in more effective pest management than traditional pesticides (Verma and Saroop 2024). Nanopesticides are site-selective nanoparticles that exhibit metal ion-mediated proteolytic activity and nanoscale chirality. They demonstrate efficacy, safety, and ecological benignity as antiviral agents against agricultural plant viruses (Zhu et al. 2024).

This study details the synthesis of poly(octyl acrylate) nanogels incorporating lambda-cyhalothrin via microemulsion polymerization. This technique enables simple preparation, allows size control (20–200 nm), and supports customized distribution, improving stability and effectiveness in agricultural applications (Wu et al. 2024a, b). The study developed nano-IMI and micro-IMI variants of imidacloprid that share equivalent compositions but differ in particle size. Nano-IMI, due to its reduced size, demonstrates improved stability, dispersibility, and foliar application, thereby enhancing its effectiveness as a nanopesticide while safeguarding nontarget organisms (Du et al. 2024).

5.2.3.2 Mode of Action and Advantages

As nanopesticides (NPts) comprises nanoscale metal materials and nanosized carriers, they improve pest management and lower dangers to the environment and human health by offering high efficiency, low toxicity, focused action, and regulated gradual release (Li et al. 2023). Nanoscale functional molecules in nanopesticides offer improved formulation stability, mobility, solubility, and biodegradability as well as providing the better leaf coverage, enhanced soil properties, and stronger pest activity at lower dosages that increase agricultural productivity and nutritional security (Verma and Saroop 2024). It ensures controlled release, tailored distribution, and increased efficacy while lowering the required amount and environmental impact and lowering the dangers to the health of nontargeted creatures (Nanda et al. 2024).

It is believed that nanopesticides encourage sustainable pest control by fulfilling several function like boosting the agrarian production attribute, extending product lifespan, minimizing damage to nontarget organisms, elevating biological response and antifeedant attributes (Adetuyi et al. 2024). As environmentally acceptable substitutes for chemical pesticides, they increase crop production, biomass, and tolerance to environmental stress while improving food security by increasing seed germination, delivering pests precisely, and providing broad-spectrum protection (Adetunji and Ugbenyen 2019). More mortality and longer residual action are caused by nanoinsecticides, which also disrupt insect hormone and enzyme levels to delay growth and provide more precise and efficient pest management (Graily Moradi and Asgari 2021). By effectively lowering insect populations and deactivating phytopathogenic bacteria and fungi, N-doped nanopesticides demonstrate multifunctionality with improved surface adhesion and a variety of modes of action, including physical, biochemical, and catalytic mechanisms, for better crop protection (Pho et al. 2020).

5.2.3.3 Ecotoxicological Considerations

Although there is potential for the use of nanomaterials in pest management, health and environmental problems still exist. For safe agricultural applications, more investigations are essential (Jasrotia et al. 2022). Compared with traditional pesticides, nanopesticides, such as nanoencapsulated imidacloprid, present greater toxicity risks to creatures that are not their intended target, such as zebrafish embryos. Assessing environmental hazards and putting safety precautions in place require an understanding of their toxicokinetics (Wu et al. 2024a, b). In *Enchytraeus crypticus*, NUCOP-M and copper oxychloride had comparable long-term toxicity, with copper oxychloride being more harmful in the short term and NUCOP-M being less lethal during the cocoon stage (Chidiamassamba et al. 2024). To ensure environmental safety, comprehensive ecotoxicological studies are necessary because tebuconazole nanoformulations are more toxic to nontarget organisms such as nematodes than are regular formulations (Eghbalinejad et al. 2024). Chen et al. (2024) observed that nanopesticides can interrupt the nitrogen cycle by two different ways: decreasing oxidative stress and blocking detoxifying enzymes along with blocking iron-based metal–organic framework. This activity leads to destruction of soil microbial communities, endangering soil creatures, including earthworms (Chen et al. 2024). The stability, interactions with organic matter, and impacts on aquatic organisms should all be considered in ecotoxicological evaluations of nanopesticides (Castanha et al. 2023).

The necessity for thorough toxicity protocols is highlighted by the fact that stability loss can reduce toxicity (Castanha et al. 2023). Though nanopesticides have huge positive impact and most importantly it increase agricultural productivity, they are not completely free from the possibilities of creating new environmental contaminants. So it is critical to understand their structures, interactions, and negative effects to diminish dangers to the environment and human health (Verma and Saroop 2024). Ecotoxicological knowledge is essential to create safer and more efficient nanopesticides as those might have unintended toxic effects on both target and nontarget species (Roy and Roy 2022). Studies on biotoxicology and ecological impacts are needed as the existing studies are insufficient to comprehensively understand the environmental impact and hazards of nanopesticides (Li et al. 2023).

5.3 Mechanism of Action of Nanomaterials in Pest Management

5.3.1 Improved Effectiveness in Pest Control

5.3.1.1 Structural Characteristics

Nanoparticles facilitate the efficient delivery of active compounds by penetrating insect exoskeletons (Belagalla et al. 2024). Nanomaterials can mitigate harmful impacts on beneficial insects and nontarget organisms by selectively targeting particular pests (Paramanik et al. 2023). Tirunagaru et al. (2025) argue that these methods facilitate controlled release and targeted action, reducing harm to nontarget organisms and enhancing surface area coverage on insect bodies, thus improving penetration and effectiveness. In contrast to conventional pesticides, these alternatives employ enhanced solubility, regulated release, and precise delivery of active components, optimizing pest control efficacy and reducing adverse impacts on environmental and human health (Belagalla et al. 2024). Nanomaterials facilitate the accurate delivery and controlled release of active compounds, reducing necessary quantities and enhancing agricultural pest management (Kannan et al. 2020).

Nanomaterials enhance the solubility and bioavailability of pesticides, reducing dependence on chemical pesticides and promoting environmental sustainability (Sabrine, 2023).

5.3.1.2 Results of the Experiment

Copper-based nanoparticles (CuNPs) enhance insect body penetration due to their small size, improve plant uptake, and exhibit strong insecticidal properties, thereby supporting sustainable agriculture (Gacem and Chaibi 2022). Research on Cu-doped ZnO nanoparticles as insecticides for *Spodoptera littoralis* has emphasized environmentally sustainable pest control in Egypt; however, it lacks a thorough examination of the mechanisms underlying the nanomaterial's action. Failure to address these processes significantly reduces *Spodoptera littoralis* populations attributed to Cu-doped ZnO-NPs, highlighting their potential insecticidal properties (Abd El-Latef et al. 2023). Silver nanoparticles and other environmentally friendly nanoparticles enhance insect body penetration and effectively target pests like *Tribolium castaneum*, offering sustainable alternatives to synthetic insecticides (Saini et al. 2023).

5.3.2 Disruption of Physiological Pathways

5.3.2.1 Targeting Neural Pathways

Nanomaterials provide us with improved pest management technique which include silver, zinc, copper, and titanium nanoparticles and helps improve biogenicity and agricultural production. Mainly their distinct physiomorphic characteristics are responsible for this efficient pest management and tailored delivery in sustainable agriculture (Jafir et al. 2023). Nanomaterials are used to promote environmentally friendly pest control through metal-, polymer-, and lipid-based pesticide formulations, which enhance stability, permeability, and target specificity (Paramanik et al. 2023). Silver nanoparticles rupture enzymatic pathways causing internal damage that ultimately results in cell death while entering into insect bodies via the cuticle membrane (Nazima et al. 2020). It also interacts with the insect cuticle or is absorbed through the digestive tract, enhancing insecticidal efficacy and prolonging exposure to the active compounds, as these are more mobile and act faster than conventional formulations.

5.3.2.2 Breaking the Metabolic Resistance

Nanomaterials are capable of altering the antioxidant and detoxifying enzymes in insect body as they interfere with protein synthesis and gene regulation, resulting in oxidative stress, problems with development and reproduction, denaturation of enzymes, and cell death in pests (Nivetha et al. 2021). According to Ghosh and Bera (2021), they efficiently manage infections and pests without causing harm to the environment by adhering to microbial membranes, penetrating cells, and causing structural changes that disrupt respiratory enzymes and DNA, limit replication, and cause cell death. Through toxicity, reactive oxygen species (ROS) generation, and physical disruption of insect physiology, especially in the midgut, nanomaterials increase insecticidal efficiency by causing cell death, immunological activation, and decreased food intake (Shahzad and Manzoor 2021). Compared with conventional dsRNA approaches, they provide sustainable pest management by effectively delivering dsRNA, improving gene silencing, interrupting key genes, and causing pest mortality. Alternative pest control is offered by transgenic RNAi, with core–shell polymeric nanostructures improving the delivery of dsRNA (Bi et al. 2023). Nanomaterials improve the effectiveness of nanopesticides and lower the chance of resistance by penetrating insects because of their small size and special characteristics, ability to stick to surfaces, and ability to alter their physiological activities (Sinha et al. 2017). This can serve as an effective pesticide as those can penetrate across the cellular membrane and pass the barrier and bind with proteins and DNA that leads to cellular damage and destruction and finally death of organism (Benelli 2018).

5.4 Eco-toxic Consequence and Safety Measures

5.4.1 Assessing Environmental Impact

5.4.1.1 Fate of Nanomaterials in Soil and Water

In soil and aquatic environments, nanomaterials' transport, bioavailability, behavior, and fate are influenced by adsorption, aggregation, sedimentation, desorption, degradation, dissolution, and mediated reactions (Bolan et al. 2024). Environmental factors and engineered nanomaterials significantly influence these materials' transformation, transport, retention, and conversion into secondary products. Thus, they affect soil contamination and agricultural health. Agriculture is influenced by engineered nanomaterials in soil and water, alongside existing knowledge gaps concerning their environmental behavior and transport (Kokilavani et al. 2023). This study highlights the need for innovative approaches to evaluate the environmental and health risks linked to the characterization of inorganic nanoparticles in natural and anthropogenic surface water and groundwater. Malakar and Snow (2020) demonstrated that the toxicological effects of different nanoparticles on living organisms vary. The fate, transport, and toxicological effects of engineered nanomaterials on organisms are affected by the aging and transformation of these materials in soils and water, highlighting environmental hazards.

Nanomaterials predominantly enter soils and water bodies via surface runoff and effluent discharges, intensifying ecological problems and influencing bioavailability. Aggregation, sedimentation, and interactions among natural colloids are significant factors. Essential processes, including dissolution and aggregation, affect bioavailability, which demonstrate distinct behaviors in natural environments compared to synthetic media (Batley et al. 2013). Nanomaterials are often transported via rainfall runoff and are released from consumer products, solid waste, and wastewater effluents (Kokilavani et al. 2023). Bolan et al. (2024) report that more than 90% of food additive nanoparticles are excreted and inadequately removed in wastewater systems, leading to soil and water contamination via food waste and recycled water disposal. Soil quality and nanoparticle characteristics affect retention and mobility via physical strain and chemical interactions (Yecheskel and Dror). Nanomaterials can enter the food chain via plant absorption, potentially harming aquatic and soil organisms (Bolan et al. 2024). Engineered nanomaterials can enhance plant growth at low concentrations; however, their effects may become harmful at elevated concentrations (Kokilavani et al. 2023). The behavior of engineered nanomaterials in soil and groundwater is determined by the physicochemical characteristics of the media, nanoparticles, and flow conditions, along with their transformation and stability, which influence retention, transport, and release (Qing et al. 2022).

5.4.1.2 Engagement with Nontarget Species

Nanomaterials can generate reactive oxygen species (ROS) in nontarget organisms, resulting in apoptosis, DNA damage, and oxidative stress (Tripathi et al. 2022). Nontarget species, such as aquatic organisms and microalgae, may absorb nanomaterials, potentially leading to biomagnification in the food chain (Avelelas et al. 2017). Avelelas et al. (2017) indicate that nanomaterials may disrupt ecosystems and biodiversity in aquatic environments, posing a threat to marine organisms. The persistence of nanomaterials in ecosystems influences soil and water quality, subsequently impacting the health of aquatic and terrestrial organisms. Enhancing the understanding of nanomaterial interactions with biological systems can create biocompatible materials, thereby reducing negative impacts on nontarget species (He et al. 2018).

The immobilization of biocides within nanostructured materials, including LDH and SiNC, represents an environmentally acceptable approach to anti-fouling. This method minimizes the toxicity of the agent to nontarget species, including *Tetraselmis chuii*. The ban on TBT resulted in heightened biofouling of submerged structures, necessitating adopting environmentally sustainable measures and raising maintenance expenses. Consequently, it is essential to examine the structures and metabolic processes of LDH in both target and nontarget species (Avelelas et al. 2017). This study primarily examines the interactions between engineered nanomaterials and biological systems, focusing on toxicity and absorption in models such as A549 cells and CD-1 rodents rather than interactions with nontarget species. Researchers aim to investigate the absorption and toxicity of nanomaterials in cellular and murine models to anticipate relationships between their biological interactions and the characteristics of these materials. This will inform the advancement of secure nanomaterials. The assessment of the environmental safety of engineered nanomaterials is complicated by their diverse characteristics, leading to intricate interactions with ecosystems. Evaluating the interconnections among nano-, bio-, and ecosystems, along with the transformation processes and nanotoxicity across various ecosystems, is essential (He et al. 2018).

Research has examined the interactions of nanomaterials with nontarget species in light of ecological concerns, focusing on their toxicity and effects on aquatic and terrestrial systems, including microorganisms and plants (Darnault 2023). The absorption of nanoparticles (NPs) by diverse organisms, including bacteria, microalgae, crustaceans, molluscs, fish, plants, and animals, may lead to toxicity affecting growth, development, and health, especially in aquatic species and flora. This raises concerns regarding environmental and human health (Tripathi et al. 2022). The interactions between the Mediterranean bivalve *Mytilus galloprovincialis* and the European sea bass *Dicentrarchus labrax* represent significant biological effects from coexposure to environmental contaminants (Canesi et al. 2022). Nasser et al. (2020) indicate that nanomaterials adsorb ambient biomolecules, forming an eco-corona that modifies the stability, identity, and toxicity of these biomolecules. This modification impacts the organism's health and necessitates the execution of targeted toxicity assessments. Nanomaterials engage with fish, aquatic plants, invertebrates, and microorganisms via adsorption and internalization processes. The toxicity of nanoparticles is influenced by the properties of nanomaterials, environmental conditions, and species characteristics (Ma and Lin 2013). The publication highlights the impact of engineered nanoparticles on nontarget aquatic animals. The effects include the generation of reactive oxygen species (ROS), the uptake of nanoparticles compared to the release of metal ions, and surface attachment, all of which influence toxicity and bioavailability (Sigg et al. 2014).

5.4.2 Human Health Concerns

5.4.2.1 Potential Toxicity of Nanopesticides

Because of the possible toxicity of nanopesticides, a thorough risk evaluation of the nanocarrier–active ingredient combination, empty nanocarrier, released active ingredient, and nanoparticle ions is needed (Kah et al. 2021). According to Demir et al. (2022), interactions with the human microbiota can cause endocrine disruption, lung cancer, and gastrointestinal disorders as a result of toxic effects such as oxidative stress and DNA damage. Owing to their designed characteristics and environmental persistence, nanopesticides must be carefully formulated, which poses health hazards (Chaud et al. 2021). Inhalation toxicity assessment approaches is a great need for the agricultural workers who could be greatly exposed to the adverse effect of neurotoxicity and chronic inhalation exposure (Hamdaoui et al. 2019). Again the toxicity measurement is also important to develop more efficient evaluation methods (John et al. 2022).

5.4.2.2 Safety Regulation and Guidelines

Standardized processes should be stabilizing in each country for assessing the hazard of nanopesticides (Graczyk et al. 2021). To ensure worker safety, clear limits on the amount of a substance in the air should be set based on its concentration and safety guidelines (Graczyk et al. 2021). Guidelines should take into consideration the outcomes for unintended species and ecological systems, for that reason careful sustainable risk analyses are essential. More research on the behavior of nanopesticides in the environment is required for efficient control (de Albuquerque et al.

2020). Policies should encourage worldwide cooperation to help define safety criteria and enable the safe commercialization of nanopesticides (Isibor 2024).

Standardized risk assessment processes for nanopesticides are badly needed considering their different reactivity and toxicity in contrast with conventional pesticides (Graczyk et al. 2021; Kookana et al. 2014). Policies should encourage worldwide cooperation for consistent safety evaluations and simplify the trade of nanomaterials (Isibor 2024). Current environmental risk assessment approaches are very critical as nanopesticides exhibit distinct properties and behaviors like ecotoxicity in diverse ecosystems, their fate, exposure (Kookana et al. 2014). Effectiveness and effects on the environment can be identified by studying on particular nanopesticides such as nanoatrazine (de Albuquerque et al. 2020). To understand risk assessment on human health, the potential health hazards of nanopesticides and their active components should be fully examined (Kah et al. 2021).

The absence of specific regulations in many countries is currently obstructing the development and commercialization of nanopesticides. A regulatory framework is crucial for protecting the interests of the public, industry, and scientific perspectives (Jain and Das 2021). The lack of specific legislation in many countries obstructs the research and commercialization of nanopesticides. A practical regulatory framework that integrates public, industry, and scientific community perspectives is crucial for ensuring safety (Jain and Das 2021). This study presents a tiered assessment method, modifications to current pesticide frameworks, and increased data requirements to effectively perform a risk assessment of human health (Kah et al. 2021). The importance of regulatory frameworks in managing health issues, impacts on nontarget organisms, and overall safety in commercializing nano-biopesticides for sustainable agriculture are highlighted (Karthik et al. 2024). Current frameworks for nanoregulation lack cohesion; the EU has established guidelines specifically for agricultural nanotechnology, whereas India has developed comprehensive guidelines for assessing nano-inspired agricultural inputs. The guidelines highlight the necessity of risk assessment, exposure evaluation, and risk categorization to ensure safety (Chowdhury et al. 2023).

5.5 Regulatory and Policy Framework for Nanopesticides

5.5.1 Current Regulatory Landscape

5.5.1.1 International Guidelines

There are concrete legal structures present in nanotechnology for manufacturing, handling, and labeling of nanomaterials and nanomedicines which are governed by organizations such as the USEPA, USFDA, and European Commission (Malik and Patil 2020). They look at the present international laws pertaining to risk assessment, public opinion, and nanofood safety in light of the commercial potential of nanotechnology in food products (Tarhan and Yardımcı 2021). The article describes the regulatory frameworks that handle the health and environmental concerns associated with nanomaterials, pointing out changes to the OECD 412 and OECD 413 recommendations for inhalation risk assessments (Rehman and Moore 2021). The need for particular nanomaterial rules that address physicochemical characteristics, environmental release, human entry points, and toxicity is emphasized, and global regulatory difficulties are covered (Kaur et al. 2020). The EU's nanotechnology regulatory frameworks include sector-specific initiatives in food, cosmetics, and biocides. These frameworks were built via a hybrid governance model that combines old tools and new governance methodologies, reflecting a case-by-case approach (Ruggiu 2015). The precautionary principle, REACH, and possible nanotechnology registries are examples of frameworks that support customized methods over general fixes for tracking nanoparticle properties (Dean and Isabelle 2016). The study highlights how difficult it is to integrate many mandatory regulatory frameworks because of their cohabitation and the wide range of stakeholder uses, which makes it more difficult to make effective policies in nanotechnologies (Auplat 2012). Additionally, it looks at how ISO, IEC, and CEN technical committees produce standards and conduct regional and national regulatory activities, such as national registrations and certification systems, to guarantee safety and compliance (Ali and Hassan 2019). Nanopesticide regulations and policies are given in Table 5.1.

5.5.1.2 Registration and Approval Process

To guarantee safety and effectiveness before nanopesticides are used in agriculture, thorough ecotoxicological analyses, regulatory compliance procedures, and environmental impact studies are needed for registration and approval (Grillo et al. 2021). It highlights how most nations lack particular legislation for the licensing of nanopesticides, which hinders research and commercialization, and it calls for the creation of a regulatory framework (Jain and Das 2021). The registration process for nanobiopesticides faces greater regulatory hurdles compared to synthetic chemicals, as it is more complex and time-consuming (Machado et al. 2023). The authorization of nanopesticides should guarantee detailed risk evaluations for health, safety, and the environment. The present policy is supposed to be versatile enough to adapt new nanomaterials and formulations (Villaverde et al. 2017).

5.5.2 Prospective Legislative Barriers

5.5.2.1 Insufficiency of Environmental Metrics

Though long-term data on nanomaterials are crucial for understanding ecosystem dynamics and serve as baseline information for developing current environmental policies, there is clear evidence that historical environmental data are often discarded or poorly archived (Lavery and Ross 2023). Again environment is a complex matrix and it is challenging to investigate and distinguish nanomaterials in complicated systems due to several specialized properties of nanomaterials like diluted concentration, structural distinctions, and continuous change, all of which make it difficult to define exposure scenarios for risk assessment (Jiang et al. 2022). Furthermore, in-depth evaluation and interpretation of the traits and effects of green

Table 5.1 Nanopesticide regulation and policy

Regulatory framework/act	Description	Region/authority
Framework Convention for Nanotechnology	Balancing regulation to avoid hindering innovation while addressing risks effectively	International
UN Framework Convention on Climate Change (UNFCCC)	Adapting lessons from environmental agreements to emerging technologies like nanotech	Global
Vienna Convention for the Protection of the Ozone Layer	Developing similar adaptive mechanisms for nanotechnology governance	International
World Health Organization (WHO) Framework Convention on Tobacco Control	Translating public health regulatory mechanisms to address nanotechnology risks	Global
European Union Water Framework Directive	Incorporating public and stakeholder engagement in nanotechnology governance	European Union
Directive 89/391/EEC	Encourages improvement in the safety and health of workers at work.	European Union
Directive 98/24/EC	Protects workers from risks related to chemical agents at work	European Union
Directive 2004/37/EC	Addresses risks related to exposure to carcinogens or mutagens at work	European Union
Council Directive 89/655/EEC	Sets safety requirements for the use of work equipment by workers	European Union
Directive 1999/92/EC	Focuses on safety in explosive atmospheres	European Union
Directive 67/548/EEC	Regulates classification, packaging, and labeling of dangerous substances and preparations	European Union
Council Regulation (EEC) No. 793/93	Evaluates and controls risks of existing substances	European Union
REACH (Registration, Evaluation, Authorization, and Restriction of Chemicals)	Addresses risks of chemical substances, including nanoscale materials.	European Union
Voluntary Codes	Examples include the European Commission's Code of Conduct for responsible nanotechnology	Various organizations
ISO/TS 27687	Defines terminology for nano-objects like nanoparticles, nanofibers, and nanoplate	International
ISO/TR 12885	Guides health and safety practices in occupational settings relevant to nanotechnologies	International
EN 14530:2004	Determines diesel particulate matter in workplace atmospheres	European Standards (EN)

(continued)

Table 5.1 (continued)

Regulatory framework/act	Description	Region/authority
EN 14175–3:2003	Assesses the performance of fume cupboards	European Standards
EN 12469:2000	Sets criteria for microbiological safety cabinets in biotechnology	European Standards
Voluntary Nanoscale Materials Stewardship Program	Encourages submission of data on engineered nanomaterials in the United States	United States
Nanoinventory	Maintains data on engineered nanomaterials in Switzerland	Switzerland
Environment (Protection) Act, 1986	Regulates hazardous substances, including nanomaterials	India
Hazardous Waste Rules, 2016	Covers nanomaterial management	India
Food Safety and Standards Act, 2006	Ensures food safety with nanotechnology applications	India
Insecticides Act, 1968	Includes nanobased insecticides	India
CEPA, 1999	Framework for nanomaterials assessment	Canada
FDA	Ensures food product safety and labeling	Canada
PCPA	Regulates pesticides, including nanobased products	Canada
CAPA	Inspects agricultural products for safety	Canada
APVMA	Oversees agrochemicals with nanotechnology	Australia
FSANZ	Regulates food products containing nanotechnology	Australia
WHS law	Ensures worker safety concerning nanomaterials	Australia
TSCA	Regulates chemical substances, including nanomaterials	United States
FIFRA	Governs pesticide registration and usage	United States
FD and C Act	Manages nanotechnology in food and cosmetics	United States
NOP	Oversees organic product standards	United States
NRDA	Coordinates nanotechnology research efforts	United States
Nanotechnology Safety Guidance** (HSE, 2011)	Provides safe handling and use practices for nanomaterials	United States
FSA	Addresses nanomaterials in food safety assessments	United States
Regulation (EC) No 1107/2009	Rules for plant protection products	Europe
Regulation (EC) No 396/2005	Maximum residue levels for pesticides	Europe
Regulation (EC) No 1935/2004	Standards for food contact materials	Europe
Regulation (EC) No 767/2009	Feed additive safety requirements	Europe
Regulation (EU) No 2019/1009	CE-marked fertilizers safety guidelines	Europe

Source: Kumari et al. (2023) and Abbott et al. (2006)

nano-objects have been limited due to the complexity and the previous labeling of these particles into broad categories (Montaño et al. 2024).

5.6 Advances in Sustainable Nanotechnology Applications

5.6.1 Biodegradable and Eco-friendly Nanomaterials

Barrier materials, flexible electronics, and coatings are all sustainable substitutes for synthetic plastics (Shanmugam 2024). Eco-materials have lower carbon footprints and include recyclable, biodegradable, and low-impact materials (Sonia and Meena 2024). They are essential in industries for waste reduction and energy efficiency through ecolabeling and life cycle assessments (Nandhini et al. 2024). Biodegradable nanocomposites enhance mechanical qualities and decrease their impact on the environment by combining polymers with nanofillers. Continuous research is necessary for sustained development since, despite their advantages, problems with mechanical performance and cost-effectiveness still exist.

Being eco-friendly, biocompatible, and efficient in terms of photocatalytic dye degradation and possible anticancer characteristics, enzyme-powered ZnO and TiO2 nanoparticles made from biowastes are appropriate for environmental remediation and medical applications (Suvathi et al. 2024). Arora et al. (2024) reported that biobased nanomaterials, which are derived from renewable resources, are environmentally friendly and biodegradable. They have a lower carbon footprint and can decompose into harmless compound, which improves sustainability and lessens their influence on the environment. Sustainability as well as bioacceptance for eco-friendly restoration is facilitated by a number of biodegradable nanoparticles which are derived from sustainable means such as plant extracts and farm-generated waste.

Application of bionanocomposites enables healthcare sector with environmentally friendly solutions involving antimicrobial substances, pharmaceutical transporters, and wound healing, which boosts strength and biological properties while maintaining biodegradability (Nandhini et al. 2024). Biodegradable compositions, such as nanocomposites and nanofillers, are highly suitable for producing various products in the packaging, construction, and biomedical subsectors of the medical field. They support sustainability while offering enhanced mechanical properties (Shobhit et al. 2024).

Increased application of biodegradable and eco-friendly produce like green manufacturing of nanomaterials promotes environmental sustainability and reduces health concerns (Ranjana et al. 2024). Natural extracts approach is very advantageous in drug delivery and production of antibacterial medicine. It is also very helpful in energy saving, with limited production of hazardous by-products (Ponduru et al. 2023). Biodegradable nanocomposite materials made from biobased polymers reduces the reliance upon non-renewable energy sources and optimizes eco-friendly practices (El-Sayed and Youssef 2023). Efficient, stable, and budget-friendly environmentally friendly synthesis of biodegradable nanomaterials using plant-mediated methods emphasizes their significance in promoting sustainability across a variety of applications (Khan et al. 2023). Natural sources of biodegradable and environmentally friendly nanomaterials, including nanochitosan, nanocellulose, and nanostarch, are noteworthy for their biodegradability, renewability, and biocompatibility, which makes them appropriate for drug delivery, tissue engineering, and sustainable packaging (Kausar and Ahmad 2023).

5.6.1.1 Advantages of Biodegradable Nanoparticles

According to Geszke-Moritz and Moritz (2024), biodegradable nanoparticles reduce toxicity and adverse effects while improving therapeutic agent stability, bioavailability, targeted distribution, and controlled release. In sectors such as packaging, construction, automotive, and biomedicine, biodegradable nanocomposites promote sustainability by enhancing mechanical qualities, lowering environmental impact, and being adaptable for a range of structural applications. Because they are made from renewable resources, they also increase safety by lowering water consumption and emissions, decreasing waste, increasing food shelf-life, and improving safety through their antibacterial qualities (Yılmaz 2024). In food packaging, these nanoparticles enhance the mechanical qualities, barrier effectiveness, and material compatibility, propelling the industry forward with environmentally beneficial methods (Mishra et al. 2024). They are stable, nontoxic, biodegradable, and biocompatible, making them efficient nanocarriers with antibacterial and anticancer properties (Kučuk et al. 2023). Because of their low toxicity and biocompatibility, they are ideal for imaging, tissue engineering, and medication delivery, which decreases their long-term environmental impact (Sreena and Nathanael 2023). They are stable, affordable, and adaptable to a range of distribution methods, combining both hydrophilic and hydrophobic materials. They are synthesized from natural resources (Ponduru et al. 2023). They improve immunotherapy, especially for triple-negative breast cancer, by increasing antigen presentation, targeted drug administration, and tumor immunosuppression (Wang et al. 2022).

5.6.2 Smart Nanopesticides with Environmental Triggers

Nanopesticides encapsulated in iron–tannic acid networks react to seven environmental stimuli (pH, H_2O_2, glutathione, phosphates, EDTA, and near-infrared light) to facilitate controlled release, biodegradability, and reduced bioaccumulation, hence improving agricultural output (Dong et al. 2021). These insecticides, which are reactive to biotic and abiotic stimuli, seek to diminish pesticide application and alleviate health and environmental issues (Camara et al. 2019). They employ regulated release tactics for extended delivery, enhancing pest control while reducing consequences for ecosystems and human health (Verma and Saroop 2024). Stimuli-responsive nanopesticides activated by pH, laccase, phosphate, and phytic acid improve the efficacy and inhibit the premature release of avermectin (Liu et al. 2023). These nanocarriers, which are responsive to pH, temperature, light, and redox conditions, regulate the release of active ingredients, improving their efficacy while minimizing their environmental impact, frequently demonstrating dual or triple responsiveness (Shen et al. 2023). There are several enzyme-responsive fluorescent nanopesticides, for example, trypsin-mediated protein which allows for intelligent delivery and environmentally friendly formulations suitable for sustainable pest management (Ma et al. 2023).

Real-time plant monitoring studies have demonstrated that dual-responsive (redox/pH) fluorescent nanoparticles enhance pesticide release against *Botrytis cinerea* (Wang et al. 2024a, b). According to Hu et al. (2023), stimulus-sensitive nano-agrochemicals, like ZIFs, discharge bioactive compounds in counteraction to pH environmental fluctuations, increasing their efficiency and assuring food safety and standards. Avermectin is unleashed by nanopesticides in response to several interactions with phytic acid, laccase, phosphate, and pH. This increases the insecticidal efficacy while protecting crops and nontarget organisms (Liu et al. 2023). The intelligent constitutive structure of nanopesticides enhances crop protection while minimizing environmental impact by utilizing environmental stimuli—such as light, temperature, and pH—to control their release (Fincheira et al. 2023).

5.6.2.1 Sensitivity of Nanomaterials Toward pH and Temperature

Chang et al. (2024) explained how nanomaterials exhibit significant reactivity to pH and temperature, which affects the electrical properties of PIL films. This conclusion was drawn from research conducted on PDMAEMA. Again research on pH- and thermoresponsive hydrogels have revealed that structural alterations, swelling characteristics, and nanoparticle transport are influenced by temperature and calcium ions, indicating the potential for drug administration (Ruiz-Virgen et al. 2024). A supramolecular assembly of hydroxyamino amide and boric acid exhibits adjustable viscosity and viscoelastic characteristics through variations in pH and temperature, hence improving material regulation (Lin et al. 2023). The nanocomposite employs pH-temperature responsiveness to induce the collapse of the polymer layer at elevated temperatures and low pH, thereby releasing doxorubicin from the mesoporous silica shell and enhancing chemo-photothermal therapy (Mustafa et al. 2023). pH-responsive nanomaterials alter their properties according to pH, facilitating drug release in acidic tumors, whereas temperature-responsive materials improve targeted drug delivery and therapeutic effectiveness in cancer by transitioning from hydrophobic to hydrophilic states at designated temperatures (Mercy et al. 2023). Stimuli-responsive nanomaterials, which modify their properties in response to pH or temperature, can be manufactured into diverse nanostructures for recyclable industrial applications, hence solving environmental issues (Mirzababaei et al. 2022). MNs are engineered for targeted drug release, optimizing chemotherapy and photothermal therapy in triple-negative breast cancer by increasing the tumor drug concentration and minimizing systemic adverse effects, hence bolstering antitumor immunity (Liu et al. 2023). Smart pesticide systems responsive to the pH of insect saliva have shown success in pest-specific targeting, reducing off-target effects.

5.6.3 Use of Nanorobots for Pest Control

According to Shivalkar et al. (2023), microbots effectively break down dangerous pesticides in water through programmable propulsion, which is regulated by temperature and pH. This makes them suitable for pest management because of their low cytotoxicity and flexibility in challenging environments. Highlighting the promise of nanorobots in agriculture, hydrogel chitosan magnetic microrobots distribute ethyl parathion to mealworm larvae, improving pesticide specificity and efficiency (Maria-Hormigos et al. 2023). Nanotechnology enhances horticultural crop pest management by concentrating on nanopesticides, related hazards, and regulations (Ambethgar 2024). Although particular pest control applications are not described, it investigates hybrid control systems for nanorobots that could be modified for pest management by maximizing resource utilization through intelligent behavior and environmental interaction.

5.6.3.1 Nanobots in Agriculture

Equipped with sensors and delivery systems, nanobots can autonomously detect and target insect populations, offering an unprecedented level of precision in pest control. Cavalcanti et al. (2006) focused on nanorobots for environmental management and hydrology monitoring, not pest control, although their work is still in the experimental stage. This method has drawn attention for use in environmental monitoring and agriculture. Mohammed and Yahya (2024) discuss the role of nanotechnology in agriculture, highlighting how it can increase productivity and lower costs. They also support research on the use of these developments in the agricultural industry.

5.7 Case Studies of Nanotechnology Applications in Insect Pest Management

5.7.1 Improving Crop Yield with Nanotechnology

Silver nanoparticles have proven to be successful in managing *Bemisia tabaci*, one of the most damaging pests in cotton production, according to field tests conducted in cotton fields. Reports have shown that crop productivity has increased by 50% after application of nanoparticles along with the significant decrease in the use of chemical pesticides (Adeel et al. 2021). The production of vegetable also increased around 20% to 40% after application of nanofertilizers in experimental vegetables (radish, cabbage, eggplant, peppers, tomatoes, celery, and leek). An increasing trends of yields of 10.29%, 10.93–16.74%, 28.81% and 13.19% were found in field crops rice, spring maize, soybeans, and soybean oil, respectively (Liu et al. 2008). The 6–17% increase in production is due to improved nutrient use efficiency provided by nanofertilizers. According to Naik et al. (2021), this improvement results from the unique properties of nanomaterials that enhance nutrient absorption and distribution. Agri-nanotechnology boost crop yields by making plants more resilient to environmental stressors and improving pesticide delivery (Singh et al. 2023).

According to reports, compared with traditional fertilizers, foliar application of nanofibers increases grain yields by 10–25%, oilseed yields by 20–30%, and pulse yields by 13–15%. When phosphorus-based nanofibers are compared with traditional synthetic fertilizers, soybean seed production can be increased by 20%. Field studies with transplanted eggplant in soil infested with *Verticillium dahliae* revealed increased crop yield potential with NP CuO treatment, resulting in a notable 17–31% increase in marketable yield compared with bulk treatments (Servin et al. 2015). Compared with the application of granular zinc sulfate, the application of nanozinc oxide particles greatly increases tomato plant output and zinc content (Khanm 2017). According to Kole et al. (2013), fullerol treatment increased the yield of bitter melon by up to 54% in terms of biomass, 128% in fruit, 20% in fruit length, 59% in fruit number, and 70% in fruit weight. This finding indicates that nanotechnology significantly increases yield. Compared with traditional fertilizers, the use of nanofertilizers increased rice plant height by 3.6%, chlorophyll content by 2.72%, and reproductive tiller count by 9.10%, and spikelet count by 15.42% (Jyothi and Hebsur 2017). Soybeans plant show improvements of 20.4% in seed yield and 32.6% in the growth rate after application of conventional phosphorus fertilizers along with the nanohydroxyapatite (nHAP) nanoparticles (Tiwari 2022).

5.7.2 Application of Nanopesticides in Rice and Vegetables

Applying nanopesticides at low to medium concentrations has shown significant benefits for growth and metabolism in rice (Hu et al. 2024). The application of tebuconazole (TEB NPs) nanopesticides on rice and vegetables improves pesticide efficiency and supports sustainable agriculture by enhancing leaf deposition and increasing fungicidal effectiveness (Tong et al. 2024). In cucumber plant, the application of glycine-functionalized nanopesticides has demonstrated effective root-knot nematode prevention (Zhong et al. 2024). Nanoparticles have insecticidal properties as ZnO nanoparticles have shown about 55% death at varying dosages against the mealybug (Agredo-Gomez et al. 2024). In rice and vegetable farming, nanopesticides effectively eliminate pests including *Plutella xylostella*; in addition, efficient dose can also stimulate plant growth (Wang et al. 2018).

5.7.2.1 Enhanced Pest Mortality and Lower Residue Levels

Nanopesticides demonstrate significant effectiveness in rice and vegetable crops compared to traditional pesticides, resulting in reduced residues and higher insect mortality rates. Minimizing pesticide residues is crucial for adherence to food safety regulations. Nanobiopesticides effectively manage insect populations in agriculture, particularly rice and vegetable crops, by acting as antifeedants, larvicides, and pesticides. Specific delivery systems enhance environmental safety and plant protection (Lade and Google 2019). Applying a nanopesticide formulation of imidacloprid using chitosan nanoparticles led to a 60% decrease in aphid populations in affected fields (Huang et al. 2023). The investigation focuses solely on the 97.33% mortality rate of *Aedes aegypti* induced by biogenic silver nanoparticles (AgNPs) derived from *Azadirachta indica.* No investigation has been conducted on nanoemulsions composed of essential oils (Ansari et al. 2024). Silver nanoparticles (AgNPs) demonstrate superior efficacy as mosquito-larvicidal agents compared to other biosynthesized nanoparticles (Mondal et al. 2022). The combination of AgNPs and Triazophos in laboratory settings achieved over 90% larval mortality against *P. gossypiella*, indicating a significant larvicidal effect. Combined with Triazophos, green-synthesized silver nanoparticles (AgNPs) significantly eliminate pink bollworm (*Pectinophora gossypiella*) infestations, achieving over 90% larval mortality. This demonstrates their efficacy as larvicidal agents in cotton crop management (Khan et al. 2024). This paper does not comprehensively analyze the targeted release of polymeric nanocapsules to control *Helicoverpa armigera*. It refers to the use of microencapsulated pesticide formulations, including chlorpyrifos. Wilson and Boucher (2010) focused on pest management formulations, specifically addressing aphids and beet armyworms. A hematoporphyrin-gold nanocomposite is utilized as a photopesticide against the cotton leafworm *Spodoptera littoralis*. Nanotechnology demonstrates the capacity to develop environmentally friendly herbicides for cotton crops, as evidenced by its effectiveness in the phototoxic destruction of second-instar larvae (Salama et al. 2024). Mahesh and Madhavi (2023) found that Zn demonstrated significant antifeedant activity against fourth instar *Papilio demoleus* larvae, with antifeedant efficacy of 84.38% and 73.92% at 250 ppm after 24 and 48 h, respectively.

Combining neonicotinoid insecticides and ZnO nanoparticles enhances insecticidal activity against *Spodoptera litura* and *Aphis gossypii* (Elmasry 2021). Zinc oxide nanoparticles exhibit antifeedant properties. This research examines the insecticidal effects of these compounds on *Spodoptera frugiperda*, specifically regarding their influence on fertility, body malformations, adult emergence, and development (Pittarate et al. 2021). Zinc oxide nanoparticles are commonly referred to as ZnO nanoparticles. Engineered nanomaterials can be antimicrobial agents in agriculture by effectively controlling soil-borne *Fusarium* pathogens, promoting beneficial microorganisms, and enhancing soil microbial diversity (Agredo-Gomez et al. 2024). This research investigated the ability to eradicate mealybugs (*Puto barberi*), resulting in a mortality rate of approximately 55% (Li et al. 2024). Nanotechnology employs various nanoparticles, such as silver, zinc oxide, and copper, to address soil-borne diseases (Vijayreddy et al. 2023). Metal and metal oxide nanoparticles play a role in preventing soil-borne diseases by improving nutritional status, enhancing nutrient availability, and exhibiting antibacterial properties (Servin et al. 2015). Applying nanomaterials, such as chitosan and silica nanoparticles, can enhance plant resistance to soil-borne diseases (Worrall et al. 2018). Silver nanoparticles and chitosan represent highly effective treatments for soil-borne diseases (Vaghasiya et al. 2022). Nanomaterials facilitate enhanced crop health monitoring through nanosensors, accelerating disease detection (Jabran et al. 2024). Solid nanopesticides, controlled-release formulations, nanoemulsions, and nanosuspensions represent effective alternatives to traditional methods for managing grain storage pests. Their small size makes them secure and effective against nontarget species, making them enjoyable for pest management in stored cereals

(Tirunagaru et al. 2025). A biononanocomposite designed to inhibit stored-grain pests is synthesized by integrating purified montmorillonite (MMT) nanoclay and cinnamaldehyde derived from cinnamon essential oil. High fumigant toxicity and significant repellency toward *Callosobruchus chinensis* (Das et al. 2024).

Silver nanoparticles (AgNPs) and silicon nanoparticles (SiNPs) have shown promise in controlling *Tribolium castaneum*, a common pest in cereal storage practices (Salah et al. 2023). Research indicates that zinc, copper, and silica green nanoparticles effectively manage the rice meal moth *Corcyra cephalonica* (Biradar et al. 2021). Acay et al. (2024) demonstrated that *Pleurotus eryngii*-lycithin-chitosan nanomaterial (PEELCN) displays promising antifungal activity against key phytopathogenic fungi, including *V. dahliae* and *B. sorokiniana*. This suggests its potential as a biocompatible and sustainable nanofungicide for effective plant disease management. The environmental impact of antifungal medicines can be minimized, and their effectiveness can be enhanced through the targeted delivery of nanoparticle-engineered bio nanocomposites (Rameez et al. 2024).

5.8 Future Directions and Recommendations

5.8.1 Addressing Research Gaps

5.8.1.1 Need for Diverse Field Trials

Verifying the safety of nanopesticides in human and animal environments is essential for effective management and addressing research and extension gaps (Tirunagaru et al. 2025). Assessing the extent of nanopesticides harm to nontarget predators is essential, along with integrating chemical and biological control strategies (Yan et al. 2024). To address the challenges and ambiguities related to pest management applications, it is essential to enhance the development, assessment, and regulation of nanotechnology (Yousef et al. 2023). Developing innovative and secure nanopesticide formulations for aquatic applications is crucial. An initial investigation into antifeedant strategies and the safety of nanopesticide formulations is essential, according to Adetuyi et al. (2024). It is crucial to clarify the mechanisms of nanoparticle uptake and translocation in plants and optimize formulations and application methods for new cultivars (Heinisch et al. 2019).

Various agricultural field trials are essential for optimizing application and toxicity management to improve productivity and sustainability, as well as for characterizing nanoparticle absorption, mobility, and crop responses across different species and conditions. Hendrickson et al. (2017) emphasize the necessity of assessing the toxicity and appropriate application levels of nanoparticles in crop tissues and characterizing their responses. This research evaluates the safety, efficacy, and environmental impact of nanoformulations and nanoparticles by optimizing practices, enhancing yields, and reducing adverse effects on humans and ecosystems. Saranya et al. (2019) emphasize the necessity of exploring nanoparticles as substitutes for traditional agrochemicals and developing nanoformulations for these substances. Agricultural practitioners must be aware of nanoparticles' environmental and safety implications in agriculture, along with the potential of nanotechnology (Nadarajah 2020). Simultaneously examining toxicology and risk management, alongside assessing the environmental impact, safety, and efficacy of nanomaterials like nanopesticides and nanofertilizers, is essential for sustainable production (Kumar et al. 2022). To assess the impact of nanotechnology on crop yields, nutritional enhancement, and environmental safety, it is crucial to perform a variety of agricultural field trials (Jha et al. 2011). González-Feijoo et al. (2024) argue that a range of experiments is essential for evaluating the effectiveness of nanofertilizers across different contexts. The environmental impact will be reduced, and nutrient inputs will be decreased without affecting quality or yield. This will guarantee ongoing crop efficiency.

Field trials are essential to evaluate the effectiveness of nanofertilizers under different agricultural conditions (González-Feijoo et al. 2024). The experiments are crucial for enhancing nutrient management, as they demonstrate the efficacy of metal-based nanofertilizers in increasing agricultural yields, reducing environmental impact, and refining formulations, dosages, and application methods for various crops. Existing understanding of nanoparticle assimilation and translocation mechanisms in plants is insufficient, indicating the necessity of additional research (Heinisch et al. 2019). More experiments are essential for improving formulations, dosages, and application methods of metal-based nanofertilizers, which have been shown to enhance productivity while minimizing negative environmental impacts and optimizing nutrient management across various crops González-Feijoo et al. (2024).

5.8.2 Policy and Regulatory Considerations

5.8.2.1 Advancing Precise Legislative Protocols

Regulatory guidelines are essential for establishing safe utilization frameworks, assessing environmental impacts, standardizing practices, promoting innovation, and protecting human health from the risks associated with nanoparticles in products (Isibor 2024). Frameworks must assess the impact of nanomaterials on ecosystems to ensure the ethical application of food and agricultural systems. A conceptual framework comprising preassessment, exposure assessment, and risk classification is essential for ensuring environmental safety and public acceptance within a strong regulatory framework for nanotechnology in agriculture. Current frameworks exhibit inconsistency and ambiguity, highlighting the need for improved nano-regulatory frameworks and more precise risk assessment methodologies (Chowdhury et al. 2023). Trade-facilitating regulations and standardized safety evaluations enhance international cooperation (Isibor 2024). Rashid and Sharvil (2020) argue that the lack of a universally accepted definition of nanomaterials results in fragmented legislation. The need for enhanced risk assessment and characterization methodologies is highlighted by the inadequate implementation of many recommendations from previous reports despite notable progress. This research advocates for the establishment of tailored regulatory frameworks for nanotechnology, the advancement of enhanced measurement instruments, the standardization of terminology, and the rigorous enforcement of current regulations. The distinct challenges posed by nanomaterials require the adaptation of current regulations. The lack of established characterization and risk assessment methodologies requires the revision of pertinent regulations and development of measurement tools (Nielsen et al. 2023). Gupta et al. (2022) assert that improving technical competency in nanotechnology necessitates incorporating feedback from stakeholders, regulatory agencies, and established frameworks from aviation and biotechnology to develop regulatory structures.

Creating a regulatory framework for nanotechnology addresses discrepancies in definitions and safety concerns. Regulatory agencies, including the European Commission and the USFDA, establish global standards for evaluating nanomaterials (Malik and Patil 2020). This study highlights the lack of regulations tailored explicitly for health products utilizing nanotechnology. This document offers further direction on testing methodologies, classification, and shared characteristics for subsequent items. Halamoda-Kenzaoui et al. (2020) highlighted the necessity of ensuring consistency in silico biodistribution methodologies and testing protocols. International organizations aim to standardize safety studies within regulatory frameworks to mitigate potential risks linked to nanoparticles in food and enhance public perception and market opportunities (Tarhan and Yardımcı, 2021). Engelmann et al. (2018) argue that resolving issues and enhancing governance require a regulatory framework integrating research investments, classification, and risk assessments across the product lifecycle. Stone et al. (2018) assert that effective governance requires tailored legislative and regulatory frameworks that involve stakeholders and adapt current systems to ensure compliance and improve proactive safety. Kaur et al. (2020) posited that the secure utilization and disposal of nanomaterials necessitate an analysis of physicochemical properties, environmental release, human exposure pathways, and toxicity within the framework of nanotechnology regulations.

5.8.3 Integration into IPM Systems

Technological innovations that limit fungal growth and prevent pest penetration, such as biopesticides and antimicrobial sheets, improve food safety and agricultural productivity. Effective antimicrobial sheets and pest control nanocomposites must be tested in a range of agricultural settings (Hamzah et al. 2023). It is currently unclear how green nanotechnology affects various pest species and stages, as well as how well it works with biological control treatments. The susceptibility of plants to species-specific nanoparticles is not well understood (Helmy et al. 2022). More research is needed to quantify and examine the effects on biological outcomes (Beggiato et al. 2024). The transmission of knowledge from the laboratory to the field and the survival of microorganisms as biocontrol agents present obstacles when nanotechnology is used to increase microbial activity. An in-depth description of the interactions between nanoparticles is needed (Pasquoto-Stigliani et al. 2023).

5.8.3.1 Synergy with Biological Control

Nanotechnology enhances biological control through the development of nanocoatings and nanoparticles for improved pest and microbiological management (Hamzah et al. 2023). In conjunction with biological control agents, green nanotechnology effectively targets stored bean pest beetles and their parasitoid wasps, indicating potential for integrated pest management (Helmy et al. 2022). The engineering of nanoscale material interfaces facilitates the precise manipulation of biomolecular interactions, thereby enhancing biomedical device functionality and controlling biological responses. This synergy improves the efficacy and delivery of biobased insecticides, reduces environmental pollution, and enhances agricultural productivity through advanced nanocomposite matrices (Mishra et al. 2019). The utilization of nanoparticles derived from endophytic fungi combats bacterial and fungal pathogens, offering a sustainable alternative to traditional methods and reducing reliance on chemical pesticides (Shankar Naik 2020).

Nanotechnology enhances biopesticides by enabling targeted delivery, improved stability, and controlled release, supporting sustainable agriculture. However, challenges like high costs, limited persistence, and complex registration remain (Ghodake et al. 2018; Şahin et al. 2023). Again slow action and poor infection rates under harsh conditions can be resolved by using entomopathogenic fungi with nanoparticles. Follow-up research are required on the mode of action of entomopathogenic fungi nanosilver and the performance of bionanosilver for practical implementations (Bihal et al. 2023).

5.9 Conclusion

Nanotechnology has enhanced insect pest management's efficacy, precision, and sustainability. Nanoemulsions, nanocapsules, and nanopesticides address pesticide resistance, environmental pollution, and disease management. The enhanced solubility, concentrated distribution, and regulated release of these chemicals minimize inputs and environmental impacts, optimizing pest management. Nanomaterials affect parasite biology, selectivity, and off-target effects, enhancing sustainability. Nonetheless, research is necessary to assess the ecotoxicological effects on soil quality, aquatic habitats, and the safety of nontarget species. Regulations facilitate the achievement of nanopesticide safety, innovation, and public confidence. International collaboration and well-defined policies are essential. Case studies and field advancements illustrate the scalability and relevance of sustainable nanotechnology within agroecosystems. Future directions encompass interdisciplinary research, safety considerations, and sustainability initiatives. Collaboration among scientists, legislators, and industry leaders is essential for optimizing agricultural nanotechnology. Nanotechnology has the potential to enhance food security, safeguard the environment, and support sustainable agricultural practices. Nanomaterials have the potential to enhance agricultural soil quality, improve crop yield, and promote sustainability. These innovations improve sustainable agriculture and mitigate environmental impact, tackling food security challenges. Advancements in self-repairing materials, smart nanodevices, nanobio interfaces, and soil health rejuvenation present significant opportunities in agricultural nanotechnology. Collaboration among businesses, educational institutions, and governmental entities is essential for effective implementation and innovation (Gautam et al. 2023). The demand for enhanced crop yields and advanced farming techniques is projected to propel the growth of nanotechnology in agriculture from \$321.1 billion in 2022 to \$868.98 billion by 2031. Nanotechnology in agriculture is anticipated to enhance yield, optimize land management, improve fertilizer distribution, and reduce environmental impact. Nanotechnology addresses soil degradation and nutrient inefficiency to ensure food security amid changing environmental and socioeconomic conditions (Kumari et al.). Nanotechnology addresses critical agricultural challenges through sustainable, viable, innovative methods, transforming insect pest management.

References

Abbas T, Zahir ZA, Naveed M, Kremer RJ (2018) Limitations of existing weed control practices necessitate development of alternative techniques based on biological approaches. Adv Agron 147:239–280

Abbott KW, Marchant GE, Sylvester DJ (2006) A framework convention for nanotechnology. Envtl L Rep News and Analysis 36:10931

Abd El-Latef EA, Wahba MN, Mousa S, El-Bassyouni GT, El-Shamy AM (2023) Cu-doped ZnO-nanoparticles as a novel eco-friendly insecticide for controlling *Spodoptera littoralis*. Biocatal Agric Biotechnol 52:102823

Acay H, Güler Güney İ, Yildirim A, Derviş S, Dereli E (2024) Green synthesis of pleurotus Eryngii-derived nanomaterials for phytopathogen control. Chem Biodiversity 2:e202401972

Adeel M, Farooq T, White JC, Hao Y, He Z, Rui Y (2021) Carbon-based nanomaterials suppress tobacco mosaic virus (TMV) infection and induce resistance in Nicotiana benthamiana. J Hazard Mater 404:124167

Adetunji CO, Ugbenyen MA (2019) Mechanism of action of nanopesticide derived from microorganism for the alleviation of abiotic and biotic stress affecting crop productivity. Nanotechnology for Agriculture: Crop Production and Protection, pp 119–142

Adetuyi BO, Olajide PA, Adetunji CO (2024) Application of plant-based nanobiopesticides that could be applied for the rejuvenation of heavily contaminated environments. Handbook of agricultural biotechnology. Nanopesticides 1:193

Agredo-Gomez AD, Molano-Molano JA, Portela-Patiño MC, Rodríguez-Páez JE (2024) Use of ZnO nanoparticles as a pesticide: In vitro evaluation of their effect on the phytophagous *Puto barberi* (mealybug). Nano-Structures and Nano-Objects 37:101095

Ali S, Hassan P (2019) Standardization and regulations of nanotechnology and recent government policies across the world on nanomaterials. Academic Press, (pp. 419–446). https://doi.org/10.1016/B978-0-12-815322-2.00020-1

Al-Rawi ASM, Mohammed AA, Al-Taweel SK, Cheyed SH (2024) The role of nanotechnology in crop improvement—a review. SABRAO J. Breed. Genet 56(5):1929–1937

Alsmaeil AW, Hammami MA, Enotiadis A, Kanj MY, Giannelis EP (2021) Encapsulation of an anionic surfactant into hollow spherical nanosized capsules: Size control, slow release, and potential use for enhanced oil recovery applications and environmental remediation. ACS omega 6(8):5689–5697

Ambethgar AAS (2024) Applications of nanotechnology in pest management of horticultural crops. In Futuristic Trends in Chemical, Material Sciences & Nano Technology (Vol. 3, Book

4, Part 1, Chapter 4, pp. 34–60). IIP Series Publishing. https://iipseries.org/assets/docupload/rsl2024D50BE6E9985677A.pdf
Ansari N, Tripathi A, Ameen S, Shaheer Akhtar M, Jabeen F, Rahman Khan A et al (2024) Green Synthesis of Nanomaterials: Properties and Their Potential Applications. Sci Adv Mater 16(8):837–854
Anton N, Vandamme TF (2011) Nano-emulsions and micro-emulsions: clarifications of the critical differences. Pharm Res 28:978–985
Araujo Robusti E, Godoy Androcioli H, Ventura MU, Hata FT, Soares Júnior D, Menezes Júnior ADO (2023) Integrated pest management versus conventional system in the common bean crop in Brazil: insecticide reduction and financial maximization. Int J Pest Manag:1–11
Arora D, Kumar M, Bhatt S, Gautam RK, Taneja Y (2024) Biodegradability and sustainability of biobased nanomaterials. In: Biobased nanomaterials: applications in biomedicine, food industry, agriculture, and environmental sustainability. Singapore, Springer Nature Singapore, pp 509–535
Atreya K, Sitaula BK, Johnsen FH, Bajracharya RM (2011) Continuing issues in the limitations of pesticide use in developing countries. J Agric Environ Ethics 24:49–62
Auplat CA (2012) The challenges of nanotechnology policy making part 1. Discussing mandatory frameworks. Glob Policy 3(4):492–500
Avelelas F, Martins R, Oliveira T, Maia F, Malheiro E, Soares AM et al (2017) Efficacy and ecotoxicity of novel anti-fouling nanomaterials in target and non-target marine species. Mar Biotechnol 19:164–174
Batley GE, Kirby JK, McLaughlin MJ (2013) Fate and risks of nanomaterials in aquatic and terrestrial environments. Acc Chem Res 46(3):854–862
Beggiato M, Payen H, Rastogi R, Sidhoum C, Singh AC, Balakrishnan D, Krishnamoorthy S (2024) Nanoengineered bio-interfaces to interrogate/control biological events on material interfaces. In: In 2024 IEEE 24th International Conference on Nanotechnology (NANO). IEEE, pp 167–170
Belagalla N, Kumar A, Badekhan A, Chaturvedi K, Kumar A, Panigrahi CK et al (2024) Enhancing agricultural efficiency and biodiversity conservation through nano pesticides; a focus on food research. J adv biol 27(5):319–335
Benelli G (2018) Mode of action of nanoparticles against insects. Environ Sci Pollut Res Int 25(13):12329–12341
Bhaskar M, Kumar A, Rani R (2023) Application of nano formulations in agriculture. Biocatal Agric. Biotechnol 102934
Bhushan I, Mehta M, Sharma M, Chopra C, Chandra R, Noor ISM et al (2024) Role of nanomaterials in modern agriculture. Zastita MaterijalaE-ISSN:2466–2585
Bi F, Yu D, Wei Z, Wei H, Ren H, Wang Y et al (2023) Core–Shell polymeric nanostructures with intracellular ATP-fueled dsRNA delivery toward genetic control of insect pests. J Agric Food Chem 71(6):2762–2772
Bihal R, Al-Khayri JM, Banu AN, Kudesia N, Ahmed FK, Sarkar R et al (2023) Entomopathogenic fungi: an eco-friendly synthesis of sustainable nanoparticles and their nanopesticide properties. Microorganisms 11(6):1617
Biradar W, Nadagouda S, Aralimarad P, Hiregoudar S (2021) Entomotoxic effect of green nanoparticle an alternate strategy for stored grain pest management. Int J Trop Insect Sci 41:2829–2840
Bolan S, Sharma S, Mukherjee S, Zhou P, Mandal J, Srivastava P et al (2024) The distribution, fate, and environmental impacts of food additive nanomaterials in soil and aquatic ecosystems. Sci Total Environ 170013
Camara MC, Campos EVR, Monteiro RA, do Espirito Santo Pereira A, de Freitas Proença PL, Fraceto LF (2019) Development of stimuli-responsive nano-based pesticides: emerging opportunities for agriculture. J Nanobiotechnology 17:1–19
Canesi L, Della Torre C, Balbi T, Corsi I (2022) Interactive effects of nanomaterials with other contaminants on aquatic organisms: nTiO 2 as a case study. Toxicology of Nanoparticles and Nanomaterials in Human, Terrestrial and Aquatic Systems:101–120

Castanha RF, Santo Pereira ADE, Villarreal GPU, Vallim JH, Pertrini FS, Jonsson CM et al (2023) Ecotoxicity studies of two atrazine nanoformulations: From the evaluation of stability in media to the effects on aquatic organisms. Environ Pollut 335:122235

Cavalcanti A, Wood WW, Kretly LC, Shirinzadeh B (2006) Computational nanomechatronics: A pathway for control and manufacturing nanorobots. In: In 2006 International Conference on Computational Inteligence for Modelling Control and Automation and International Conference on Intelligent Agents Web Technologies and International Commerce (CIMCA'06). IEEE, pp 185–185

Chang CW, Lemich SB, Schütz P, Weerathaworn S, Weißpflog M, Wu CT et al (2024) Dual-responsive PIL films with gold nanoparticles: tailoring electrical responses to pH and thermal stimuli. ACS Appl Mater Interfaces 16(44):61096–61104

Charpe AM (2022) Nano-encapsulation of neem based pesticide formulations. J Pl Dis Sci 17(2):82–92

Chaud M, Souto EB, Zielinska A, Severino P, Batain F, Oliveira-Junior J, Alves T (2021) Nanopesticides in agriculture: benefits and challenge in agricultural productivity, toxicological risks to human health and environment. Toxics 9(6):131

Chavda VP, Balar PC, Bezbaruah R, Vaghela DA, Rynjah D, Bhattacharjee B et al (2024) Nanoemulsions: summary of a decade of research and recent advances. Nanomedicine 19(6):519–536

Chen Y, Dong J, Sun L, Chen W, Li X, Wang C et al (2023) Multidimensional stimuli-responsive and degradable pesticide nanocapsules through on site coordination assembly for enhanced pest and pathogen control. ACS Sustain Chem Eng 11(34):12809–12820

Chen H, Yang L, Zhao S, Xu H, Zhang Z (2024) Long-term toxic effects of iron-based metal-organic framework nanopesticides on earthworm-soil microorganism interactions in the soil environment. Sci Total Environ 917:170146

Chidiamassamba SB, Gomes SI, Amorim MJ, Scott-Fordsmand JJ (2024) Considering safe and sustainable by design alternatives–Environmental hazards of an agriculture nano-enabled pesticide to non-target species. Chemosphere 367:143582

Chowdhury AR, Sarkar DJ, Singh BK, Sarkar B (2023) Nanotoxicological issues in agriculture and related regulatory framework. In: Biological applications of nanoparticles. Singapore, Springer Nature Singapore, pp 217–231

Cui J, Sun C, Wang A, Wang Y, Zhu H, Shen Y et al (2020) Dual-functionalized pesticide nanocapsule delivery system with improved spreading behavior and enhanced bioactivity. Nanomaterials 10(2):220

Culliney TW (2014) Crop losses to arthropods. In: Integrated pest management: pesticide problems, vol 3. Springer, Netherlands, pp 201–225

Darnault C (2023) Nanomaterial interactions with the environment and biological systems: implications for soil, water, plants, and microorganisms. In EGU General Assembly Conference Abstracts (pp. EGU-10830)

Das R, Borthakur S, Arokiyaraj C, Hazarika M, Borgohain P, Borah TR et al (2024) A novel montmorillonite clay based bionanocomposite as an emerging biocontrol agent against stored grain pulse beetle. Hybrid Adv 5:100138

Dasgupta N, Ranjan S, Gandhi M (2019) Nanoemulsion ingredients and components. Environ Chem Lett 17:917–928

de Albuquerque FP, Preisler AC, Fraceto LF, Oliveira HC, de Castro VLS (2020) Overview of nanopesticide environmental safety aspects and regulatory issues: the case of nanoatrazine. From Research and Development to Mechanisms of Action and Sustainable Use in Agriculture, Nanopesticides, pp 281–298

Dean S, Isabelle S (2016) Regulating nanoparticles: a complex balancing act. (Doctoral dissertation, Durham University) Durham E-Theses

Demir E, Kansız S, Doğan M, Topel Ö, Akkoyunlu G, Kandur MY, Turna Demir F (2022) Hazard assessment of the effects of acute and chronic exposure to permethrin, copper hydroxide, acephate, and validamycin nanopesticides on the physiology of Drosophila: novel insights into the cellular internalization and biological effects. Int J Mol Sci 23(16):9121

Dhananjayan V, Jayakumar S, Ravichandran B (2020) Conventional methods of pesticide application in agricultural field and fate of the pesticides in the environment and human health. Controlled release of pesticides for sustainable agriculture 1–39
Dhumal N, Yadav V, Borkar S (2022) Nanoemulsion as novel drug delivery system: development, characterization and application. Asian J Pharm Sci 10(6):120–127
Dong J, Chen W, Feng J, Liu X, Xu Y, Wang C et al (2021) Facile, smart, and degradable metal–organic framework nanopesticides gated with FeIII-tannic acid networks in response to seven biological and environmental stimuli. ACS Appl Mater Interfaces 13(16):19507–19520
Dong W, Ren Y, Xue H (2024) Fabrication and application of carrier-free and carrier-based nanopesticides in pest management. Arch Insect Biochem Physiol 116(2):e22124
Du Q, Gao F, Cui B, Wang T, Chen F, Zeng Z et al (2024) Improving the stability, foliar utilization and biological activity of imidacloprid delivery systems: size effect of nanoparticles. Environ Res 119386
Dwivedi SK, Tomer A (2020) Applications for nanotechnology in the polyphagous destructive insect pest management of agricultural crops. Nanotechnology for Food, Agriculture, and Environment 181–188
Eghbalinejad M, Hofman J, Kotouček J, Grillo R, Hochmanová Bílková Z, Reiff N, Höss S (2024) Nanoenabled pesticides: a comprehensive toxicity assessment of tebuconazole nanoformulations with nematodes at single species and community level. Environ Sci Eur 36(1):51
Elmasry NS (2021) Efficacy of zinc oxide nanoparticle adding to two neonicotinoids pesticides against *Spodoptera litura* (Lepidoptera: *Noctuidae*) and *Aphis gossypii* Glover. Environ Sci Eur F Toxicology and Pest Control 13(1):209–215
El-Sayed SM, Youssef AM (2023) Eco-friendly biodegradable nanocomposite materials and their recent use in food packaging applications: a review. Sustain Food Technol 1(2):215–227
Engelmann W, Gaymard S, von Hohendorff R (2018) The legal perception of the risks of nanotechnologies in the environment: challenges and possibilities in the construction of a framework. Veredas do Direito 15:243
Faizan M, Karabulut F, Khan I, Akhtar MS, Alam P (2024) Emergence of nanotechnology in efficient fertilizer management in soil. S Afr J Bot 164:242–249
Fincheira P, Hoffmann N, Tortella G, Ruiz A, Cornejo P, Diez MC et al (2023) Eco-efficient systems based on nanocarriers for the controlled release of fertilizers and pesticides: Toward smart agriculture. Nanomaterials 13(13):1978
Gacem MA, Chaibi R (2022) Cu-based nanoparticles as pesticides: applications and mechanism of management of insect pests. In: Copper nanostructures: next-generation of agrochemicals for sustainable agroecosystems. Elsevier, pp 203–218
Gao Y, Xiao W, Gu C, Yu S, Li S, Zou A (2024) Layer-by-layer assembled decomposable nanocapsules for light-responsive release of pesticide imidacloprid on *Aphis craccivora* Koch. Pest Manag Sci 80(7):3207–3214
Gaur N, Chaudhary R, Diwan B (2022) Nanoencapsulation for chemical intermediate, biocides, and biobased binder. In: Smart nanomaterials for bioencapsulation. Elsevier, pp 267–279
Gautam R, Manisankar G, Bhuyan S, Kurdekar AK (2023) Exploring the future of agriculture through nanotechnology: a review. Int J Environ Clim 13(10):215–228
George S (2015) Nanomaterial properties: implications for safe medical applications of nanotechnology. In Nanotechnology in endodontics: current and potential clinical applications. Cham: Springer International Publishing, (pp. 45–69)
Geszke-Moritz M, Moritz M (2024) Biodegradable polymeric nanoparticle-based drug delivery systems: comprehensive overview, perspectives and challenges. Polymers 16(17):2536
Ghodake VN, Naik SV, Bhukhanwala KN, Kande KV, Bhor NJ, Patravale VB (2018) Nanoengineered systems for biopesticides. In: Handbook of Nanomaterials for Industrial Applications. Elsevier, pp 243–259
Ghosh SK, Bera T (2021) Unraveling the mechanism of nanoparticles for controlling plant pathogens and pests. In Advances in nano-fertilizers and nano-pesticides in agriculture. Woodhead Publishing, (pp. 415–436)

González-Feijoo R, Martinez-Castillo C, Rodríguez-Seijo A, Pérez-Rodríguez P, Arenas-Lago D (2024) Nanoagrochemicals versus conventional fertilizers: a field case study with tailor-made nanofertilizers for sustainable crop efficiency of *Brassica oleracea* L. convar. Capitata var. Sabauda. Agronomy 14(9):1885

Graczyk H, Fontana L, Younes M, Iavicoli I (2021) Safety, regulation, and policy. In: Health and environmental safety of nanomaterials. Woodhead Publishing, pp 83–95

Graily Moradi F, Asgari Lajayer B (2021) Nanoinsecticides: preparation, application, and mode of action. Nanotechnology applications in health and environmental sciences:385–404

Grillo R, Fraceto LF, Amorim MJ, Scott-Fordsmand JJ, Schoonjans R, Chaudhry Q (2021) Ecotoxicological and regulatory aspects of environmental sustainability of nanopesticides. J Hazard Mater 404:124148

Gupta A, Gupta S, Mukherjee G (2022) Nanotechnology and Regulatory Issues. In: Handbook of Nanotechnology in Nutraceuticals. CRC Press, pp 441–460

Gupta R, Malik P, Rani R, Solanki R, Ameta RK, Malik V, Mukherjee TK (2024) Recent progress on nanoemulsions mediated pesticides delivery: insights for agricultural sustainability. Plant Nano Biol 100073

Halamoda-Kenzaoui B, Box H, van Elk M, Gaitan S, Geertsma RE, Lafuente EG et al (2020) Launching stakeholder discussions on identified regulatory needs for nanotechnology-enabled health products. Precision Nanomedicine 3(2):608–621

Hamdaoui Q, Gaie-Levrel F, Macé T, Vaslin-Reimann S, Flamant F, Bencsik A (2019) Development and metrological characterization of an aerosol generation device dedicated to inhalation toxicology studies: the nanopesticide case. In: In 19th International Congress of Metrology (CIM2019). EDP Sciences, p 07002

Hamed R, Jodeh S, Alkowni R (2024) Nano bio fertilizer capsules for sustainable agriculture. Scientific Report 14(1):13646

Hamzah WM, Reyes IGC, Mendoza Pérez JA (2023) Nanotechnology for pest and microbiological control In Agricultural and Environmental Nanotechnology: Novel Technologies and their Ecological Impact. Springer Nature Singapore, Singapore, pp 393–409

He X, Fu P, Aker WG, Hwang HM (2018) Toxicity of engineered nanomaterials mediated by nano–bio–eco interactions. J Environ Sci Health C 36(1):21–42

Heinisch M, Jácome J, Miricescu D (2019) Current experience with application of metal-based nanofertilizers. In: MATEC Web of Conferences, vol 290. EDP Sciences, p 03006

Helmy EAM, San PP, Zhang YZ, Adarkwah C, Tuda M (2022) Fungus-synthesized nanoparticles and their target and nontarget effects on stored bean pest beetles and their parasitoid. Research Square (preprint)

Hendrickson C, Huffstutler G, Bunderson L (2017) Emerging applications and future roles of nanotechnologies in agriculture. Agric Res Technol 11(1)

Hu S, Yan C, Fei Q, Zhang B, Wu W (2023) MOF-based stimuli-responsive controlled release nanopesticide: mini review. Front Chem 11:1272725

Hu Z, Huang W, Wang M, Yang T, Pei H, Zhang F (2024) Regulation of nanofungicide based on ZIF-8 on the growth and metabolism of rice seedlings. Mater Today Commun 38:107873

Huang Y, Yang Y, Liang B, Lu S, Yuan X, Jia Z et al (2023) Green nanopesticide: pH-responsive eco-friendly pillar [5] arene-modified selenium nanoparticles for smart delivery of carbendazim to suppress sclerotinia diseases. ACS Appl Mater Interfaces 15(13):16448–16459

Isibor PO (2024) Regulations and policy considerations for nanoparticle safety. In: Environmental nanotoxicology: combatting the minute contaminants. Cham: Springer Nature Switzerland, pp 295–316

Isman MB (2020) Botanical insecticides in the twenty-first century—fulfilling their promise? Ann Rev Entomol 65(1):233–249

Jabran M, Ali MA, Muzammil S, Zahoor A, Ali F, Hussain S et al (2024) Exploring the potential of nanomaterials (NMs) as diagnostic tools and disease resistance for crop pathogens. Chem biol technol agric 11(1):75

Jacob S, Kather FS, Boddu SH, Shah J, Nair AB (2024) Innovations in nanoemulsion technology: enhancing drug delivery for oral, parenteral, and ophthalmic applications. Pharmaceutics 16(10):1333

Jafir M, Irfan M, Zia-ur-Rehman M, Hafeez F, Ahmad JN, Sabir MA et al (2023) The global trend of nanomaterial usage to control the important agricultural arthropod pests: a comprehensive review. Plant stress 10:100208

Jain A, Das S (2021) Regulatory requirements for nanopesticides and nanofertilizers. In: Advances in nano-fertilizers and nano-pesticides in agriculture. Woodhead Publishing, pp 145–152

Jasrotia P, Nagpal M, Mishra CN, Sharma AK, Kumar S, Kamble U et al (2022) Nanomaterials for postharvest management of insect pests: current state and future perspectives. Front Nanotechnol 3:811056

Jha Z, Behar N, Sharma SN, Chandel G, Sharma DK, Pandey MP (2011) Nanotechnology: prospects of agricultural advancement. Nano Vision 1(2):88–100

Jiang C, Liu S, Zhang T, Liu Q, Alvarez PJ, Chen W (2022) Current methods and prospects for analysis and characterization of nanomaterials in the environment. Environ Sci Technol 56(12):7426–7447

John AT, Wadhwa S, Mathur A (2022) Nanotoxicology: exposure, mechanism, and effects on human health. Front Toxicol:35–77

Jyothi TV, Hebsur NS (2017) Effect of nanofertilizers on growth and yield of selected cereals-A review. Agric Rev 38(2):112–120

Kah M, Johnston LJ, Kookana RS, Bruce W, Haase A, Ritz V, ..., Gubala V (2021) Comprehensive framework for human health risk assessment of nanopesticides. Nature nanotechnology 16(9), 955–964

Kannan M, Elango K, Tamilnayagan T, Preetha S, Kasivelu G (2020) Impact of nanomaterials on beneficial insects in agricultural ecosystems. In Nanotechnology for food, agriculture, and environment. Cham: Springer International Publishing, (pp. 379–393)

Karthik P, Saravanaraj A, Vijayalakshmi V, Ragavan KV, Vaidyanathan VK (2024) Nanobiopesticides: Significance, preparation technologies, safety aspects, and commercialization for sustainable agriculture. In: Nanohybrid Fungicides. Elsevier, pp 367–393

Kaur R, Watson JA (2024) A scoping review of postharvest losses, supply chain management, and technology: implications for produce quality in developing countries. J ASABE 67(5):1103–1131

Kaur J, Singh H, Khatri M (2020) Regulatory considerations for safety of nanomaterials. Nanomedicine for Bioactives: Healthcare applications, pp 431–450

Kausar A, Ahmad I (2023) Degradable green polymers, green nanopolymers and green nanocomposites derived from natural systems: statistics and headways. Nano-Horizons: J Nanosci Nanotechnol 2:36

Khan S, Qi K, Khan I, Wang A, Liu J, Humayun M et al (2023) Eco-friendly graphitic carbon nitride nanomaterials for the development of innovative biomaterials: Preparation, properties, opportunities, current trends, and future outlook. J Saudi Chem Soc 27(6):101753

Khan MN, Zhuo Z, Qayyum MA, Ali H, Xu D, Niaz Y, Yasin M (2024) Insecticidal efficacy of green synthesized silver nanoparticles, *Bacillus thuringiensis* and Triazophos against Pectinophoragossypiella. Emir J Food Agric 36:1–7

Khanm H (2017) Nano Zinc Oxide Boosting Growth and Yield in Tomato: The Rise of "Nano Fertilizer Era." Int J Agric Sci Res 7(3):197–206. https://doi.org/10.24247/IJASRJUN201724

Kokilavani S, Janani B, Balasurya S, Khan SS (2023) Accumulation of engineered nanomaterials in soil, water, and air. In: Agricultural and environmental nanotechnology: novel technologies and their ecological impact. Singapore, Springer Nature Singapore, pp 551–582

Kole C, Kole P, Randunu KM, Choudhary P, Podila R, Ke PC, Rao AM, Marcus RK (2013) Nanobiotechnology can boost crop production and quality: first evidence from increased plant biomass, fruit yield and phytomedicine content in bitter melon (Momordica charantia). BMC Biotechnol 13(1):37. https://doi.org/10.1186/1472-6750-13-37

Kookana RS, Boxall AB, Reeves PT, Ashauer R, Beulke S, Chaudhry Q, Van den Brink PJ (2014) Nanopesticides: guiding principles for regulatory evaluation of environmental risks. J Agric Food Chem 62(19):4227–4240

Kučuk N, Primožič M, Knez Ž, Leitgeb M (2023) Sustainable biodegradable biopolymer-based nanoparticles for healthcare applications. Int J Mol Sci 24(4):3188

Kumar S, Spandanagowda ND, Tirole R, Dave V (2022) Nanotechnology in the agriculture industry. In: Nanotechnology. CRC Press, pp 157–174

Kumar S, Bhuvaneshwari R, Jain S, Nirwan S, Fatima Z, Kumar D, … Poonam (2024) A systematic review on pesticide-loaded nanocapsules: a sustainable route for pesticide management to enhance crop productivity. Curr Nanosci 20(3), 280–297

Kumari R, Suman K, Karmakar S, Mishra V, Lakra SG, Saurav GK, Mahto BK (2023) Regulation and safety measures for nanotechnology-based agri-products. Front Genome Ed 5:1200987

Lade BD, Gogle DP (2019) Nano-biopesticides: synthesis and applications in plant safety. In Nanobiotechnology Applications in Plant Protection: Volume 2 (pp. 169–189). Cham: Springer International Publishing

Lavery H, Ross H (2023) Archives–an important requirement in environmental management. Australas J Environ Manag 30(2):141–147

Li X, Chen Y, Xu J, Lynch I, Guo Z, Xie C, Zhang P (2023) Advanced nanopesticides: advantage and action mechanisms. Plant Physiol. Biochem 108051

Li H, Guo Y, Liang A, Xu X, Hao Y, Shang H et al (2024) Commonly used engineered nanomaterials improve soil health via suppressing soil-borne fusarium and positively altering soil microbiome. ACS ESandT Engineering 4(4):915–927

Lin YT, Liu S, Bhat B, Kuan KY, Zhou W, Cobos IJ et al (2023) pH-and temperature-responsive supramolecular assemblies with highly adjustable viscoelasticity: a multi-stimuli binary system. Soft Matter 19(29):5609–5621

Liu J, Zhang YD, Zhang ZM (2008) Application study on nanobiotechnology in increasing yield benefit of rice, maize and soybean. J Anhui Agric Sci 36:15814–15816

Liu J, Xu D, Xu G, Li X, Dong J, Luan X, Du X (2023) Smart controlled-release avermectin-nanopesticides based on metal–organic frameworks with large pores for enhanced insecticidal efficacy. Chem Eng J 475:146312

Luo J, Gao Y, Liu Y, Du J, Zhang DX, Cao H et al (2021) Using a reactive emulsifier to construct simple and convenient nanocapsules loaded with lambda-cyhalothrin to achieve efficient foliar delivery and insecticidal synergies. Nanoscale 13(37):15647–15658

Ma S, Lin D (2013) The biophysicochemical interactions at the interfaces between nanoparticles and aquatic organisms: adsorption and internalization. Environ Sci-Proc Imp 15(1):145–160

Ma E, Fu Z, Chen K, Sun L, Zhang Y, Liu Z, Li L, Guo X (2023) Smart Protein-Based Fluorescent Nanoparticles Prepared by a Continuous Nanoprecipitation Method for Pesticides' Precise Delivery and Tracing. J Agric Food Chem 71(22):8391–8399. https://doi.org/10.1021/acs.jafc.3c01267

Machado S, Pereira R, Sousa RMO (2023) Nanobiopesticides: are they the future of phytosanitary treatments in modern agriculture? Sci Total Environ 166401

Mahesh L, Madhavi M (2023) Antifeedant efficacy of green synthesized zinc oxide nanoparticles against the larvae of *Papilio demoleus* L. J Agric Res ANGRAU 51(2):05–14

Malakar A, Snow DD (2020) Nanoparticles as sources of inorganic water pollutants. In: Inorganic pollutants in water. Elsevier, pp 337–370

Malik R, Patil S (2020) Nanotechnology: Regulatory outlook on nanomaterials and nanomedicines in United States, Europe and India. Appl Clin Res Clin Trials Regul Aff 7(3):225–236

Maria-Hormigos R, Mayorga-Martinez CC, Pumera M (2023) Magnetic hydrogel microrobots as insecticide carriers for in vivo insect pest control in plants. Small 19(51):2204887

Mei Z, Chen H, Weng T, Yang Y, Yang X (2003) Solid lipid nanoparticle and microemulsion for topical delivery of triptolide. Eur J Pharm Biopharm 56(2):189–196

Mercy DJ, Harini K, Madhumitha S, Anitha C, Iswariya J, Girigoswami K, Girigoswami A (2023) pH-responsive polymeric nanostructures for cancer theranostics. J Met Mater Miner 33(2):1–15

Mirzababaei S, Saeedian K, Navaei-Nigjeh M, Abdollahi M (2022) Using pH/thermal responsive materials. In: Nanomaterials Recycling. Elsevier, pp 223–247

Mishra P, Seenivasan R, Mukherjee A, Chandrasekaran N (2019) Polymer/layered silicate nanocomposites as matrix for bioinsecticide formulations. In: Nano-biopesticides today and future perspectives. Academic Press, pp 161–178

Mishra M, Mittal A, Negi GS, Singh P, Kaur S, Prasanna JL, Sharma P (2024) Biodegradable nanomaterials for sustainable food packaging applications. In: E3S Web of Conferences, vol 547. EDP Sciences, p 01014

Mohammed AAH, Yahya RT (2024) The role of nanotechnology in the treatment of agricultural problems. In achieving food security through sustainable agriculture. IGI Global, (pp. 268–282)

Mondal A, Debnath P, Kumar MN (2022) Multidimensional controlling properties of biofabricated silver-nanoparticles on different mosquito species. In: Cognitive Data Models for Sustainable Environment. Academic Press, pp 1–18

Mondéjar-López M, García-Simarro MP, Navarro-Simarro P, Gómez-Gómez L, Ahrazem O, Niza E (2024) A review on the encapsulation of "eco-friendly" compounds in natural polymer-based nanoparticles as next generation nano-agrochemicals for sustainable agriculture and crop management. Int J Biol Macromol 136030

Montaño MD, Goodman AJ, Ranville JF (2024) Past progress in environmental nanoanalysis and a future trajectory for atomic mass-spectrometry methods. Nano Impact 100518

Mustafa RA, Ran M, Wang Y, Yan J, Zhang Y, Rosenholm JM, Zhang H (2023) A pH/temperature responsive nanocomposite for chemo-photothermal synergistic cancer therapy. Smart Mater Med 4:199–211

Nadarajah K (2020) Advances in agronanotechnology and future prospects. Nanotechnol Food Agric Environ:85–104

Naik BSS, Mahawar N, Rupesh T, Dhegavath S, Singh Meena R (2021) Nanotechnology based nanofertilizer: a sustainable approach for enhancing crop productivity under climate changing situations. Curr Rese Agri Far 2(1):21–29

Nanda S, Ganguly A, Sarkar S, Mandi M, Das K, Ghanty S et al (2024) Nanopesticides in agriculture: scopes and limitations. In: Technical and technological solutions towards a sustainable society and circular economy. Cham: Springer Nature Switzerland, pp 179–193

Nandhini J, Karthikeyan E, Rajeshkumar S (2024) Eco-friendly bio-nanocomposites: pioneering sustainable biomedical advancements in engineering. Discover Nano 19(1):86

Nasser F, Constantinou J, Lynch I (2020) Nanomaterials in the environment acquire an "Eco-Corona" impacting their toxicity to Daphnia magna—a call for updating toxicity testing policies. Proteomics 20(9):1800412

Nazima S, Raul PK, Goswami D, Das D, Islam S, Tyagi V et al (2020) Bio-nanoparticle assembly: a potent on-site biolarvicidal agent against mosquito vectors. RSC Adv 10(16):9356–9368. https://doi.org/10.1039/C9RA09972G

Ngangom L, Shabaaz Begum JP, Gautam S, Venugopal D, Joshi S (2024) Nanotechnology in food crop production and food processing industry. In: Food production, diversity, and safety under climate change. Springer Nature Switzerland, Cham, pp 235–247

Nielsen MB, Skjolding L, Baun A, Hansen SF (2023) European nanomaterial legislation in the past 20 years—closing the final gaps. Nano Impact 100487

Nivetha N, Asha AD, Bandeppa Thakur, JK, Dukare AS, Paul B, Paul S (2021) Microbial-based nanoparticles as potential approach of insect pest management. Microbes for Sustainable Insect Pest Management: Hydrolytic Enzyme and Secondary Metabolite–Volume 2, 135–157

Olusola AB, Adetunji CO (2024) The influence of nanopesticides on the social economy, its bioeconomy perspectives in attaining sustainable development goals. handbook of agricultural biotechnology. Nanopesticides 1:279

Padrilah SN, Samsudin NIP, Shukor MYA, Masdor NA (2024) Nanoemulsion strategies in controlling fungal contamination and toxin production on grain corn using essential oils. Green Chem Lett Rev 17(1):2315138

Paramanik S, Ganguly A, Jana K, Rajak P (2023) 14 Nanotechnology: a novel weapon for insect pest and vector management. Nanocomposite and Nanohybrid Materials: Processing and Applications 17:277

Pasquoto-Stigliani T, Guilger-Casagrande M, Campos EV, Germano-Costa T, Bilesky-José N, Migliorini BB et al (2023) Titanium biogenic nanoparticles to help the growth of *Trichoderma harzianum* to be used in biological control. J Nanobiotechnology 21(1):166

Patil SR, Patil Rahul B (2022) Global scenario of pesticides and benefits from pesticide usage: a review. Int J Zool Invest 8(2):805–813

Patir K (2022) Nanoemulsion and its application in pesticide formulation. In: Handbook of research on nanoemulsion applications in agriculture, food, health, and biomedical sciences. IGI Global, pp 401–424

Pho QH, Losic D, Ostrikov KK, Tran NN, Hessel V (2020) Perspectives on plasma-assisted synthesis of N-doped nanoparticles as nanopesticides for pest control in crops. React Chem Eng 5(8):1374–1396

Pimentel D, Hart K (2001) Pesticide use: ethical, environmental, and public health implications. In New dimensions in bioethics: science, ethics and the formulation of public policy (pp. 79–108). Boston, MA: Springer US

Pittarate S, Rajula J, Rahman A, Vivekanandhan P, Thungrabeab M, Mekchay S, Krutmuang P (2021) Insecticidal effect of zinc oxide nanoparticles against *Spodoptera frugiperda* under laboratory conditions. Insects 12(11):1017

Ponduru HK, Gurubilli CSR, Mudunuru S, Banisetti DK (2023) Green synthesis of nano particles from biodegradable waste extracts and their applications. Int J Pharm Chem:46–54

Pu J, Luo ML, Wang K, Li X, Wu JB, Zhang N et al (2024) Fabrication and release mechanism study of a nanocapsule breaker for controlling degradation rate of insoluble residue in slickwater at moderate-temperature reservoirs. SPE J 1:1–12

Qing HS, Zhou S, Jiang Y, Xi X, Tan Y, Zhang G, Jiang NH et al (2022) Fate and transport of engineered nanoparticles in soils and groundwater (pp. 205–251). https://doi.org/10.1016/b978-0-12-824088-5.00003-3

Qu H, Wu S, Gong J (2023) A sustainable and smart fungicide release platform through cocrystal nanocapsules for improved utilization rate and environmental safety. Chem Eng J 473:145284

Rajić D, Vuković S, Vuković A, Cvjetković VG, Pelemiš S (2022) Nanocapsules based on plant proteins. Contemp. Mater 13(2)

Rameez M, Khan N, Ahmad S, Ahmad MM (2024) Bionanocomposites: a new approach for fungal disease management. Biocatal Agric Biotechnol 103115

Ranjana D, Gangopadhyay S, Das A, Das A, Bhattacharjee C (2024) Environment friendly green synthesis of nanomaterials. Elsevier BV, (pp. 320–333) https://doi.org/10.1016/b978-0-323-96020-5.00160-6

Rashid MR, Sharvil P (2020) Nanotechnology: Regulatory Outlook on Nanomaterials, and Nanomedicines in United States, Europe and India 7(3):225–236. https://doi.org/10.2174/2213476X06666191129094236

Rehman N, Moore S (2021) An overview of the state of the regulatory and preclinical requirements for nanomaterials including medical devices. Integ Envir Asses and Manag 17(6):1098–1104

Riyaz SM, Kathiravan K (2019) Integrated pest management approaches. Geminiviruses: Impact, Challenges and Approaches, pp 219–235

Roy J, Roy K (2022) Nano-read-across predictions of toxicity of metal oxide engineered nanoparticles (MeOx ENPS) used in nanopesticides to BEAS-2B and RAW 264.7 cells. Nanotoxicology 16(5):629–644

Ruggiu D (2015) The consolidation process of the EU regulatory framework on nanotechnologies: within and beyond the EU case-by-case approach. Eu J Law Tech 6(3)

Ruiz-Virgen L, Hernandez-Martinez MA, Martínez-Mejía G, Caro-Briones R, Herbert-Pucheta E, Río JMD, Corea M (2024) Analysis of structural changes of pH–thermo-responsive nanoparticles in polymeric hydrogels. Gels 10(8):541

Sabrine S (2023) Nano Green Pesticide, an innovative approach. All Sciences Abstracts 1. https://doi.org/10.59287/as-abstracts.925

Şahin EÇ, Aydın Y, Utkan G, Uncuoğlu AA (2023) Nanotechnology in agriculture for plant control and as biofertilizer. In: Synthesis of bionanomaterials for biomedical applications. Elsevier, pp 469–492

Saini H, Meena S, Mali PC, Kumari V, Kachhwaha N, Kaushik P, … Jangir RN (2023) Green nanoparticles application in the management of stored grain pests, *Tribolium* sp. 29(2), 195–202. ISSN 2456–9364
Salah S, Alyousuf AA, Abass MH (2023) Efficiency of silicon and silver nanoparticles against the infestation of *Triboliumcastanium* (Herbst)(Coleoptera: *Tenebrionidae*) on wheat grains under laboratory conditions. Basrah J Agric Sci 36(2):175–184
Salama SM, Al-Sherbini ASA, El-Ghannam G, Imam AI, Darwish WM (2024) Hematoporphyrin-gold nanocomposite as a smart generation of photopesticides to control cotton leafworm stages, *Spodoptera Littoralis*. Egypt J Chem 67(12):307–317
Santos PA, Biraku X, Nielsen E, Ozketen AC, Ozketen AA, Hakki EE (2024) Agricultural nanotechnology for a safe and sustainable future: current status, challenges, and beyond. J Sci Food Agric
Saranya S, Aswani R, Remakanthan A, Radhakrishnan EK (2019) Nanotechnology in agriculture. Nanotechnology for agriculture: advances for sustainable agriculture:1–17
Schmitz A, Badgujar C, Mansur H, Flippo D, McCornack B, Sharda A (2022) Design of a reconfigurable crop scouting vehicle for row crop navigation: a proof-of-concept study. Sensors 22(16):6203. e solutions against these pests
Servin A, Elmer W, Mukherjee A, De la Torre-Roche R, Hamdi H, White JC et al (2015) A review of the use of engineered nanomaterials to suppress plant disease and enhance crop yield. J Nanopart Res 17:1–21
Shafiq S, Shakeel F, Talegaonkar S, Ahmad FJ, Khar RK, Ali M (2007) Development and bioavailability assessment of ramipril nanoemulsion formulation. Eur J Pharm Biopharm 66(2):227–243
Shahzad K, Manzoor F (2021) Nanoformulations and their mode of action in insects: a review of biological interactions. Drug Chem Toxicol 44(1):1–11
Shankar Naik B (2020) Nanoparticles from endophytic fungi and their efficacy in biological control. Nanotechnology for Food, Agriculture, and Environment 161–179
Shanmugam K (2024) Recyclability of Spray Coated Smooth Nanocellu-lose Films as a Potential Sustainable Alternative to Synthetic Food Packaging-A Perspective
Sharma HC (2012) Conventional and biotechnological approaches for pest management: Potential and limitations. International Crops Research Institute for the Semi-Arid Tropics (ICRISAT) Patancheru-502 324, 3–25
Sharma S, Kooner R, Arora R (2017) Insect pests and crop losses. In: Breeding insect resistant crops for sustainable agriculture. Springer, Singapore, pp 45–66
Shen M, Liu S, Jiang C, Zhang T, Chen W (2023) Recent advances in stimuli-response mechanisms of nano-enabled controlled-release fertilizers and pesticides. Eco-Environmental Health 2(3):161–175
Shivalkar S, Yadav S, Tauheed A, Tariq M, Samanta SK, Sk MP, Sahoo AK (2023) Programmed logic operation-based controlled movement of self-destructive microbots for toxic pesticide degradation. ACS Appl Eng Mater 1(11):2858–2867
Shobhit D, Sawood GM, Mishra A, Singh N, Gautam SB, Gupta SK (2024) A biodegradable nanocomposite for structural applications (pp. 157–172). https://doi.org/10.58532/v3becs9p6ch1
Sigg L, Yue Y, Schug H, Röhder L, Piccapietra F, Odzak N et al (2014) Chemical aspects of nanoparticle ecotoxicology. Chimia 68(11):806–806
Singh U, Pramanik A, Sinha A, Pal AK, Rajput P, Chaubey KK et al (2023) Safety regulation of nanoemulsion in the food, agriculture and biomedical sector. In: Current trends in green nanoemulsions: food, agriculture and biomedical sectors. Singapore, Springer Nature Singapore, pp 299–311
Singh A, Tomar B, Harutyunyan A, Rajput P, Minkina T, Mandzhieva S et al (2024a) Utilizing nanotechnology in agriculture: a balancing approach between environmental health and risks. In: Nanotechnology applications and innovations for improved soil health. IGI Global, pp 380–399
Singh S, Angurana R, Kaur H, Katoch V, Dhanjal DS, Aguilar-Marcelino L et al (2024b) Nanoencapsulation of fertilizers: Delivery and applications. In: Nanofertilizer delivery, effects and application methods. Elsevier, pp 97–110

Sinha K, Ghosh J, Sil PC (2017) New pesticides: a cutting-edge view of contributions from nanotechnology for the development of sustainable agricultural pest control. In: New pesticides and soil sensors. Academic Press, pp 47–79

Sonia YK, Meena S (2024) Eco-materials: opportunity, applications, and challenges. Eco-Materials and Green Energy for a Sustainable Future 75–92

Sparks TC, Nauen R (2015) IRAC: mode of action classification and insecticide resistance management. Pestic Biochem Physiol 121:122–128

Sreena R, Nathanael AJ (2023) Biodegradable biopolymeric nanoparticles for biomedical applications-challenges and future outlook. Materials 16(6):2364

Stone V, Führ M, Feindt PH, Bouwmeester H, Linkov I, Sabella S et al (2018) The essential elements of a risk governance framework for current and future nanotechnologies. Risk Analysis 38(7):1321–1331

Suvathi S, Ravichandran K, Karunakaran M, Praseetha PK, Ayyanar M, Gobalakrishnan S (2024) Biowastes-derived enzyme-powered zinc oxide and titanium oxide nanomaterials synthesis for anticancer and eco-friendly photocatalytic activity. Appl Mater Today 36:102024

Tamta S, Vimal V, Verma S, Gupta D, Verma D, Nangan S (2024) Recent development of nanobiomaterials in sustainable agriculture and agrowaste management. Biocatal Agric. Biotechnol 103050

Tang FH, Lenzen M, McBratney A, Maggi F (2021) Risk of pesticide pollution at the global scale. Nat. Geosci 14(4):206–210

Tarhan Ö, Yardımcı AI (2021) Regulatory principles on food nano-particles legislated by international organizations. Safety and Regulatory Issues of Nanoencapsulated Food Ingredients, 251–290

Tirunagaru KC, Kolluru R, Singh IR (2024) A review on nanotechnological approaches for the management of stored insect-pests: present scenario and future prospects. J Sci Res Rep 30(10):756–769

Tirunagaru KC, Desai S, Singh IR (2025) Nano Technological Approaches for the Management of Stored Insect-Pests: Present Scenario and Future Prospects. In: Sustainable Development Goals Towards Environmental Toxicity and Green Chemistry. Springer, Cham, pp 249–267

Tiwari KN (2022) Nano technology based P fertilizers for higher efficiency and agriculture sustainability. Ann Plant Soil Res 24(2):198–207. https://doi.org/10.47815/apsr.2022.10149

Tong X, Tang J, Wang J, Wu Y, Zheng Z, Yuan C et al (2024) Engineered biodegradable nanopesticides for efficient leaf deposition and plant fungicides. AIChE J 70(6):e18414

Tripathi C, Sharma P, Singh IK, Singh A (2022) Integrative behavioral and ecotoxicological effects of nanoparticles. In: Emerging Contaminants in the Environment. Elsevier, pp 311–333

Vaghasiya TP, Kumar A, Nakum K (2022) A review on wide range application of nanoparticles in agriculture and its implications in plant disease management. J Nanoworld 8:55–65

Verma R, Saroop S (2024) Impact of nanopesticides in the environment: Solutions, threats, and opportunities. In: Pesticides in a changing environment. Elsevier, pp 251–292

Vijayreddy D, Dutta P, Puzari KR (2023) Nanotechnology in plant disease management. Research Biotica 5(2):56–62

Villaverde JJ, Sevilla-Morán B, López-Goti C, Sandín-España P, Alonso-Prados JL (2017) An overview of nanopesticides in the framework of European legislation. New Pesticides and Soil Sensors:227–271

Wang G, Xiao Y, Xu H, Hu P, Liang W, Xie L, Jia J (2018) Development of multifunctional avermectin poly (succinimide) nanoparticles to improve bioactivity and transportation in rice. J Agric Food Chem 66(43):11244–11253

Wang X, Ye L, He W, Teng C, Sun S, Lu H et al (2022) In situ targeting nanoparticles-hydrogel hybrid system for combined chemo-immunotherapy of glioma. J Control Release 345:786–797

Wang K, Li JQ, Lu J, Wang D, He S, Wang JX, Chen JF (2024a) Redox/pH Dual-Responsive Fluorescent Nanoparticles for Intelligent Pesticide Release and Visualization in Gray Mold Disease Synergistic Control. Langmuir 40(31):16511–16520

Wang Y, Qian M, Xie Y, Zhang X, Qin Y, Huang R (2024b) Biodegradable nanoparticles-mediated targeted drug delivery achieves trans-spatial immunotherapy. Fundamental Research 4(6):1639–1649
Wilson SL, Boucher RE (2010) Microencapsulated insecticide formulations. https://patents.google.com/patent/EP2403333A1/en
Wisnujati NS (2023) Unveiling the epoch: exploring pesticide utilization and trade trends globally and regionally from 1990 to 2020. Sci Res Jr Multidisc 2(2):108–121
Worrall EA, Hamid A, Mody KT, Mitter N, Pappu HR (2018) Nanotechnology for plant disease management. Agron 8(12):285
Wu S, Li G, Qu H, Xu W, Xu Y, Kovaleva EG et al (2023) Ecofriendly nanoemulsion to promote the wetting and spreading of droplet on soybean leaves for regulating the sheath blight. ACS Agric Sci Technol 3(11):1025–1033
Wu F, Zhang S, Li H, Liu P, Su H, Zhang Y et al (2024a) Toxicokinetics explain differential freshwater ecotoxicity of nanoencapsulated imidacloprid compared to its conventional active ingredient. Environ Sci Technol 58(22):9548–9558
Wu T, Wang Y, Liu S, Zheng Z, He C, Yao W et al (2024b) Facile-prepared size-controllable nanogels for enhancing bidirectional translocation, control efficiency, and security of nanopesticide. Adv Funct Mater 2410555
Xu C, Cao L, Chen H, Liu T, Liang W, Li Y (2024) Copper-driven formation of prothioconazole nanocomplex: an innovative strategy to prepare nanopesticide with improved bioactivity and reduced environmental impacts. Small 2406419
Yan X, Shan Y, Ma Y, Wang Y, Wu C, Ren X et al (2024) Thiacloprid-silica nano-delivery system enhances toxicity against *Aphis gossypii* and improves non-target biosafety. Chemosphere 367:143596
Yılmaz BB (2024) Perspectives of biodegradable nanocoatings in food packaging. In: Sustainable approach to protective nanocoatings. IGI Global, pp 113–169
Yousef HA, Fahmy HM, Arafa FN, Abd Allah MY, Tawfik YM, El Halwany KK et al (2023) Nanotechnology in pest management: advantages, applications, and challenges. Int J Trop Insect Sci 43(5):1387–1399
Zhong X, Su G, Hao L, Chen H, Li C, Xu H et al (2024) Foliar application of glycine-functionalized nanopesticides for effective prevention and control of root-knot nematodes via a targeted delivery strategy. Pest Manag Sci 80(4):2120–2130
Zhou Z, Tang G, Liu Y, Zhang X, Huang Y, Wang J et al (2024) Facile fabrication of pesticide nanocapsules using cinnamaldehyde derived imide polymer as wall material for pH-responsive and ultraviolet shielding properties. Chem Eng J 489:151430
Zhu F, Wang XW, Chen H, Wen J (2024) Chiral nanopesticides: the invincible opponent of plant viruses. Trends Plant Sci 29(2):120–122

Nanotechnology and Agribusiness: Part-I

6

Andersen E. Schlogl, Patrícia Avelar Ribeiro, Marcos Sousa Silva, Cintia Maria Sousa Ferreira, Millena Alves Máximo Vaz, Elton Barreto, Ana Luíza Schuenck Alberice Monteiro, and Igor J. B. Santos

Abstract

Nanotechnology has emerged as a revolutionary field with the potential to transform agribusiness by addressing some of the most pressing challenges in sustainable agriculture and environmental restoration. This chapter provides a comprehensive overview of the integration of nanotechnology in agribusiness, exploring new possibilities for enhancing productivity, reducing environmental impact, and improving resource efficiency. The introduction sets the stage by discussing the global need for innovative solutions in agriculture. Next, the chapter discusses applications of nanotechnology, including smart delivery systems for agrochemicals, sensors for real-time monitoring, and nanomaterials for improving soil health and water retention. In the section on startups, innovation, and scale-up, the chapter highlights how nanotechnology is catalyzing the growth of new enterprises and fostering innovation through scalable technologies. This

A. E. Schlogl · P. A. Ribeiro
Postgraduate of the Physics and Chemistry of Materials Program (FQMAT), Federal University of São João del-Rei (UFSJ), São João del-Rei, Brazil

M. S. Silva
Postgraduate of the Chemical Engineering Program (PPGEQ), Federal University of São João del-Rei (UFSJ), São João del-Rei, Brazil

C. M. S. Ferreira
Postgraduate of the Bioprocess Engineering and Biotechnology Program (PPGEnBio), Federal University of São João del-Rei (UFSJ), São João del-Rei, Brazil

M. M. Vaz · E. Barreto · A. L. S. A. Monteiro · I. J. B. Santos (✉)
Department of Chemical Engineering (DEQUI), Federal University of São João del-Rei (UFSJ), São João del-Rei, Brazil
e-mail: igorboggione@ufsj.edu.br

S. S. Purewal et al. (eds.), *Agri-Nanotechnology: Innovations for Sustainable Agriculture and Environmental Restoration*,
https://doi.org/10.1007/978-981-96-9756-4_6

is followed by an analysis of patents and market trends, showing how intellectua-Gproperty is shaping the competitive landscape and the commercialization of nanotechnology-based solutions. The strategies section outlines critical approaches for successfully integrating nanotechnology into agribusiness, from R&D investment to collaborative partnerships with industry stakeholders. Finally, the chapter concludes with a future outlook, discussing the potential long-term impacts of nanotechnology on global food security, environmental sustainability, and the development of smart farming systems. Through these lenses, the chapter underscores the vital role of nanotechnology in driving the next wave of innovation in agriculture and environmental restoration.

Keywords

Agribusiness · Agrochemicals · Environmental restoration · Nanotechnology · Productivity · Sensors · Sustainable agriculture

6.1 Introduction

Conventional agriculture, based on the intensive use of chemical inputs and monocultures, has caused significant negative impacts on the environment. The excessive use of fertilizers and pesticides contaminates soil and water, while also contributing to the loss of environmental biodiversity (Patel et al. 2020).Large-scale agricultural production also requires vast amounts of water, a resource that is becoming increasingly scarce in various regions of the world (Ungureanu et al. 2020). Alarming data from the United Nations reveals that the number of people experiencing hunger is rising exponentially, with approximately 800 million facing food shortages (United Nations 2022). The projection is even more concerning: by 2050, the global population is expected to reach 9 billion, requiring a 50% increase in food production (FAO 2017). The current production, characterized by traditional and unsustainable agricultural practices, is proving insufficient to meet this growing demand. To reverse this scenario, it is essential to adopt more sustainable agricultural practices that promote the conservation of natural resources and the resilience of agricultural systems (Usman et al. 2020).Nanotechnology can play a significant role in achieving the zero-hunger goal, one of the 17 Sustainable Development Goals proposed by the United Nations, by providing innovative solutions that enhance food production efficiency and environmental preservation (Kah et al. 2019; Usman et al. 2020). In this context, nanotechnology-based solutions are gaining prominence for their ability to improve seed germination and stimulate plant growth through the controlled release of agrochemicals. Controlled release significantly reduces the use of chemical inputs and minimizes nutrient losses during fertilization. Furthermore, nanotechnology in agriculture offers innovative approaches for the protection and remediation of water and soil resources, contributing to the sustainable production of high-quality food (Fu et al. 2020; Iavicoli et al. 2017; Machado et al. 2022). In this study, we will explore how nanotechnology is revolutionizing agribusiness by providing innovative solutions to the challenges of food production. Throughout the following sections, we will discuss specific applications of nanotechnology in agriculture, such as smart delivery systems for agrochemicals, sensors for real-time monitoring, and nanomaterials to enhance soil health and water retention. Additionally, we will analyze the market landscape and public policies driving this field and present an overview of prospects. Our goal is to demonstrate how nanotechnology can contribute to a more sustainable and secure future, ensuring the production of high-quality food to meet the needs of a growing global population.

6.2 New Possibilities of Nanotechnology in Agribusiness

The global agricultural sector is currently grappling with major challenges such as low productivity, declining soil fertility, excessive use of agrochemicals, limited arable land, and dwindling water resources for irrigation (Shahzad et al. 2019). These issues are further worsened by climate change, driven by rising CO_2 levels and increasing temperatures (Zhang et al. 2021). The finite availability of natural resources and the rapidly growing global population highlight the need for innovations in conventional agriculture (Mittal et al. 2020). With limited access to essential resources like water, land, soil, and forests, alongside the plateauing of crop productivity, there is a demand for effective agricultural approaches that are both economically viable and environmentally sustainable (Sharma et al. 2022). During the first Green Revolution, various agrochemical products including insecticides, herbicides, fungicides, and fertilizers were developed, significantly boosting agricultural production (Kumar et al. 2021). However, conventional agrochemicals are no longer as effective in improving crop productivity due to their negative impact on soil microbial health and nutrient balance. This, combined with growing environmental pressures and pollution, blocks progress toward sustainable agriculture (Husen 2021). To address these challenges, the shift toward more efficient and sustainable agricultural systems has become essential. Nanotechnology offers a promising solution, driving innovations in pest and disease management, fertilizers, agrochemicals, biofertilizers, biostimulants, and post-harvest storage, paving way for a sustainable and economic agricultural future (Santos et al. 2024).

6.2.1 Smart Delivery Systems for Agrochemicals

Traditional agrochemical methods for pest control and fertilization often lead to rapid, uncontrolled release of substances into the environment, reducing resource efficiency and posing health risks. The emergence of smart delivery systems has introduced a more regulated approach to agrochemical release (Grillo et al. 2021). Controlled-release systems (CRS) allow for the gradual and targeted delivery of active ingredients, maintaining consistent concentrations over extended periods. This innovation offers benefits such as enhanced durability, reduced toxicity, and increased efficacy, aligning with the goals of sustainable agriculture (Guha et al. 2020; Shao et al. 2022). Incorporating nanotechnology into CRS has further

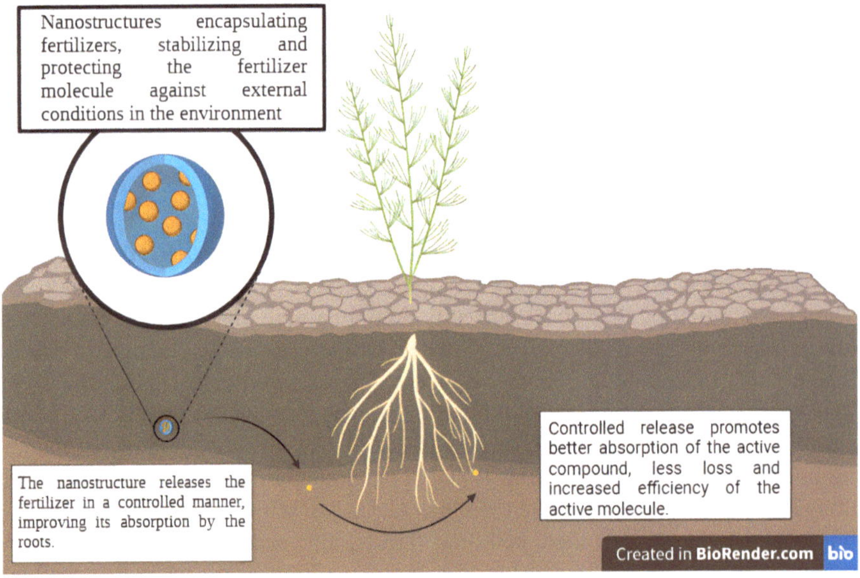

Fig. 6.1 Schematic showing a controlled release process for a nanofertilizer

advanced nutrient delivery to plants and soil. Nanoagrochemicals, including nanofungicides, nanobacteriocides, nanoherbicides, nanoinsecticides, and nanofertilizers, provide notable advantages such as improved crop protection, enhanced nutrient efficiency, and enriched soil fertility (Fincheira et al. 2023; Sharma et al. 2022). These solutions are projected to surpass traditional agro chemicals, potentially boosting agricultural productivity by 20–30%. Among them, nanofertilizers are particularly impactful in supporting crop growth and development, positioning nanoagrochemicals as essential tools for advancing sustainable farming practices (Rajput et al. 2021). Fig. 6.1 shows a schematic of a controlled release process for a nanofertilizer.

6.2.2 Nanopesticides

Nanopesticides utilize nanomaterials (NMs) to create innovative pesticide formulations for agriculture (Usman et al. 2020). These formulations include nanostructures (NPs) with natural antimicrobial, insecticidal, or herbicidal properties, as well as NPs specifically designed to encapsulate synthetic molecules for slow and sustained release (Wang et al. 2022). Nanoscale particles bring unique advantages to nanocarrier complexes, employing materials such as silica, lipids, polymers, and metals. They enable efficient loading of active ingredients and support gradual release, enhancing water solubility and environmental stability (Yadav et al. 2020). This prolongs exposure to targets like pests and weeds while reducing the amount of pesticide needed (Ali et al. 2021). By optimizing delivery, this approach lowers the required pesticide dose, mitigates toxicity, and minimizes undesirable side effects, offering a more sustainable and efficient solution for pest management in agriculture (da Silva et al. 2024). Nanopesticides can be developed through two main approaches: one involves directly processing the active ingredients into nanostructures, while the other relies on using nanocarriers to encapsulate the substances (Ding et al. 2023). Various nanomaterials have been explored as nanopesticides due to their insecticidal, bactericidal, and fungicidal properties (Table 6.1). Metallic nanoparticles, in particular, are primarily utilized to combat fungi and bacteria because of their well-known antimicrobial properties (Correa et al. 2020). Several studies have focused on the direct transformation of metallic nanoparticles for pesticide production (Abdallah et al. 2020; Abdelaziz et al. 2023; Ahmed et al. 2020; Akpinar et al. 2021; Al-Otibi et al. 2021; Kashyap et al. 2023; Li et al. 2020; Mali et al. 2020; Manimegalai et al. 2020; Mishra et al. 2020; Oussou-Azo et al. 2020; Rizwana and Alwhibi 2021; Vanti et al. 2020). Additionally, the use of nanocarriers or nanoencapsulation has become a widely adopted strategy for achieving controlled pesticide release (Artusio et al. 2021; Dong et al. 2021; Elabasy et al. 2020; Kumar et al. 2022; Maluin et al. 2020). This approach enhances the dispersibility and wettability of pesticides while shielding them from environmental factors (Ding et al. 2023).

6.2.3 Nanofertilizers

Nanofertilizers act as "smart" carriers for macronutrients (e.g., nitrogen, potassium, phosphorus, calcium, magnesium) and micronutrients (e.g., zinc, copper, iron, manganese), enhancing nutrient delivery to plant parts, which boosts crop yields and improves vegetation (Singh et al. 2021). They increase nutrient uptake by roots, optimize soil nutrient management, reduce nutrient waste and environmental losses, and minimize immobilization (Fincheira et al. 2023). Kumar et al. (2021) presented various nanoformulations in fertilizers designed to enhance plant growth phases, enzymatic activities, and nutrient absorption and release. Nanomaterials also serve as targeted biofertilizers, such as micronutrient formulations containing elements like iron (Fe) and zinc (Zn) (Haydar et al. 2021), or macronutrient fertilizers with nitrogen (N), phosphorus (P), and potassium (K) (Alimohammadi et al. 2020). Recent studies have demonstrated the benefits of using nanomaterials, such as bio-nano-CuO, bio-nano-Se, and ZnO nanostructures, to improve the growth of crop such as strawberries and corn. These materials allow for the slow release of nutrients and more efficient absorption by plants, highlighting their potential to overcome the limitations of conventional fertilizers and contribute to more sustainable agriculture (Azam et al. 2022; El-Bialy et al. 2023).

Table 6.1 New nanomaterial-based pesticides identified in recent literature

Type of pesticides/nature	Nanotechnology	Size (nm)	Pathogen	References
Bactericide	AgNPs derived from *Solanum torvum* fruit extract	27	*Xanthomonas axonopodis* pv. *Punicae* *Ralstonia solanacearum*	Vanti et al. (2020)
Bactericide	AgNPs biosynthesized from *Stenotrophomonas* sp. BHU-S7	~12	*Xanthomonas oryzae* pv. *oryzae* (Xoo)	Mishra et al. (2020)
Bactericide	AgNPs synthesized from *Bacillus cereus*	18–39	*Xanthomonas oryzae* pv. *oryzae (Xoo)*	Ahmed et al. (2020)
Bactericide	Nanocomposite that MoS2 sheets loaded with CuNPs (MoS2-CuNPs)	<10	*Xanthomonas oryzae* pv. *oryzae* (Xoo)	Li et al. (2020)
Bactericide	Chitosan nanostructures (CSNPs) and zinc oxide nanoparticles (ZnONPs) biosynthesized from dried green tomato extract	23.8–91.9 31.3–88.9	*Xanthomonas oryzae* pv. *oryzae* (Xoo)	Abdallah et al. (2020)
Bactericide/fungicide	Copper nanoparticles produced from the strain *Stenotrophomonas maltophilia*	18–32	*Agrobacterium tumefaciens, Xanthomonas campestris, Xanthomonas axonopodis, Ralstonia solanacearum/ Alternaria pori, Fusarium roseum, Macrophaminaphaseolina, Sclerotinia sclerotiorum, Fusarium solani, Rhizoctonia solani, Alternaria brassicicola, Fusarium oxysporum* sp. *Lycopersicum*, and *Pencillium digitatum*	Kashyap et al. (2023)
Fungicide	Silver nanoparticles (AgNPs) synthesized from *Menthapulegium* leaves	61.81	*Alternaria alternata, Colletotrichum musae, Fusarium solani, F. oxysporum, Helminthosporium sativum*, and *Macrophomina phaseolina*	Rizwana and Alwhibi (2021)
Fungicide	AgNPs	3–10	*Fusarium oxysporum*	Akpinar et al. (2021)
Fungicide	AgNPs synthesized from Ziziphus spina-christi leaves (ZSCE)	11.25	*Fusarium oxysporum*	Abdelaziz et al. (2023)

(continued)

Table 6.1 (continued)

Type of pesticides/ nature	Nanotechnology	Size (nm)	Pathogen	References
Fungicide	AgNPs synthesized from *Malva parviflora* L	50.6	*Helminthosporium rostratum*, *Fusarium solani*, *Fusarium oxysporum*, and *Alternaria alternata*	Al-Otibi et al. (2021)
Fungicide	Copper nanoparticles (Cu-NPs) and copper oxide nanoparticles (CuO-NPs)	40–60	*Colletotrichum gloeoesporioides*	Oussou-Azo et al. (2020)
Fungicide	Cu-NPs biosynthesized using *Celastruspani culatus* leaf extract	2–10	*Fusarium oxysporum*	Mali et al. (2020)
Insecticide	AgNPs biosynthesized using *Leonotis nepetifolia*	8.5	*Spodoptera litura*, *Helicoverpa armigera*	Manimegalai et al. (2020)
Fungicide	Nanoencapsulation of fungicides in chitosan	5–58	*Ganoderma boninense*	Maluin et al. (2020)
Fungicide	Nanoencapsulation of fungicides in biopolymeric chitosan-carrageenan nanocomposites	66.6–231.82	*A. solani, A. alternata, S. lycopersici, SclerotiniaSclerotiorum*	Kumar et al. (2022)
Fungicide/ herbicide	Nanoencapsulation of pesticides in carboxylated porous carbon nanoparticles (PCNs)	90.8	*Candida albicans*	Dong et al. (2021)
Herbicide	Alginate nanoparticles using an inverse mini emulsion template in sunflower oil and used them to encapsulate a herbicide	20	–	Artusio et al. (2021)
Insecticide	Nanoformulation (NF) ofthiamethoxam (TMX) in cellulose nanocrystals (CNCs)	124,5 +/−0.5	*Phenacoccus solenopsis*	Elabasy et al. (2020)

6.3 Sensors for Real-Time Monitoring

Traditional farming methods significantly contribute to water body eutrophication and soil quality degradation. Additionally, improper agricultural techniques harm beneficial insect habitats and other wild life. To address these challenges, precision agriculture techniques should be implemented for real-time monitoring (Bharti et al. 2024). Integrating nanomaterials with biosensors—referred to as nanobiosensors—has enhanced detection capabilities for numerous environmental applications. Nanobiosensors are specifically designed, noninvasive, and highly sensitive tools that combine the unique properties of nanomaterials with the specificity of biological components (Kalyani et al. 2020). These sensors provide real-time response signals that can be easily collected and analyzed, offering efficient detection with higher precision, continuous monitoring, and reduced time and cost requirements (Bhandari et al. 2023).

Overall, nanobiosensors have become integral to the advancement of “smart agriculture” or “precision agriculture” by enabling sensing, processing, and detecting environmental changes (Thakur et al. 2022). They serve various purposes in agriculture, including soil condition assessment, crop protection, and pest diagnostics. When used appropriately and responsibly, nanosensors support sustainable agriculture and help increase crop yields (Hazarika et al. 2022). Nanosensors can be categorized into optical sensors, including colorimetric, fluorescence, and surface plasmon resonance sensors; electrochemical sensors, such as amperometric, potentiometric, conductometric, and impedimetric sensors; and piezoelectric sensors (Lata et al. 2024). A wide range of nanomaterials is currently used to develop biosensors, including metallic nanoparticles, metal oxides, quantum dots, and carbon-based materials (Aoko et al. 2021; Foroughi and Jahani 2022; Jayeoye et al. 2022; Jebril et al. 2021; Liang et al. 2021; Liu et al. 2022; Moarefdoust et al. 2021; Mohammadi et al. 2020; Wang et al. 2022).

6.4 Nanomaterials for Improving Soil Health and Water Retention

Soil health plays a pivotal role in agriculture and ecosystem balance. It supports plant growth by supplying nutrients, hosting diverse organisms, and breaking down organic matter (Yang et al. 2020). Described as a dynamic, complex, and living system, soil underpins sustainable agricultural development (van Es and Karlen 2019). Recognized as one of the most influential factors for crop growth and sustainable farming, soil health shows promising improvement through the application of engineered nanomaterials (ENMs) (Khan 2020). Recent studies highlight a strong connection between ENMs and enhanced soil quality (Arora et al. 2019; Baig et al. 2022; Sharma and Kumar 2024; Suazo-Hernández et al. 2023). Nanotechnologies directly impact physical soil properties, including porosity, bulk density, structure, and texture. By creating nanoscale pores or channels between soil aggregates, ENMs improve water drainage and aeration, fostering better soil performance (Hasan et al. 2020; Yang et al. 2020). Water retention capacity varies between different types of soil (García et al. 2021) and is influenced by structural, organic, textural, and topographical factors. This property is essential for holding water available for plants, increasing soil fertility, protecting against erosion, and mitigating the impacts of periods of drought. Soils with low water retention present challenges for agriculture, as they cannot store enough moisture to sustain crops. In addition, the excessive use of fertilizers in agricultural intensification contributes to erosion and reduced soil fertility (Nassaj-Bokharaei et al. 2021). Nanostructured hydrogels are emerging as a sustainable solution for soils with low water retention. Studies have shown that these materials can significantly improve the soil's water and microbiological properties. For example, carboxylated nanocellulose polymers increased water absorption capacity and microbial respiration, as well as promoting greater activity of microorganisms due to the availability of carbon (Barajas et al. 2021). Similarly, starch-based hydrogels with natural carbon nanostructures proved to be effective in growing tomatoes, reducing the effects of water stress and improving growth and nutrient concentration in the plants (Nassaj-Bokharaei et al. 2021). In addition to hydrogels, other nanomaterials help with water-holding capacity, such as nanocarbon and nanobiochar, due to their carbon retention properties (Nazir et al. 2020).

6.5 Startups, Innovation, and Scaleup

The agribusiness sector is highly susceptible to the effects triggered by political-economic shifts, climate change, and food security challenges. Meeting the demands of a high-demand market presents a significant challenge to achieving the goals outlined by the Sustainable Development Goal (SDG) of eradicating hunger, achieving food security, and promoting sustainable agriculture (United Nations Development Programme 2019) (Giller et al. 2021). The inequality in food input production does not correspond proportionally to global demand. Factors such as wars, excessive consumption, food waste, and climate changes contribute to this disparity (Daszkiewicz 2022). In response to these fluctuations, the sector increasingly relies on science and technological innovations to identify operational errors, market gaps, and implement new technologies at various stages of the process (Li et al. 2020). Within this context, startups have gained prominence on the global stage. Startups are innovative enterprises, often at an early stage, aiming to explore and develop innovative, scalable products, services, or business models with the potential for rapid growth (Zawislak et al. 2023). Unlike traditional companies, startups operate in highly uncertain conditions, progressing through stages such as pre-startup, startup, and growth/exit. Additionally, they face the so-called "Valley of Death" (VoD), characterized by an inability to advance, often due to insufficient funding or technologies failing to achieve market adoption (Gbadegeshin et al. 2022). Consequently, these innovative companies tend to adopt an agile experimental approach, testing hypotheses and quickly adapting to market feedback.

AgTechs, a category of startups, provide technologies applicable to the agribusiness sector, such as production monitoring applications, georeferencing drones, and biotechnology solutions (Mendes et al. n.d.). Investments in AgTechs have been driven primarily by the decline in arable land and the growing population, which demands rapid, large-scale production (Wilson et al. 2022). In the United States, investments in startups are predominantly sourced from private capital such as AgFunder, a venture capital firm that in 2021 accounted for $51.7 billion in venture investments. In the European Union (EU), the European Innovation Council (EIC) plans to allocate €1.4 billion in 2025 to support technological innovation companies (European Innovation Council and SMEs Executive Agency 2024). In Brazil, programs have been developed to facilitate the dissemination and adoption of technological innovations in agribusiness, leading to the expansion of research and development (R&D) and the reduction of operational costs (Romani et al. 2021; Xue et al. 2024). Government investment initiatives include the Inova Startup Program by the Brazilian Micro and Small Business Support Service (Sebrae) (Sebrae Startup, 2024) and the Finep Startup Program ("Programa Finep Startup," 2024), offering up to R$750,000 and R$2 million, respectively, to companies. Other initiatives, such as the Supera program, a partnership between the government and the University of São Paulo (USP), provide business incubation, attract innovative companies, offer technological services, support local production arrangements, encourage internationalization, assist with intellectual property protection, facilitate technology transfer, and resource mobilization. In 2019, over 60 startups were assisted, generating a total revenue of R$6.4 million (20200409-100,416-Relatorio de atividades dez n.d.). However, despite current funding and investments in startups, reaching the end consumer requires technological scaling. Technological scaling in the agribusiness sector refers to adapting and expanding technological solutions developed at a laboratory scale to larger scales, such as pilot plants or industrial production. This process is essential to transform laboratory innovations into viable and competitive products for the agribusiness market (Thorpe and Ridgman 2016; Vasconcellos et al. 2021). To assess the maturity of developing technologies and facilitate communication across various application areas, the Technology Readiness Level (TRL) model, developed by the National Aeronautics and Space Administration (NASA), is commonly used. This model evaluates technology maturity based on budgetary and temporal constraints, comprising nine distinct levels (Ferreira et al. 2021).

In technological scaling, the stages can be mapped to the levels of technological readiness:

- Applied research development in the laboratory (TRL 1–3): Direct connection between the proposed solution and agribusiness market demands.
- Product development in the laboratory (TRL 4–5): Focus on properties and functionalities for end-use.

- Process development in the laboratory (TRL 6): Validation of small-scale production feasibility and adjustments to critical variables.
- Implementation in a pilot plant (TRL 7–8): Transition to larger scale, ensuring product quality and characteristics.

Achieving TRL 9 indicates that the technology is ready for large-scale industrial application and commercialization to meet sector demands (Escalab 2023; Ferreira et al. 2021; Vasconcellos et al. 2021).

The Technology Readiness Level (TRL) model is a comprehensive, versatile system designed to evaluate technology development across multiple sectors, including the agroindustrial sector, where innovation plays a critical role. To address manufacturing risks and assess process maturity, the US Department of Defense (DOD) developed a complementary methodology similar to TRL, known as Manufacturing Readiness Levels (MRL), which is divided into ten levels (Ozcan et al. 2023; Wang 2018). In the context of agroindustry, these frameworks gain additional relevance as they guide the integration of emerging technologies, such as precision agriculture, bio-based materials, and smart packaging, into production systems. Simultaneously, as the sector seeks sustainable growth and enhanced competitiveness, addressing business-related challenges becomes paramount. Insights into revenue streams, technology costs, profitability, and market dynamics are crucial to justify investments in innovative solutions and drive economic development. To meet this demand, the Business Case Readiness Level (BcRL) was established, providing structured criteria to align technological advancement with market viability and scalability (Bigliardi and Filippelli 2022; Kucher et al. 2023). These models also play a crucial role in preventing and identifying potential risks during the scaling-up process, which can be classified into three categories: technological, production (or manufacturing), and economic. Technological risks pertain to threats to the functionality and quality of the product. Production risks are related to manufacturing challenges, such as chemical reactions that produce gas or release heat. While these phenomena may be manageable at a laboratory scale, where only a few milliliters of reaction are involved, they can become significant hazards at industrial scales involving tens or hundreds of liters, posing risks to both equipment and personnel. Economic risks encompass factors that may hinder product commercialization, such as competition, selling price, market conditions, and marketing strategies (Ferreira et al. 2021; Ozcan et al. 2023). By categorizing and addressing these risks, TRL, MRL, and BcRL collectively provide a roadmap for optimizing the scaling-up process in the agroindustry, ensuring safety, efficiency, and economic viability. This systematic approach enhances the sector's ability to transition innovative technologies from the research stage to full-scale production and market implementation, driving sustainable growth and fostering resilience in global agricultural markets. The criteria for each level of TRL, MRL, and BcRL are summarized in Table 6.2.

Table 6.2 Levels and charges of different innovation management methods

Level/ model	Technology readiness level (TRL)	Manufacturing readiness levels (MRL)	Business case readiness levels (BcRL)
1	Basic physical principle	Assessment of manufacturing feasibility	Research concept proven in the laboratory; principal investigator defines potential market value
2	Technical-scientific concept	Definition of manufacturing concepts	Analysis and validation by the university team of the research potential and market insertion concept
3	Proof of concept	Developed manufacturing concepts	Research concept presented to an external entity (incubator board, venture capital investors, etc.) for analysis
4	Laboratory validation	Capability of producing the technology in a laboratory environment	Research concept/target markets presented to industrial partners; alignment with strategic plan objectives
5	Prototype validation	Capability of producing prototype components in a production-relevant environment	Financial aspects defined; return on investment required; margins, funding sources (internal, external, or both)
6	Field testing	Capability of producing a prototype system or subsystem in a production-relevant environment	Research prototype ready for market, approved by an external entity and key customers
7	Scaleup testing	Capability of producing systems, subsystems, or components in a representative production environment	Product insertion into a target market; positive market focus response from the group
8	Tested and qualified product	Pilot line capability demonstrated. Ready to begin low-rate production	Full-scale production in the local market; financial confirmation
9	Technology incorporation	Low-rate production demonstrated. Installed capacity for full production	Full-scale production in national markets; future improvements planned
10	–	Full production demonstrated and optimized manufacturing practices applied	–

6.6 Patents and Market Trend

According to the Basic Patent Protection Manual, a patent is an official title of temporary ownership granted by the State, under the law, to its holder or successors (either an individual or a legal entity), who then possess exclusive rights over the asset. This asset may pertain to a product, a manufacturing process, or improvements to existing products and processes that are the subject of the patent. Third parties may only exploit the patent with the permission of the holder (via a license)

(Ministério da Economia Instituto Nacional da Propriedade Industrial Diretoria de Patentes 2021). To claim a patent, the invention must be novel, useful, and non-obvious in terms of the "state of the art." The patent must also meet other legal requirements, such as deadlines concerning how long the invention has been disclosed to the public, if it has been, prior to filing the patent application. In some countries, such as the United States and Brazil, there is a one-year grace period during which the application can still be filed; in most other countries, public disclosure of the invention before the patent application is filed is generally sufficient to cause its denial (Hall 2007; Nard 2019). In certain jurisdictions, patents cannot be granted for inventions that are contrary to "public order" or considered "immoral." For example, inventions related to devices for the consumption of illegal drugs may not be patentable due to moral concerns. However, this restriction may sometimes be more a matter of how the patent claim is structured than an absolute prohibition. Not all jurisdictions share the same definitions of morality, and some choose not to implement patent laws addressing moral issues (Hall 2007; Nard 2019).

In recent years, nanotechnology has emerged as an innovative field in the agricultural sector, offering more efficient and sustainable solutions to many challenges faced by modern agriculture. Intellectual property, particularly patents, has played a crucial role in protecting and advancing these innovations, enabling companies to safeguard their rights and position themselves competitively in the market. Patent analysis reveals how nanotechnology-based technologies are shaping the future of agriculture (Ponta et al. 2022; Wu et al. 2022). Among the patents analyzed, US 2019/0202750 A1 stands out for describing the synthesis of NPK nanocomposites for plant nutrition, using nanoparticles such as ZnO and TiO_2. This patent involves the creation of controlled-release fertilizers that enhance nutrient use efficiency—a growing trend in the agricultural market. Additionally, EP3205637A1 (aqueous fertilizer with metallic nanoparticles) also focuses on improving nutrient absorption by plants, utilizing metallic nanoparticles such as copper and silver through an aqueous fertilizer application method. Both patents represent significant advancements in the pursuit of high-precision agricultural solutions and reduced input waste, which are increasingly valued in a context of sustainability and resource-use efficiency. Another relevant patent is WO2024144660A1, which describes the use of graphene oxide modified with iron oxide to combat plastic pollution in agricultural soils and waters. Growing awareness of the environmental impacts of agriculture and the pollution caused by plastics and microplastics has driven the development of solutions that not only enhance crop productivity but also reduce environmental harm. This patent exemplifies how nanotechnology can improve soil and water health, reflecting a rising trend toward more eco-friendly solutions in the agricultural sector.

Furthermore, patents such as US 20170367325A1 and US 20210292253A1 demonstrate the use of nanoparticles, such as nanoceria and iron oxide, to improve plant photosynthesis under abiotic stress. Adapting plants to extreme climatic conditions, such as drought and high temperatures, is an increasing priority due to climate change. These innovations are essential for enhancing the resilience of agricultural crops to such challenges, enabling stable production even under adverse conditions. The market is decisively shaping the future of the agricultural sector by offering solutions that are more efficient, sustainable, and precise. The growing number of patents related to the use of nanotechnology in fertilizers and agricultural solutions reflects emerging market trends, such as precision agriculture and sustainability. Intellectual property, through patents, ensures the protection of technological innovations that meet these demands, offering a competitive edge to companies leading the development of such solutions. Finally, patents US 11639319 B2 and CN 113735656 B, which address encapsulated fertilizers, illustrate the growing trend toward solutions that enable controlled nutrient release. The use of modified chitosan and potassium nanosilicate to regulate fertilizer release offers an innovative solution to increase nutrient use efficiency, reduce waste, and mitigate environmental impacts. The market for controlled-release fertilizers has been expanding as farmers seek more sustainable and effective alternatives. Table 6.3 summarizes the selected patents and their respective applications in the agricultural sector.

When analyzing the market and the products utilizing nanotechnology currently being commercialized worldwide, the StatNano database indicates that approximately 260 nanotechnology-based products are produced by 96 companies across 29 countries. Nanotechnology products in the agro-industrial sector are divided into

Table 6.3 Patents filed by countries in the agro-industrial sector

Patent number	Country of origin	Description	Application in agribusiness	Year
US 2019/0202750 A1	USA	NPK nanocomposites with nanoparticles and aerosol-based foliar application	Efficient and sustainable fertilization, controlled nutrient release	2019
EP3205637A1	Spain	Aqueous fertilizer with metallic nanoparticles	Efficient fertilization, pathogen protection	2019
CN113735656B	China	Potassium nanosilicate fertilizer with controlled release	Increased fertilizer efficiency, reduced leaching	2022
US 11639319 B2	USA	Fertilizers encapsulated in modified chitosan	Gradual nutrient release, enhanced plant growth	2023
WO2015150903A1	France	Nanomanufacturing with clay minerals for sustainable fertilizers	Biodegradable fertilizers, controlled nutrient release	2015
WO2024144660A1	France	Graphene oxide nanomaterial with iron oxide for plastic pollution mitigation	Nano–/microplastic removal, agricultural ecosystem protection	2024
US 20170367325 A1	USA	Use of nanoceria to enhance photosynthesis under abiotic stress	Improved photosynthesis, resistance to environmental stress	2020
US 20210292253 A1	USA	Seed pre-soaking with hematite nanoparticles	Increased plant growth and productivity	2021

five main segments: fertilizers, animal husbandry, plant cultivation, plant protection, and soil improvement. Comparatively, the agro-industrial sector still represents a relatively low percentage of nanotechnology-based products compared to other sectors such as electronics, medical, construction, and cosmetics. Figure 6.2 illustrates the percentage of nanotechnology products commercialized in the agro-industrial sector, broken down by its main segments and the producing countries, as well as the active properties of nanostructured products in each segment. Specifically, "a" corresponds to the fertilizer segment, "b" to the animal husbandry segment, "c" to plant protection, and "d" to plant cultivation (StatNano 2023a, b). Additionally, Table 6.4 lists commercialized products, detailing the nanostructures used, their country of manufacture, and the respective segment they belong to.

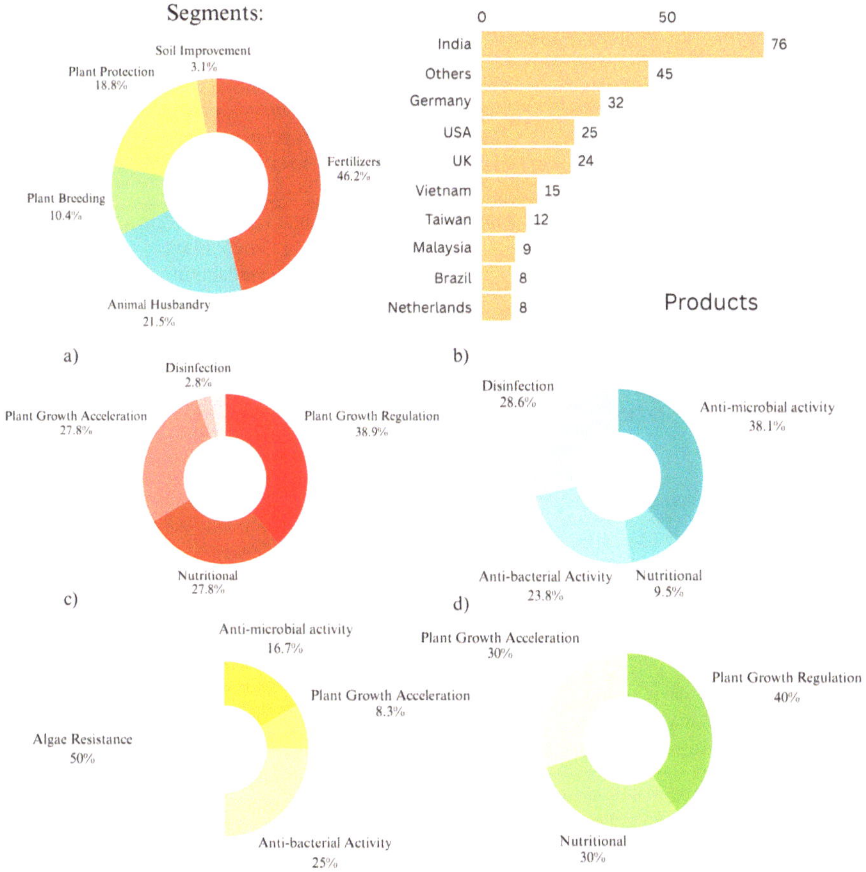

Fig. 6.2 Percentage of nanotechnology products commercialized in the agro-industrial sector, broken down by its main segments and the producing countries, as well as the active properties of nanostructured products in each segment. Specifically, (**a**) corresponds to the fertilizer segment, (**b**) to the animal husbandry segment, (**c**) to plant protection, and (**d**) to plant cultivation

Table 6.4 Products marketed in the agro-industry sector and StatNano's database

Applications	Product name	Nanotechnology/ nanomaterial	Properties	Country
Fertilizers	Lithocal	Calcium nanoparticle/ nanopowder Manganese nanoparticle/ nanopowder	Nutritional; absorption enhancement	Brazil
	Fértil calmag Fertilecalcium 25	Magnesium nanocapsule Calcium nanoparticle/ nanopowder	Plant growth regulation; absorption enhancement Nutritional; absorption enhancement	Brazil
	Nanovec TSS 80	Manganese, molybdenum, and zinc nanocapsule	Seed treatment	Brazil
	AgroCyprum®	Copper nanoparticle/ nanopowder	Plant growth regulation; plant stress reduction	Netherlands
	AgroArgentum® Bor	Silver nanoparticle/ nanopowder	Pesticide removal; plant growth regulation; plant stress reduction	Netherlands
	High-tech Nano ChelatedZinc	Zinc nanocapsule	Plant growth regulation; seed germination promotion	United Arab Emirates
	NANOZINC	Zinc oxide nanoparticle/ nanopowder	Plant growth regulation	Argentina
	Nano potash	Potassium nanoparticle/ nanopowder	Plant growth acceleration; soil nutritious	India
	NanoK™ NanoZn™	Potassium nanocapsule Zinc nanocapsule	Crop yield enhancement; potassium/zinc delivery to crops	United States
Animal husbandry	Arcavit amino Arcavit l forte	Nano-encapsulation technology	Nutritional; vitamin; antibiotic	Italy
	Nano posheshfelez	Silver nanoparticle /nanopowder	Anti-bacterial activity; disinfection	Iran
	Super betamax™ Betamax™	Nano-locked Beta Glucans Nano-carriers of vitamin B complex	Immunity improvement; vitamin Active growth promoter; nervous system ailments	United States

(continued)

Table 6.4 (continued)

Applications	Product name	Nanotechnology/ nanomaterial	Properties	Country
	Somguard	Silver nanoparticle /nanopowder	Anti-bacterial activity; non-toxic; environmentally friendly; anti-microbial activity	India
Plant protection	Nano shield	Silver nanoparticle /nanopowder	Antibiotic; virucidal; fungicide	India
	AZterknotfungicide	Nano-polymertechnology	Plant health improvement; fungicide; plant stress reduction	Canada
	Nanoscaled biocides	Biocide nanoencapsulation	Anti-microbial activity	Germany
	Pesticide synergist	Nanoscale catalytic materials	Anti-bacterial activity; non-toxic; inorganic	China
Plant breeding	Nano agrocrop	Biodegradable and biocompatible nanoparticles of plant and natural origin	Anti-bacterial activity; anti-fungal activity	Brazil
	Plant growth promoter	Variety of nano-oxides	Anti-bacterial activity; non-toxic; inorganic	China
	Nano-boron Nano-molybdenum Nano-silver Nano-magnesium Nano-iron Nano-potassium Nano-phosphorous Nano-zinc	Boron, molybdenum, silver, iron, potassium, phosphorous, and zinc oxide nanoparticle/ nanopowder; magnesium nanocapsule	Plant growth regulation; nutritional; plant growth acceleration	India
Soil improvement	Hydro: Flower Coco: Flower Coco: Grow Soil: Flower Soil: Grow	Nanoparticles from plant vitality's hydro	Plant growth regulation; root activity improvement; seed germination promotion	United Kingdom

6.7 Barrier and Strategies

Although nanotechnology has the potential to significantly impact the agricultural industry, its full implementation faces several barriers. These obstacles may stem from various domains, including technological, legal, and acceptance barriers, as illustrated in Fig. 6.3 (Martínez-Azúa and Sama-Berrocal 2022; Martínez-Azúa and Seoane 2019):

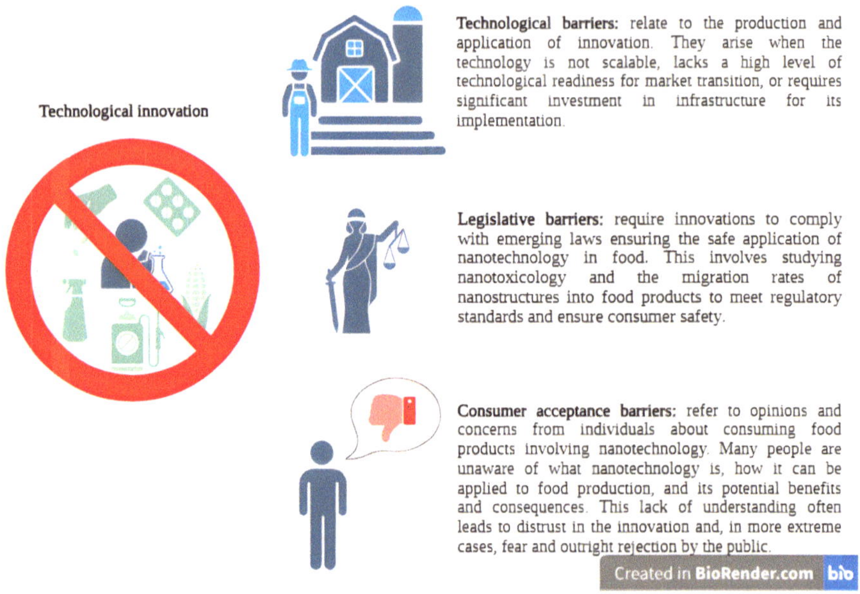

Fig. 6.3 Types of barriers encountered when implementing innovation in agribusiness

Technological barriers pertain to challenges in the production and application of innovations. These include whether the technology is scalable, whether it has achieved a high technology readiness level (TRL) to transition to the market, and whether the developed product is easily applicable within the industry or market. Often, producers may need to invest in infrastructure to adopt the new technology. These barriers should be addressed and resolved during the scaling-up process through proof-of-concept trials, testing in realistic environments and gathering consumer feedback to align prototypes with industry and market contexts (Vasconcellos et al. 2021).

Consumer acceptance barriers encompass opinions and concerns from the general population consuming foods associated with nanotechnology. Despite decades of academic research in nanotechnology, much of the non-scientific population remains unaware of what nanotechnology entails, how it applies to food production, and its potential consequences and benefits. This information gap fosters mistrust in innovations, potentially leading to fear and outright rejection among consumers. A common concern with food-related nanotechnology is its toxicity. Nanomaterials can be toxic depending on the exposure duration, potentially causing DNA damage, increased production of reactive oxygen species (ROS), and carcinogenesis, contributing to negative perceptions of nanoproduced goods (Hasmin et al. 2021; Sodano et al. 2016). To mitigate these barriers, regulatory agencies such as the Food and Drug Administration (FDA), Brazil's National Health Surveillance Agency (ANVISA), and the European Food Safety Authority (EFSA), which oversee product commercialization approvals, must conduct further analyses to demonstrate the safety of nanotechnology applications, thereby improving public perception (He et al. 2019; Cushen et al. 2012). In response to these concerns, countries are establishing parameters and analyses to determine the safety of new products. Using disciplines such as mathematics, physics, bioinformatics, and nanotoxicology, advancements are being made to ensure the safety of nanomaterials (NM) before they reach the market. These analyses evaluate the structural behavior of applied materials, thermodynamic behavior in various matrices, bioaccumulation propensity, and potential interactions (Hofmann et al. 2020; Buzby 2010).

Given that the toxicological and migratory behavior of NMs is not yet fully understood, migration testing of nanostructures is critical to prevent bioaccumulation or adverse outcomes. When NMs are introduced into foods, a transfer rate occurs, which may or may not be safe. Consequently, these analyses are vital. NMs must be stable and precisely targeted, with selective multi-target mechanisms to avoid instability or accumulation of unwanted particles in the synthesized matrix (Chausali et al. 2022). The migration rates associated with different nanostructure syntheses demand further research and improvements in current legislation by organizations like the FDA, the European Commission (EC), and ANVISA. As nanotechnology advances, regulations adapt to specify requirements. Legislations such as EC 1935/2004, EC 450/2009, EC 10/2011, Novel Food Regulation EC 258/97, and EEC 89/109 govern the nanoproduct (NP) market within the European Union. In the USA, NPs are regulated based on safety assessments by the FDA. In Brazil, ANVISA enforces limits on active ingredient concentrations to ensure safety based on existing studies (Nanostructured Materials for Food Packaging Applications, 2024). For innovation to reach the market, it must comply with these updated standards, undergo rigorous testing and analyses, and adhere to the regulations of the respective countries.

6.8 Future Outlook

Nanotechnology is revolutionizing agriculture by offering innovative solutions to increase productivity and sustainability. Through nanoformulations, nanosensors, and other applications, this technology enables more precise control of pests and diseases, optimizes the use of agricultural inputs, and promotes the development of more resistant and nutritious crops. Nanotechnology spans various areas, from the development of nanofertilizers and smart packaging to the genetic manipulation of crops and the creation of nanosensors for disease monitoring. This precision and efficiency contribute to environmental preservation, reducing food waste, pesticide use, and promoting more sustainable farming practices. Additionally, it generates opportunities for the emergence of startups and technology-based companies, contributing to job creation.

Economically, this technology has the potential to revolutionize various sectors, including agriculture. Nanotechnology promises to optimize processes, reduce production costs, and introduce materials with unprecedented properties. However, the development of this technology requires significant investment in research and development, as well as assessing potential liability risks. From a legal perspective, nanotechnology raises essential issues regarding intellectual property, product liability, and regulatory oversight. The regulation of nanotechnology in agriculture is crucial to ensuring food and environmental safety. Organizations such as the Nanotechnology Initiative Division, Ministry of Law and Justice, CDC (Center for Disease Control and Prevention), and CKMNT (Center for Knowledge Management of Nanoscience and Technology) are adopting a series of principles to establish standards and guidelines for the safe and responsible use of these technologies. The risk and benefit assessment of nanotechnology is an ongoing process involving scientists, regulators, and civil society.

References

Abdallah Y, Liu M, Ogunyemi SO, Ahmed T, Fouad H, Abdelazez A, Yan C, Yang Y, Chen J, Li B (2020) Bioinspired green synthesis of chitosan and zinc oxide nanoparticles with strong antibacterial activity against rice pathogen *Xanthomonas oryzae* pv. oryzae. Molecules 25(20):4795. https://doi.org/10.3390/MOLECULES25204795

Abdelaziz AM, Elshaer MA, Abd-Elraheem MA, Ali OMOM, Haggag MI, El-Sayyad GS, Attia MS (2023) Ziziphus spina-christi extract-stabilized novel silver nanoparticle synthesis for combating Fusarium oxysporum-causing pepper wilt disease: in vitro and in vivo studies. Arch Microbiol 205(2):1–17. https://doi.org/10.1007/S00203-023-03400-7

Ahmed T, Shahid M, Noman M, Niazi MBK, Mahmood F, Manzoor I, Zhang Y, Li B, Yang Y, Yan C, Chen J (2020) Silver nanoparticles synthesized by using Bacillus cereus SZT1 ameliorated the damage of bacterial leaf blight pathogen in rice. Pathogens 9(3):160. https://doi.org/10.3390/PATHOGENS9030160

Akpinar I, Unal M, Sar T (2021) Potential antifungal effects of silver nanoparticles (AgNPs) of different sizes against phytopathogenic Fusarium oxysporum f. sp. radicis-lycopersici (FORL) strains. *SN*. Appl Sci 3(4):1–9. https://doi.org/10.1007/S42452-021-04524-5/TABLES/2

Ali SS, Al-Tohamy R, Koutra E, Moawad MS, Kornaros M, Mustafa AM, Mahmoud YAG, Badr A, Osman MEH, Elsamahy T, Jiao H, Sun J (2021) Nanobiotechnological advancements in agriculture and food industry: applications, nanotoxicity, and future perspectives. Sci Total Environ 792:148359. https://doi.org/10.1016/J.SCITOTENV.2021.148359

Alimohammadi M, Panahpour E, Naseri A (2020) Assessing the effects of urea and nano-nitrogen chelate fertilizers on sugarcane yield and dynamic of nitrate in soil. Soil Sci Plant Nutr 66(2):352–359. https://doi.org/10.1080/00380768.2020.1727298

Al-Otibi F, Perveen K, Al-Saif NA, Alharbi RI, Bokhari NA, Albasher G, Al-Otaibi RM, Al-Mosa MA (2021) Biosynthesis of silver nanoparticles using Malva parvi flora and their antifungal activity. Saudi J Biol Sci 28(4):2229–2235. https://doi.org/10.1016/J.SJBS.2021.01.012

Aoko IL, Ondigo D, Kavoo AM, Wainaina C, Kiirika L (2021) A gold nanoparticle-based colorimetric probe for detection of gibberellic acid exuded by *Ralstonia solanacearum* pathogen in tomato (*Solanum lycopersicum* L.). Int J Hortic Sci Technol 8(3):203–214. https://doi.org/10.22059/IJHST.2021.314572.423

Arora A, Singh B, Kaur P (2019) Performance of nano-particles in stabilization of soil: a comprehensive review. Mater Today Proc 17:124–130. https://doi.org/10.1016/J.MATPR.2019.06.409

Artusio F, Casà D, Granetto M, Tosco T, Pisano R (2021) Alginate nanohydrogels as a biocompatible platform for the controlled release of a hydrophilic herbicide. PRO 9(9):1641. https://doi.org/10.3390/pr9091641/s1

Azam M, Bhatti HN, Khan A, Zafar L, Iqbal M (2022) Zinc oxide nano-fertilizer application (foliar and soil) effect on the growth, photosynthetic pigments and antioxidant system of maize cultivar. Biocatal Agric Biotechnol 42:102343. https://doi.org/10.1016/j.bcab.2022.102343

Baig N, Alowaid AM, Abdulazeez I, Salhi B, Sajid M, Kammakakam I (2022) Designing of nano-textured inorganic-organic hybrid PVDF membrane for efficient separation of the oil-in-water emulsions. Chemosphere 308:136531. https://doi.org/10.1016/j.chemosphere.2022.136531

Barajas RM, Wong V, Little K, Patti AF, Garnier G (2021) Carboxylated nanocellulose superabsorbent: biodegradation and soil water retention properties. Research Square. https://doi.org/10.21203/rs.3.rs-474606/v1

Bhandari G, Dhasmana A, Chaudhary P, Gupta S, Gangola S, Gupta A, Rustagi S, Shende SS, Rajput VD, Minkina T, Malik S, Slama P (2023) A perspective review on green nanotechnology in agro-ecosystems: opportunities for sustainable agricultural practices & environmental remediation. Agriculture 13(3):668. https://doi.org/10.3390/agriculture13030668

Bharti A, Jain U, Chauhan N (2024) From lab to field: nano-biosensors for real-time plant nutrient tracking. Plant Nano Biol 9:100079. https://doi.org/10.1016/j.plana.2024.100079

Bigliardi B, Filippelli S (2022) A review of the literature on innovation in the agro-food industry: sustainability, smartness and health. Eur J Innov Manag 25(6):589–611

Buzby JC (2010) Nanotechnology for food applications: more questions than answers. J Consum Aff 44(3):528–545

Chausali N, Saxena J, Prasad R (2022) Recent trends in nanotechnology applications of bio-based packaging. J Agr Food Res 7:100257

Correa MG, Martínez FB, Vidal CP, Streitt C, Escrig J, de Dicastillo CL (2020) Antimicrobial metal-based nanoparticles: a review on their synthesis, types, and antimicrobial action. Beilstein J Nanotechnol 11(129):1450–1469. https://doi.org/10.3762/bjnano.11.129

Cushen M, Kerry J, Morris M, Cruz-Romero M, Cummins E (2012) Nanotechnologies in the food industry—recent developments, risks, and regulation. Trends Food Sci Technol 24(1):30–46

da Silva R, Viana VE, Avila LA, Zotti MJ, Smagghe G, Junior AM, Camargo ER, Fajardo AR (2024) Advances on polymeric nanocarriers for sustainable agriculture: enhancing dsRNA/siRNA delivery to combat agricultural pests. Int J Biol Macromol 282:137000. https://doi.org/10.1016/j.ijbiomac.2024.137000

Daszkiewicz T (2022) Food production in the context of global developmental challenges. Agriculture 12(6):832

Ding Y, Wang Q, Zhu G, Zhang P, Rui Y (2023) Application and perspectives of nanopesticides in agriculture. J Nanopart Res 25(8):1–15. https://doi.org/10.1007/s11051-023-05811-5

Dong J, Liu X, Chen Y, Yang W, Du X (2021) User-safe and efficient chitosan-gated porous carbon nanopesticides and nanoherbicides. J Colloid Interface Sci 594:20–34. https://doi.org/10.1016/j.jcis.2021.03.001

Elabasy A, Shoaib A, Waqas M, Shi Z, Jiang M (2020) Cellulose nanocrystals loaded with thiamethoxam: fabrication, characterization, and evaluation of insecticidal activity against *Phenacoccus solenopsis Tinsley* (Hemiptera: Pseudococcidae). Nano 10(4):788. https://doi.org/10.3390/nano10040788

El-Bialy SM, El-Mahrouk ME, Elesawy T, Omara AED, Elbehiry F, El-Ramady H, Áron B, Prokisch J, Brevik EC, Solberg S (2023) Biological nanofertilizers to enhance growth potential of strawberry seedlings by boosting photosynthetic pigments, plant enzymatic antioxidants, and nutritional status. Plants 12(2). https://doi.org/10.3390/plants12020302

Escalab (2023, August 1). Estudo de viabilidade: O passo a passo para um EVTE completo. https://escalab.com.br/estudo-de-viabilidade-passo-a-passo-evte-completo/

European Innovation Council and SMEs Executive Agency (2024, October 29) European Commission Press release: European Innovation Council to invest €1.4 billion in deep tech and scale-up of strategic technologies in 2025. https://eic.ec.europa.eu/news/european-innovation-council-invest-eu14-billion-deep-tech-and-scale-strategic-technologies-2025-2024-10-29_en

FAO (2017) The future of food and agriculture: trends and challenges | Policy support and governance. Food and Agriculture Organization of the United Nations. https://www.fao.org/policy-support/tools-and-publications/resources-details/en/c/472484/

Ferreira CV, Biesek FL, Scalice RK (2021) Product innovation management model based on manufacturing readiness level (MRL), design for manufacturing and assembly (DFMA), and technology readiness level (TRL). J Braz Soc Mech Sci Eng 43(7):360

Fincheira P, Hoffmann N, Tortella G, Ruiz A, Cornejo P, Diez MC, Seabra AB, Benavides-Mendoza A, Rubilar O (2023) Eco-efficient systems based on nanocarriers for the controlled release of fertilizers and pesticides: toward smart agriculture. Nano 13(13):1978. https://doi.org/10.3390/nano13131978

Foroughi MM, Jahani S (2022) Investigation of a high-sensitive electrochemical DNA biosensor for determination of idarubicin and studies of DNA-binding properties. Microchem J 179:107546. https://doi.org/10.1016/j.microc.2022.107546

Fu L, Wang Z, Dhankher OP, Xing B (2020) Nanotechnology as a new sustainable approach for controlling crop diseases and increasing agricultural production. J Exp Bot 71(2):507–519. https://doi.org/10.1093/jxb/erz314

García CR, Brown S, Krzic M, Lavkulich L, Roa-García MC (2021) Relationship of soil water retention characteristics and soil properties: a case study from the Colombian Andes. Can J Soil Sci 101(1):144–156. https://doi.org/10.1139/cjss-2020-0066

Gbadegeshin SA, Al Natsheh A, Ghafel K, Mohammed O, Koskela A, Rimpiläinen A, Tikkanen J, Kuoppala A (2022) Overcoming the valley of death: a new model for high technology startups. Sustain Futures 4:100077

Giller KE, Delaune T, Silva JV, Descheemaeker K, van de Ven G, Schut AGT, van Wijk M, Hammond J, Hochman Z, Taulya G (2021) The future of farming: who will produce our food? Food Secur 13(5):1073–1099

Grillo R, Mattos BD, Antunes DR, Forini MML, Monikh FA, Rojas OJ (2021) Foliage adhesion and interactions with particulate delivery systems for plant nanobionics and intelligent agriculture. Nano Today 37:101078. https://doi.org/10.1016/j.nantod.2021.101078

Guha T, Gopal G, Kundu R, Mukherjee A (2020) Nanocomposites for delivering agrochemicals: a comprehensive review. J Agric Food Chem 68(12):3691–3702

Hall BH (2007) Patents and patent policy. Oxf Rev Econ Policy 23(4):568–587

Hasan MK, Shopan J, Ahammed GJ (2020) Nanomaterials and soil health for agricultural crop production: current status and future prospects. In: Nanomaterials for agriculture and forestry applications, pp 289–312. https://doi.org/10.1016/B978-0-12-817852-2.00012-3

Hasmin NA, Mahmood A, Azizun N, Abd Kadir NH, Matmin J (2021) Consumers' awareness and knowledge of nanofood in Malaysia. SMRJ 18(2):23–46

Haydar MS, Ghosh S, Mandal P (2021) Application of iron oxide nanoparticles as micronutrient fertilizer in mulberry propagation. J Plant Growth Regul 41(4):1726–1746. https://doi.org/10.1007/s00344-021-10413-3

Hazarika A, Yadav M, Yadav DK, Yadav HS (2022) An overview of the role of nanoparticles in sustainable agriculture. Biocatal Agric Biotechnol 43:102399. https://doi.org/10.1016/j.bcab.2022.102399

He X, Deng H, Hwang HM (2019) The current application of nanotechnology in food and agriculture. J Food Drug Anal 27(1):1–21

Hofmann T, Lowry GV, Ghoshal S et al (2020) Technology readiness and overcoming barriers to sustainably implement nanotechnology-enabled plant agriculture. Nat Food 1(7):416–425

Husen A (2021) The harsh environment and resilient plants: an overview. In: Harsh environment and plant resilience: molecular and functional aspects, pp 1–23. https://doi.org/10.1007/978-3-030-65912-7_1

Iavicoli I, Leso V, Beezhold DH, Shvedova AA (2017) Nanotechnology in agriculture: opportunities, toxicological implications, and occupational risks. Toxicol Appl Pharmacol 329:96–111. https://doi.org/10.1016/j.taap.2017.05.025

Jayeoye TJ, Eze FN, Olatunji OJ, Tyopine AA (2022) Synthesis of biocompatible Konjac glucomannan stabilized silver nanoparticles, with *Asystasia gangetica* phenolic extract for colorimetric detection of mercury (II) ion. Sci Rep 12(1):1–15. https://doi.org/10.1038/s41598-022-13384-x

Jebril S, Fdhila A, Dridi C (2021) Nanoengineering of eco-friendly silver nanoparticles using five different plant extracts and development of cost-effective phenol nanosensor. Sci Rep 11(1):1–11. https://doi.org/10.1038/s41598-021-01609-4

Kah M, Tufenkji N, White JC (2019) Nano-enabled strategies to enhance crop nutrition and protection. Nat Nanotechnol 14(6):532–540. https://doi.org/10.1038/s41565-019-0439-5

Kalyani N, Goel S, Jaiswal S (2020) On-site sensing of pesticides using point of care biosensors: a review. Environ Chem Lett 19(1):345–354. https://doi.org/10.1007/s10311-020-01070-1

Kashyap P, Shirkot P, Das R, Pandey H, Singh D (2023) Biosynthesis and characterization of copper nanoparticles from *Stenotrophomonas maltophilia* and its effect on plant pathogens and pesticide degradation. J Agric Food Res 13:100654. https://doi.org/10.1016/j.jafr.2023.100654

Khan ST (2020) Interaction of engineered nanomaterials with soil microbiome and plants: their impact on plant and soil health. In: Nanomaterials for agriculture and food applications, pp 181–199. https://doi.org/10.1007/978-3-030-33996-8_10

Kucher L, Hełdak M, Orochovska L (2023) Assessment of the readiness of agrarian enterprises to implement innovative projects. Agric Resour Econ Int Scient E-J 9(1):224–259

Kumar A, Choudhary A, Kaur H, Mehta S, Husen A (2021) Smart nanomaterial and nanocomposite with advanced agrochemical activities. Nanoscale Res Lett 16(1):1–26. https://doi.org/10.1186/S11671-021-03612-0

Kumar R, Najda A, Duhan JS, Kumar B, Chawla P, Klepacka J, Malawski S, Sadh PK, Poonia AK (2022) Assessment of antifungal efficacy and release behavior of fungicide-loaded chitosan-carrageenan nanoparticles against phytopathogenic fungi. Polymers 14(1):41. https://doi.org/10.3390/POLYM14010041/S1

Lata S, Bhardwaj S, Garg R (2024) Nanomaterials for sensing and biosensing: applications in agri-food diagnostics. Int J Environ Anal Chem. https://doi.org/10.1080/03067319.2022.2115895

Li Y, Liu Y, Yang D, Jin Q, Wu C, Cui J (2020) Multifunctional molybdenum disulfide-copper nanocomposite that enhances the antibacterial activity, promotes rice growth and induces rice resistance. J Hazard Mater 394:122551. https://doi.org/10.1016/J.JHAZMAT.2020.122551

Liang N, Hu X, Li W, Mwakosya AW, Guo Z, Xu Y, Huang X, Li Z, Zhang X, Zou X, Shi J (2021) Fluorescence and colorimetric dual-mode sensor for visual detection of malathion in cabbage based on carbon quantum dots and gold nanoparticles. Food Chem 343:128494. https://doi.org/10.1016/J.FOODCHEM.2020.128494

Liu P, Zhao M, Zhu H, Zhang M, Li X, Wang M, Liu B, Pan J, Niu X (2022) Dual-mode fluorescence and colorimetric detection of pesticides realized by integrating stimulus-responsive luminescence with oxidase-mimetic activity into cerium-based coordination polymer nanoparticles. J Hazard Mater 423:127077. https://doi.org/10.1016/J.JHAZMAT.2021.127077

Machado TO, Grabow J, Sayer C, de Araújo PHH, Ehrenhard ML, Wurm FR (2022) Biopolymer-based nanocarriers for sustained release of agrochemicals: a review on materials and social science perspectives for a sustainable future of agri- and horticulture. Adv Colloid Interf Sci 303:102645. https://doi.org/10.1016/J.CIS.2022.102645

Mali SC, Dhaka A, Githala CK, Trivedi R (2020) Green synthesis of copper nanoparticles using Celastrus paniculatus Willd. leaf extract and their photocatalytic and antifungal properties. Biotechnol Rep 27:e00518. https://doi.org/10.1016/J.BTRE.2020.E00518

Maluin FN, Hussein MZ, Azah Yusof N, Fakurazi S, Idris AS, Zainol Hilmi NH, Jeffery Daim LD (2020) Chitosan-based agro nanofungicides as a sustainable alternative in the basal stem rot disease management. J Agric Food Chem 68(15):4305–4314. https://doi.org/10.1021/ACS.JAFC.9B08060/ASSET/IMAGES/MEDIUM/JF9B08060_0008.GIF

Manimegalai T, Raguvaran K, Kalpana M, Maheswaran R (2020) Green synthesis of silver nanoparticle using Leonotis nepetifolia and their toxicity against vector mosquitoes of Aedes aegypti and Culex quinquefasciatus and agricultural pests of Spodoptera litura and Helicoverpa armigera. Environ Sci Pollut Res 27(34):43103–43116. https://doi.org/10.1007/S11356-020-10127-1/METRICS

Martínez-Azúa BC, Sama-Berrocal C (2022) Objectives of and barriers to innovation: how do they influence the decision to innovate? J Open Innov: Technol Mark Complex 8(3):134

Martínez-Azúa BC, Seoane FJF (2019) Agri-food industry in Extremadura: obstacles to innovation, willingness to innovate, and demanded public actions. Investigaciones Regionales 45:181–199

Ministério da Economia Instituto Nacional da Propriedade Industrial Diretoria de Patentes, P. de C. e T. de C. I (2021, August) Manual Básico para Proteção por Patentes de Invenções, Modelos de Utilidade e Certificados de Adição. https://www.gov.br/Inpi/Pt-Br/Servicos/Patentes/Guia-Basico/ManualdePatentes20210706.Pdf

Mishra S, Yang X, Ray S, Fraceto LF, Singh HB (2020) Antibacterial and biofilm inhibition activity of biofabricated silver nanoparticles against Xanthomonas oryzae pv. oryzae causing blight disease of rice instigates disease suppression. World J Microbiol Biotechnol 36(4):1–10. https://doi.org/10.1007/S11274-020-02826-1/FIGURES/5

Mittal D, Kaur G, Singh P, Yadav K, Ali SA (2020) Nanoparticle-based sustainable agriculture and food science: recent advances and future outlook. Front Nanotechnol 2:579954. https://doi.org/10.3389/FNANO.2020.579954/BIBTEX

Moarefdoust MM, Jahani S, Moradalizadeh M, Motaghi MM, Foroughi MM (2021) An electrochemical sensor based on hierarchical nickel oxide nanostructures doped with indium ions for voltammetric simultaneous determination of sunset yellow and tartrazine colorants in soft drink powders. Anal Methods 13(21):2396–2404. https://doi.org/10.1039/D1AY00306B

Mohammadi A, Heydari-Bafrooei E, Foroughi MM, Mohammadi M (2020) Electrochemical aptasensor for ultra-sensitive detection of PCB77 using thionine-functionalized MoS2-rGO nanohybrid. Microchem J 155:104747. https://doi.org/10.1016/J.MICROC.2020.104747

Nard CA (2019) The law of patents. Aspen Publishing

Nassaj-Bokharaei S, Motesharezedeh B, Etesami H, Motamedi E (2021) Effect of hydrogel composite reinforced with natural char nanoparticles on improvement of soil biological properties and the growth of water deficit-stressed tomato plant. Ecotoxicol Environ Saf 223. https://doi.org/10.1016/j.ecoenv.2021.112576

Nazir R, Ayub Y, Tahir L (2020) Green-nanotechnology for precision and sustainable agriculture. In: Biogenic nano-particles and their use in agro-ecosystems. Springer, pp 317–357

Oussou-Azo AF, Nakama T, Nakamura M, Futagami T, Vestergaard MCM (2020) Antifungal potential of nanostructured crystalline copper and its oxide forms. Nano 10(5):1003. https://doi.org/10.3390/NANO10051003

Ozcan S, Stornelli A, Simms C (2023) A product innovation readiness level framework. IEEE Trans Eng Manag

Patel SK, Sharma A, Singh GS (2020) Traditional agricultural practices in India: an approach for environmental sustainability and food security. Energy Ecol Environ 5(4):253–271. https://doi.org/10.1007/S40974-020-00158-2/FIGURES/3

Ponta L, Puliga G, Manzini R, Cincotti S (2022) Sustainability-oriented innovation and co-patenting role in agri-food sector: empirical analysis with patents. Technol Forecast Soc Chang 178:121595

Rajput VD, Singh A, Minkina T, Rawat S, Mandzhieva S, Sushkova S, Shuvaeva V, Nazarenko O, Rajput P et al (2021) Nano-enabled products: challenges and opportunities for sustainable agriculture. Plants 10(12):2727. https://doi.org/10.3390/PLANTS10122727

Rizwana H, Alwhibi MS (2021) Biosynthesis of silver nanoparticles using leaves of Mentha pulegium, their characterization, and antifungal properties. Green Process Synth 10(1):824–834

Romani LAS, Barianiet JM, Drucker DP et al (2021) Role of Research and Development Institutions and AgTechs in the digital transformation of Agriculture in Brazil. Rev Ciênc Agron 51:e20207800

Santos PA, Biraku X, Nielsen E, Ozketen AC, Ozketen AA, Hakki EE (2024) Agricultural nanotechnology for a safe and sustainable future: current status, challenges, and beyond. J Sci Food Agric. https://doi.org/10.1002/JSFA.13922

Shahzad AN, Qureshi MK, Wakeel A, Misselbrook T (2019) Crop production in Pakistan and low nitrogen use efficiencies. Nat Sustain 2(12):1106–1114. https://doi.org/10.1038/s41893-019-0429-5

Shao C, Zhao H, Wang P (2022) Recent development in functional nanomaterials for sustainable and smart agricultural chemical technologies. Nano Converg 9(1):1–17. https://doi.org/10.1186/S40580-022-00302-0/FIGURES/6

Sharma R, Kumar V (2024) Nano-enabled agriculture for sustainable soil. Waste Manag Bull 2(1):152–161. https://doi.org/10.1016/J.WMB.2024.01.002

Sharma S, Kumar A, Choudhary A, Harish BM, Karmakar P, Sharma P, Singh J, Pandey V, Mehta S (2022) Recent developments in smart nano-agrochemicals: a promise for revolutionizing present-day agriculture. Mater Today Proc 69:530–534. https://doi.org/10.1016/J.MATPR.2022.09.306

Silva Donda MM, Santini Pigatto GA, Satolo EG, Sommaggio Coletta LF (2021) Analysis of the innovative profile of agribusiness startups. Rev Humanid Inov 8(49):231–250

Singh H, Sharma A, Bhardwaj SK, Arya SK, Bhardwaj N, Khatri M (2021) Recent advances in the applications of nano-agrochemicals for sustainable agricultural development. Environ Sci: Processes Impacts 23(2):213–239. https://doi.org/10.1039/D0EM00404A

Sodano V, Gorgitano MT, Verneau F, Vitale CD (2016) Consumer acceptance of food nanotechnology in Italy. Br Food J 118(3):714–733

StatNano (2023a, August 1) Nanotechnology Companies. https://Statnano.Com/Companies

StatNano (2023b, August 1) Nanotechnology products database. https://Product.Statnano.Com/

Suazo-Hernández J, Arancibia-Miranda N, Mlih R, Cáceres-Jensen L, Bolan N, Mora MDLL (2023) Impact on some soil physical and chemical properties caused by metal and metallic oxide engineered nanoparticles: a review. Nano 13(3):572. https://doi.org/10.3390/NANO13030572

Thakur M, Wang B, Verma ML (2022) Development and applications of nanobiosensors for sustainable agricultural and food industries: recent developments, challenges and perspectives. Environ Technol Innov 26:102371. https://doi.org/10.1016/J.ETI.2022.102371

Thorpe M, Ridgman T (2016) Identifying and sharing the knowledge required for scaling up innovative technologies in the process industry. Int J Innov Technol Manag 13(06):1750006

Ungureanu N, Vlăduţ V, Voicu G (2020) Water scarcity and wastewater reuse in crop irrigation. Sustainability 12(21):9055. https://doi.org/10.3390/SU12219055

United Nations (2022) Food | United Nations. https://www.un.org/en/global-issues/food

United Nations Development Programme (2019, March 29) Agenda 2030—Transformando Nosso Mundo. https://Www.Undp.Org/Pt/Brazil/Publications/Agenda-2030-Transformando-Nosso-Mundo

Usman M, Farooq M, Wakeel A, Nawaz A, Cheema SA et al (2020) Nanotechnology in agriculture: current status, challenges and future opportunities. Sci Total Environ 721:137778. https://doi.org/10.1016/J.SCITOTENV.2020.137778

van Es HM, Karlen DL (2019) Reanalysis validates soil health indicator sensitivity and correlation with long-term crop yields. Soil Sci Soc Am J 83(3):721–732. https://doi.org/10.2136/sssaj2018.09.0338

Vanti GL, Kurjogi M, Basavesha KN, Teradal NL, Masaphy S, Nargund VB (2020) Synthesis and antibacterial activity of *Solanum torvum* mediated silver nanoparticle against *Xanthomonas axonopodis* pv. *punicae* and *Ralstonia solanacearum*. J Biotechnol 309:20–28. https://doi.org/10.1016/J.JBIOTEC.2019.12.009

Vasconcellos EP, de Souza PMTG, Franco MR, de Castro VG, Souza LV, Lago RM, Speziali MG (2021) Escalonamento de tecnologias: Desenvolvimento de produto e processo do laboratório à escala piloto conectado ao mercado (parte 1). Química Nova 44:377–384

Wang B (2018) The future of manufacturing: a new perspective. Engineering 4(5):722–728

Wang D, Saleh NB, Byro A, Zepp R, Sahle-Demessie E, Luxton TP, Ho KT, Burgess RM, Flury M, White JC, Su C (2022) Nano-enabled pesticides for sustainable agriculture and global food security. Nat Nanotechnol 17(4):347–360. https://doi.org/10.1038/S41565-022-01082-8

Wilson WW, Vetsch L, Bullock DW (2022) Valuing an agricultural technology startup using real options. Agribusiness 38(4):771–785

Wu L, Hu K, Lyulyov O, Pimonenko T, Hamid I (2022) The impact of government subsidies on technological innovation in agribusiness: the case for China. Sustainability 14(21):14003

Xue Z, Hou Y, Cao G, Sun G (2024) How does digital transformation drive innovation in Chinese agribusiness: mechanism and micro evidence. J Innov Knowl 9(2):100489

Yadav RK, Singh NB, Singh A, Yadav V, Bano C, Khare S, Niharika. (2020) Expanding the horizons of nanotechnology in agriculture: recent advances, challenges and future perspectives. Vegetos 33(2):203–221. https://doi.org/10.1007/S42535-019-00090-9/METRICS

Yang T, Siddique KHM, Liu K (2020) Cropping systems in agriculture and their impacts on soil health—a review. In: Global ecology and conservation, vol 23. Elsevier B.V. https://doi.org/10.1016/j.gecco.2020.e01118

Zawislak PA, Reichert FM, Barbieux D, Avila AMS, Pufal N (2023) The dynamic chain of innovation: bounded capabilities and complementarity in agribusiness. J Agribus Dev Emerg Econ 13(5):657–670

Zhang P, Guo Z, Ullah S, Melagraki G, Afantitis A, Lynch I (2021) Nanotechnology and artificial intelligence to enable sustainable and precision agriculture. Nat Plants 7(7):864–876. https://doi.org/10.1038/s41477-021-00946-6

Nanotechnology in Crop Protection: Innovations for Sustainable Agriculture

7

Kalaivani K. Nadarajah

Abstract

Nanotechnology, the manipulation of materials at the nanoscale, has revolutionized agriculture, particularly in crop protection, by offering sustainable solutions to enhance productivity and mitigate environmental impacts. This chapter explores the application of nanotechnology in developing nanopesticides, nanofertilizers, and nanosensors, emphasizing their role in improving the efficiency and precision of agrochemicals while minimizing ecological harm. Nanopesticides ensure targeted pest control with reduced chemical loads, while nanofertilizers enhance nutrient delivery and plant resilience to stress. Nanosensors provide real-time pest and stress detection, enabling timely interventions. The chapter highlights the mechanisms through which nanoparticles interact with pests and plants, including improved bioavailability, controlled release, and elicitation of plant defense responses. It also discusses the integration of these innovations into sustainable agricultural practices, addressing challenges such as pesticide resistance, environmental toxicity, and regulatory hurdles. By advancing precision agriculture and reducing dependency on conventional agrochemicals, nanotechnology offers a transformative pathway to achieving global food security while preserving environmental integrity. With continued research and responsible implementation, nanotechnology can play a pivotal role in fostering resilient and sustainable agricultural systems.

K. K. Nadarajah (✉)
School of Biological Sciences and Biotechnology, Faculty Science and Technology, Universiti Kebangsaan Malaysia, Bangi, Selangor, Malaysia
e-mail: vani@ukm.edu.my

S. S. Purewal et al. (eds.), *Agri-Nanotechnology: Innovations for Sustainable Agriculture and Environmental Restoration*,
https://doi.org/10.1007/978-981-96-9756-4_7

Keywords

Agriculture · Agrochemicals · Crop protection · Nanofertilizers · Nanopesticides · Nanosensors · Nanotechnology · Pests · Productivity

7.1 Introduction

In recent years, nanotechnology, which is the manipulation and application of materials at the nanoscale, usually between 1 and 100 nanometers, has become a field that is revolutionizing many different fields, including medicine, electronics, environmental science, and agriculture (Malik et al. 2023). The distinct physical, chemical, and biological characteristics of the nanoscale are markedly different from those seen in bulk materials. These qualities result from the large surface area-to-volume ratio, boosted reactivity, and unique optical and electrical properties that nanomaterials display (Altammar 2023). Consequently, nanotechnology has created new opportunities for the agricultural revolution by offering solutions that boost output while reducing negative environmental effects. Nanotechnology has special potential in the agricultural sector in a number of important areas, including food safety, nutrition delivery, and crop protection. Researchers are looking at using nanomaterials and nanoscale formulations to increase the efficiency of fertilizers and pesticides, which could result in lower dosages and less contamination of the environment (Malik et al. 2023; Abdul et al. 2024). For example, nanoencapsulation methods can advance the bioavailability of active chemicals and shield them against deterioration, enabling tailored distribution to particular plant tissues or insect populations. This strategy reduces the risk of environmental runoff and non-target exposure while simultaneously growing the effectiveness of agrochemicals and minimizing the amount of chemicals used. Another crucial use of nanotechnology in agriculture is in the form of nanofertilizers. By utilizing nanoparticles, these fertilizers improve crop nutrient delivery, increase nutrient usage efficiency, and decrease waste. Nanofertilizers have the ability to release nutrients in a regulated and sustained manner, in contrast to conventional fertilizers, which frequently experience large losses as a result of volatilization, leaching, or immobilization in the soil. The environmental effects of nutrient runoff into water bodies are lessened as a result of the reduced frequency and amount of fertilizer applications that are necessary. Additionally, certain nutrients can be delivered directly to plant roots or leaves using nanofertilizers, which will improve plant uptake and growth in general. This focused strategy is in line with precision agriculture's tenets and delivers a sustainable way to meet the world's expanding food demand while protecting the environment (Arora et al. 2024; Yadav et al. 2023a, b).

Sustainable crop protection presents a pressing challenge in modern agriculture. Conventional methods of pest and disease management heavily rely on chemical pesticides, which, despite their efficacy, have led to significant unintended consequences. These include the emergence of pesticide resistance among target pests, soil deprivation, and adverse health effects on non-target organisms, including humans. The extensive use of chemical pesticides also contributes to environmental pollution and disrupts ecological balance, posing long-term risks to biodiversity (Aktar et al. 2009; Kalogiannidis et al. 2022). As pests and pathogens increasingly adapt to conventional chemical controls, the agricultural sector faces an urgent need for alternative solutions that are both effective and sustainable. Sustainable crop protection aims to reduce the environmental footprint of pest management practices while maintaining or enhancing crop yields. This need has catalyzed interest in novel technologies, including nanotechnology, to offer safer and more targeted pest management approaches (Karlsson et al. 2020). Nanotechnology-based interventions can enhance the delivery of active ingredients, increase the precision of pest targeting, and potentially decrease the required chemical loads in agricultural fields, contributing to a more environmentally friendly and efficient crop protection system (Ahmad et al. 2024; Gangwar et al. 2023). In particular, nanopesticides are a revolutionary development in pest management. These insecticides are nanoscale-engineered to enhance the active components' solubility, stability, and bioavailability, enabling more efficient pest control at lower treatment rates. Active chemicals can be encapsulated in nanopesticides to avoid environmental deterioration and provide regulated release to the proposed pests (Tang et al. 2024; Ahmad et al. 2024). Furthermore, active compounds can be distributed using nanotechnology-enabled smart delivery systems in response to particular stimuli, pH, temperature, or the presence of specific pests. This accuracy minimizes the expansion of pesticide resistance, lessens the impact on non-target organisms, and lowers environmental contamination (Gangwar et al. 2023). Moreover, by integrating chemical, biological, and physical control techniques, nanocarriers can support integrated pest management (IPM) tactics. Both synthetic and biopesticides can be delivered by nanoparticles, increasing the efficacy of biological control agents and lowering dependency on traditional chemicals. These advancements provide a more environmentally responsible and balanced approach to managing pests and diseases, which is in line with the objectives of sustainable agriculture (Tang et al. 2024). Notwithstanding the technology's encouraging uses in agriculture, there are possible hazards and legal issues with its deployment. When nanomaterials are released into the environment, concerns are raised regarding their behavior, permanence, and possible toxicity to ecosystems and non-target animals. Therefore, thorough risk evaluations are necessary to examine the long-term impacts of goods enabled by nanotechnology on the environment and human health. Although regulatory frameworks are changing to address these issues, the rate of technology improvement frequently surpasses the creation of suitable rules and safety requirements (Iavicoli et al. 2017; Ray et al. 2009). Finding a balance between encouraging innovation and guaranteeing safety is essential. The development of evidence-based regulations that support the suitable application of nanotechnology in agriculture while preserving ecological health and food safety necessitates cooperation from stakeholders, including researchers, regulatory agencies, and industry representatives. This chapter examines the ways in which nanotechnology might support sustainable farming methods and solve the present crop protection issues. We demonstrate how nanotechnology enhances the efficiency of agrochemicals and lessens their environmental impact by looking at the application of nanomaterials for managing pests and diseases. Nanotechnology has the potential to help sustainable agriculture since it can increase crop yields while preserving the environment, making sure that farming methods can satisfy the needs of a growing world population without endangering ecological integrity (Gari et al. 2022). Nanotechnology can be a potent ally in the quest for sustainable food production systems with further study and prudent application.

7.2 Nanotechnology Applications in Crop Protection

The use of nanotechnology in crop protection has promises for improving fertilizer and pest control, with a focus on accuracy, efficiency, and environmental safety. This section looks at important uses like nanopesticides, nanosensors and nanofertilizers for detecting pests, and nano-biopesticide delivery systems.

7.2.1 Nanopesticides

A family of sophisticated formulations known as "nanopesticides" use nanoparticles and nanoscale carriers to improve the effectiveness and safety of traditional pesticides for the environment. By enhancing pesticide efficacy, stability, and targeted distribution, these formulations which include encapsulated insecticides, nanocapsules, and nanoemulsions represent a promising step toward precision agriculture. Conventional pesticides frequently have off-target effects, poor solubility, and quick degradation, all of which can reduce their efficacy and raise ecological concerns. These problems are addressed by nano-pesticides, which use nanotechnology to provide increased bioavailability, controlled release, and protection against environmental deterioration (Ali et al. 2023; Zainab et al. 2024).

7.2.1.1 Types and Formulation of Nanopesticides

Encapsulated pesticides, nanocapsules, and nanoemulsions are some of the different types of nanopesticides that are intended to increase pesticide efficiency and lessen their negative effects on the environment. Encapsulated pesticides protect active chemicals against deterioration by moisture, UV light, and microbiological endeavor by encasing them in nanocarriers such as polymeric, lipid, or mesoporous silica nanoparticles. For example, pesticides like imidacloprid have been encapsulated in polymeric nanoparticles manufactured from biodegradable polymers like poly(lactic-co-glycolic acid) (PLGA), which improves the pesticide's durability and allows for controlled release. The distribution of hydrophobic pesticides has also been improved by the use of lipid-based nanoparticles, such as solid lipid nanoparticles (SLNs), which provide targeted proceedings and extended release against pests (Campos et al. 2015; Chaud et al. 2021; Mittal et al. 2020). A subclass of encapsulated pesticides known as nanocapsules has a core-shell structure in which the active component is encased in a tiny shell, frequently composed of alginate or chitosan. As an illustration, chitosan nanocapsules are used to encapsulate the fungicide carbendazim in order to improve adherence to plant surfaces and enable controlled delivery, which lowers environmental runoff (Wang et al. 2017). Another type is nano-emulsions, which are water and oil mixes at the nanoscale that are stabilized by surfactants. These formulations enhance the bioavailability and solubility of insecticides that are not very soluble in water. A natural insecticide called neem oil has been made into nanoemulsions to enhance its environmental impact and boost pest control effectiveness while using less oil. When combined, these cutting-edge nanopesticide formulations improve pest control efficiency while lowering chemical use and environmental hazards, marking important breakthroughs in sustainable agriculture (Mustafa and Hussein 2020). The capacity of nanocapsules to transport both hydrophilic and hydrophobic pesticide chemicals makes them widely employed. Nanocapsules are submicron-sized particles with a liquid or solid core encased in a protective polymer shell.

Their adaptability enables them to stabilize and solubilize a broad range of insecticides, enhancing their uptake by both pests and plants. By using ultrafine oil or water-based emulsions, nanoemulsions, on the other hand, make hydrophobic pesticides more soluble and dispersible, increasing their availability for plant or pest absorption. In order to ensure consistency and extended activity in a collection of environmental conditions, these nanoscale emulsions are frequently stabilized with surfactants or polymers (Vrignaud et al. 2011) (Table 7.1).

7.2.1.2 Mechanism of Action at the Molecular Level

Nanopesticides function at the molecular level by increasing the active components' surface area and bioavailability, which facilitates more efficient interaction with the biological systems of their target pests. Since nanoparticles have a larger surface area than typical formulations, the pesticides can attach, penetrate, and absorb into pest organisms more effectively, frequently enabling effective control at shorter doses (Yin et al. 2023). Additionally, by shielding active chemicals from deterioration brought on by environmental elements like UV light and microbial activity, nanoencapsulation and nanostructuring help increase their usefullness and lifespan in the field (Kah and Hofmann 2014).

Another important benefit of nanopesticides is their controlled release. Numerous nanoformulations are designed to publish active chemicals gradually, frequently in reaction to temperature or pH changes in the environment. A longer-lasting pest control effect is made possible by this controlled release capability, which lessens the need for repeat applications and the environmental burden of pesticides overall (Huang et al. 2018). The insecticide functions exactly where it is needed since pH-responsive nanoparticles can only release their payload when they come into contact with the right pH (slightly acidic) environments found on insect exoskeletons (Kah and Hofmann 2014).

7.2.1.3 Advantages of Nanopesticides

Compared to traditional pesticides, nanopesticides have a number of clear benefits, such as decreased environmental toxicity, increased efficacy, and the possibility of targeted administration. Nanopesticides help limit pesticide runoff, leaching, and

Table 7.1 Some nanopesticides used in agriculture

Type of nanopesticide	Examples available commercially	Target agricultural pests	Reference
Encapsulated pesticides	Polymeric nanoparticles (e.g., PLGA-encapsulated imidacloprid)	Various insect pests (e.g., termites, aphids)	Chaud et al. (2021), Mittal et al. (2020)
Lipid-based nanoparticle	Solid lipid nanoparticles (SLNs) (e.g., encapsulated hydrophobic pesticides like imidacloprid)	Insect pests, specifically those resistant to hydrophobic pesticides	Campos et al. (2015)
Nanocapsules	Chitosan nanocapsules (e.g., carbendazim for fungicide delivery)	Fungal pathogens (e.g., *Fusarium, aspergillus*)	Wang et al. (2017)
Nanoemulsions	Neem oil nanoemulsions, e.g., EmulTech (natural neem oil encapsulated in nanoemulsion)	Various insect pests (e.g., aphids, whiteflies), fungi	Mustafa and Hussein (2020)
Nanoemulsion	Nanoemulsions of pyrethroids (e.g., NanoPyre)	Insect pests	Vrignaud et al. (2011)
Nanocapsules (other)	Polymeric nanocapsules (e.g., curcumin encapsulated pesticides for antifungal action)	Fungal pathogens, plant diseases	Liu et al. (2021)
Nanoemulsions (other)	Nanoemulsions of *Bacillus thuringiensis* (Bt)	Lepidopteran larvae (e.g., cotton bollworm, corn earworm)	John et al. (2017)
Nanoemulsion	Abamectin (PLGA)	Nematocidal activity	Fu et al. (2018), Ismail et al. (2022)

contamination of non-target areas, all of which are serious environmental issues linked to traditional pesticides. This is achieved by lowering the total amount of pesticide required for effective pest control (Wang et al. 2022). According to studies, nanoformulated pesticides can dramatically lower chemical loads on ecosystems by achieving comparable or better pest control with up to 50% fewer active ingredients than traditional formulations (Tang et al. 2024).

Further contributing to better pest control results are nanopesticides' increased stability and bioavailability. By shielding active chemicals from initial deterioration, encapsulation increases the amount of pesticide that reaches the target insect population in an active state (Manna et al. 2023). In field tests, pesticides encapsulated in lipid-based nanocarriers or chitosan have shown extended stability and bioavailability, resulting in more efficient insect suppression (Iavicoli et al. 2017). Furthermore, by altering their surface, nanopesticides can be made to attach more specifically to the exoskeletons of pests or the surfaces of plant leaves, boosting their effectiveness and minimizing their adversative effects on pollinators, beneficial insects, and other non-target creatures (Chaud et al. 2021).

7.2.1.4 Targeted Delivery and Reduced Non-target Effects

The potential for targeted distribution of nanopesticides is among their most exciting features. By designing nanoparticles to release their payload in response to certain pests or environmental circumstances, the pesticide can be applied where it is most required. The danger of exposure to non-target species is significantly decreased by surface modifications that enable nanopesticides to be "programmed" to attach specifically to pest organisms, such as the inclusion of ligands that recognize pest-specific markers (Kah and Hofmann 2014). Therefore, we can improve selectivity and decrease negative effects on beneficial insects like bees and natural pest predators by using nanopesticides that target certain enzymes or receptors that are unique to pest species. Nanopesticides' capacity for targeted and sustained release also opens up possibilities for their incorporation into integrated pest management (IPM) systems, where their accuracy can support biological control approaches. By working in tandem with biological agents like natural predators or pest-resistant crop varieties, nanopesticides can be employed in integrated pest management (IPM) to reduce the need for broad-spectrum chemical pesticides and enable lower pesticide dosages (Khartik et al. 2024).

7.2.2 Nanofertilizers

A revolutionary development in nutrient management, nanofertilizers provide precise nutrient delivery that boosts crop output while lessening environmental impact. These cutting-edge fertilizers use nanoscale transporters to solve issues with nutrient leaching, runoff, and volatilization that occur with traditional fertilizers. By increasing nutrient bioavailability and uptake efficiency, nanofertilizers reduce ecological footprints and increase yields (Arora et al. 2024; Anurag et al. 2023; Easwaran et al. 2024). Following are some examples of nanofertilizers that have been studied (Table 7.2).

7.2.2.1 Formulation and Mechanisms of Delivery

The purpose of nanofertilizers is to provide micronutrients like zinc (Zn) and iron (Fe) in nanoscale forms, along with macronutrients like nitrogen (N), phosphorus (P), and potassium (K). The high surface-area-to-volume ratios of these formulations increase their solubility, reactivity, and plant absorption efficiency. By reducing volatilization and leaching losses, nano-urea created by IFFCO in India was shown to improve nitrogen use efficiency. According to field tests, nano-urea reduces the necessary application rate as assessed to regular urea and boosts yields by 8–10% for crops like rice and wheat (Nongbet et al. 2022; Upadhyay et al. 2023). Likewise, phosphorus-loaded hydroxyapatite nanoparticles (HAPs) have been utilized to promote maize growth by releasing nutrients in response to root exudates, which is in line with plant need (Zhang et al. 2024). Nitrogen availability is improved via controlled-release processes, like those found in zeolite nanoparticles. As shown in wheat cultivation, these nanoparticles gradually release urea, encouraging plant uptake and decreasing nitrate leakage into water bodies (Saurabh et al. 2024).

Table 7.2 List of nanofertilizer and the performance in agriculture

Nanofertilizer	Carrier system	Performance	Reference
Nano-biochar fertilizer	Biochar nanoparticle	Possesses exceptional capacity to absorb toxins, nutrients, and pollutants	Sireesha et al. (2018), Udomkun et al. (2024)
Nano-phosphorus	Poly-beta-amino-esters (PBAE), graphene oxide (GO), poly lactic-co-glycolic acid (PLGA)	Stable compounds reduce the burden on water supplies and prevent nutritional imbalances in the soil by improving phosphorus uptake	Li et al. (2022)
Nano-hydroxyapatite	Calcium phosphate nanoparticle	Enhances bioavailablity of phosphorus and soil organic matter.	Elsayed et al. (2022), Jia et al. (2022)
Nano-nitrogen fertilizers (NNFs)	Nano-zeolite particles	Lowers nitrogen losses and increases the efficiency of nitrogen consumption to reduce environmental contamination	Abhiram (2023)
Nano-zinc oxide	Polymer nanocomposite	Exhibits gradual release of fertilizer over time and improves plants' bioavailability of zinc	Abbas et al. (2019), Adil et al. (2022)

7.2.2.2 Enhanced Nutrient Uptake and Bioavailability

Nanofertilizers improve the efficiency of nutrient uptake by directly absorbing nutrients through the roots or foliar routes. Due to their nanoscale size, they can pass through cell walls and stomata, avoiding problems with soil fixation. It has been demonstrated that zinc oxide nanoparticles greatly increase wheat's zinc bioavailability, which in turn improves the activity of enzymes necessary for photosynthesis and metabolic processes. Compared to traditional zinc sulfate applications, ZnO nanoparticle-treated plants showed a 30–40% increase in grain zinc concentration (Gangwar et al. 2023). When applied to rice, silicon nanoparticles raised photosynthetic rates and crop yield. These nanoparticles enhanced resilience to abiotic stressors like salt by strengthening cell walls with silica (Ali and Bijay-Singh 2024).

7.2.2.3 Role in Enhancing Crop Resilience to Pests

Nanofertilizers enhance plant health and strengthen defenses, which helps them indirectly to aid in pest management. In order to provide a physical barrier against pests such as the brown planthopper in rice, nanosilicon fortifies cell walls. Increased defense enzymes synthesized, including phenylalanine ammonia-lyase, is a factor in biochemical resistance (Wu et al. 2017). Fungal infections can be effectively mitigated by iron nanoparticles. Fe nanoparticles improved tomato plants' tolerance to Fusarium infections by increasing their chlorophyll content and lowering oxidative stress (Elbasuney et al. 2022).

7.2.2.4 Systemic Resistance and Induced Defense Mechanisms

Systemic resistance mechanisms are another way that nanofertilizers might prepare plants for improved immune responses. Nanochitosan, a chitin-derived nanoparticle, substantially reduces fungal infections from *Fusarium oxysporum* by inducing the enhanced levels of chitinase and β-1,3-glucanase enzymes in tomatoes. This dual function of disease resistance and nutrient provision demonstrates its multifunctionality (Filyushin et al. 2022). In potatoes, nano-calcium phosphate causes systemic acquired resistance by triggering the pathways for salicylic acid and jasmonic acid, which are critical for pathogen defense. This method lessens the requirement for chemical interventions by increasing resistance to bacterial wilt (Wilson et al. 2023).

7.2.3 Nanosensors for Pest Detection

In precision agriculture, nanosensors have become game-changing instruments that allow for thorough crop monitoring and early pest detection. In order to identify particular biochemical markers or signals that imply the presence of pests, plant stress, or environmental changes, these sensors make use of nanoscale materials. Nanosensors help minimize crop losses, maximize pesticide use, and promote sustainable farming methods by fusing sophisticated detecting mechanisms with real-time monitoring capabilities.

7.2.3.1 Types of Nanosensors

The three types of nanosensors: chemical, optical, and biosensors are each intended to identify different signals or markers. Biosensors identify pest-specific biomarkers by using biological components such as enzymes, antibodies, or nucleic acids. Early warning signs for cotton bollworm infestations have been afforded by biosensors that use gold nanoparticles functionalized with lectins to detect glycoproteins secreted by *Helicoverpa armigera* during feeding (Wang et al. 2022). Similarly, *Bemisia tabaci* (whitefly) RNA sequences have been successfully targeted by DNA-based biosensors, allowing for accurate pest species identification (Rosli et al. 2024).

Optical sensors use light-based processes to identify pest activity or plant stress. For example, reactive oxygen species (ROS) in tomato plants impacted by *Tuta absoluta* (tomato leafminer) have been monitored using fluorescent carbon dots, which provide stress signals at the cellular level (Rodríguez-Varillas et al. 2022). Drones equipped with hyperspectral nanosensors have been able to identify pest-induced variations in light reflectance in maize fields attacked by *Spodoptera frugiperda* (fall armyworm), allowing for extensive surveillance (Feng et al. 2022).

Plants under pest stress release volatile organic compounds (VOCs), which are detected by chemical sensors. VOCs such as methyl jasmonate have been found in rice under stress from *Nilaparvata lugens* (brown planthopper), thanks to zinc oxide (ZnO) nanoparticles coated with molecularly imprinted polymers (MIPs) (Hua et al. 2022). Additionally, nanoarrays with gold nanoparticles have differentiated VOC signatures in wheat infested with *Sitophilus oryzae* (rice weevil) (Liu et al.

Table 7.3 List of nanosensors and their mechanism of action

Nanosensors	Mechanism of action	Reference
Chitosan-based electrochemical sensors and biosensor	Biomolecules such as enzymes, DNA, and antibodies are immobilized by the free amine and hydroxyl groups of chitosan. Chitosan builds composites with high electron transfer rates and specific surface areas and when combined with conductive polymers and nanoparticles, improve the sensitivity and selectivity of electrochemical sensors. Transforms physical, chemical, or biological data into signals that can be measured	Karrat and Amine (2020)
Carbon nanotube (CNT)-based biosensors	CNT-based enzymatic electrochemical biosensors exploit the biospecificity of the enzymatic process to direct electrical transport between biomolecules, particularly between enzymes and CNT bulk electrode materials	Sireesha et al. (2018), Ferrier and Honeychurch (2021)
Gold nanoparticle (AuNP)-based biosensor	Electromagnetic enhancement is caused by the simultaneous oscillation of electrons in the nanostructure and the incoming light field. This phenomenon results in a higher localized electric field on the nanoparticle's surface. Consequently, a range of SERS-related components seen in living cells can be imaged and detected using AuNPs	Kumalasari et al. (2024)
Graphene quantum dots (GQDs)-based fluorescent biosensor	A more stable detection of elements like nitrogen may be achievable because to fluorescent sensors' ability to attach selectively to the target analytes and perceive the targets in real time	Rodkate et al. (2023)
Polyaniline–magnesia nanocomposites	Even at extremely low concentrations, they detect ammonia with improved sensitivity and selectivity	Ganachari et al. (2024)

2024). Table 7.3 lists some of the nanosensors and their mechanism of action against pests.

7.2.3.2 Mechanisms of Action

The extremely sensitive processes that nanosensors use to function improve their accuracy. By selectively binding to pest biomarkers, biosensors' biological recognition elements generate quantifiable signals like variations in electrical conductivity or fluorescence intensity. For example, early pest identification was made possible by graphene-based biosensors that identified particular salivary enzymes from insects that were feeding on soybeans (Bai et al. 2020). Optical sensors detect changes in light characteristics, including changed reflectance or fluorescence, brought on by pest stress. Plasmonic nanosensors have identified UV-induced stress signals in rice plants that are being attacked by thrips (Kashyap and Kumar 2021). Chemical sensors use selective relations between volatile organic compounds (VOCs) and the detecting elements. For example, titanium dioxide (TiO_2) nanoparticles can detect linalool and cis-jasmone released by *Ostrinia nubilalis*-infested maize (European corn borer).

7.2.3.3 Advantages of Nanosensors

Nanosensors have many benefits, including revolutionizing pest control methods and promoting sustainable farming. Early detection of pest presence is made possible by nanosensors, which enables prompt treatment. Within 24 h of an infection, carbon nanotube biosensors identified insect activity in sugarcane, minimizing crop losses (Miguel-Rojas and Pérez-de-Luque 2023). Targeted interventions identify certain pest risks to enable targeted pesticide applications. VOC-detecting nanosensors in greenhouse experiments decreased tomato pesticide consumption by 30% without sacrificing production quality (Pretty and Bharucha 2015). Data-driven decision-making and real-time monitoring are made possible by integration with IoT-connected technologies. An IoT-enabled ZnO nanosensor device in rice fields achieved 85% accuracy in predicting insect outbreaks (Malik 2024).

Advantages for the environment nanosensors assist preserve beneficial organisms and lessen environmental contamination by limiting the use of pesticides and chemical discharge. By using nanosensors in soybean fields, pesticide expenditure was cut by 25%, resulting in less pollution of the soil and water (Yadav et al. 2023a, b).

Conclusively, nanosensors are a significant development in agricultural pest control, providing instruments for early detection, improved accuracy, and environmentally friendly farming methods. Through the utilization of their sensitivity, specificity, and integration capabilities, nanosensors improve resource efficiency, minimize crop losses, and safeguard the environment.

7.2.4 Nanodelivery Systems for Biopesticides

In integrated pest management (IPM), the use of biopesticides which involves natural pesticides made from microbial agents, plant extracts, or organic materials has increased because of their low toxicity and little environmental effect. Their traditional formulations, however, have drawbacks such as quick disintegration, poor performance, and trouble with specific distribution. In order to overcome these obstacles, nanodelivery technologies improve the biopesticides' stability, bioavailability, and accuracy.

7.2.4.1 Types of Nanodelivery Systems

Nanoparticles, nanoemulsions, nanosuspensions, and hydrogel-based systems are examples of nanodelivery systems for biopesticides, and each has special benefits. *Beauveria bassiana*, an entomopathogenic fungus, has been encapsulated in biodegradable polymer nanoparticles, such as those derived from poly(lactic-co-glycolic acid) (PLGA), increasing its stability against UV degradation and extending its effectiveness as a pest control agent (Makadia and Siegel 2011). Similarly, when compared to conventional neem oil formulations, chitosan nanoparticles loaded with neem oil demonstrated higher mortality against *Spodoptera litura* (armyworm) (Kilani-Morakchi et al. 2021). Aphids and whiteflies have been successfully controlled with neem oil nanoemulsions; a study showed that neem-based nanoemulsions increased penetration and pest death by enhancing bioavailability (Anjali et al. 2012). In comparison to traditional formulations, *Metarhizium anisopliae* solid particle nanosuspensions enhanced fungal spore dispersion, resulting in greater efficacy against *Plutella xylostella* (diamondback moth) (Santos et al. 2022). *Bacillus thuringiensis* (Bt)-containing hydrogels demonstrated prolonged release over a number of days, preserving insecticidal effectiveness against *Helicoverpa armigera* (cotton bollworm) and lowering the frequency of treatment (Gujar et al. 2007).

7.2.4.2 Mechanisms of Action

Through a number of ways, nanodelivery devices improve the effectiveness of biopesticides. The progressive release of active substances is made possible by controlled release of nanocarriers, which sustain effective concentrations for extended periods of time. Crop protection was greatly enhanced by silica nanoparticles encasing *Paecilomyces lilacinus*, which allowed for prolonged nematode endeavor against *Meloidogyne incognita* (root-knot nematode) (Tavallali and Karimi 2023). Pests can be precisely targeted, thanks to targeted delivery of nanoparticles. In order to improve targeting and lessen non-target impacts, magnetic nanoparticles functionalized with biopesticides were directed to particular insect locations under magnetic fields (Rathore et al. 2024). Biopesticides are shielded from environmental deterioration by being encapsulated in nanoparticles, which shield them from heat, moisture, and ultraviolet light. *Trichoderma harzianum* spore-carrying PLGA nanoparticles showed improved survival rates in harsh environmental settings, guaranteeing steady effectiveness in the field (Alghuthaymi et al. 2002).

7.2.4.3 Advantages of Nanodelivery Systems

Encapsulation increases the biopesticides' resistance to environmental conditions, extending their shelf life and field performance. Nanodelivery systems also offer increased stability and efficacy. Comparing nano-encapsulated azadirachtin to non-encapsulated formulations, the former demonstrated remarkably longer bioactivity against pests (Melanie et al. 2022). Sustained and targeted pest management where nanoemulsions of plant extracts were able to lower pest population with fewer sprayings. A case study on garlic oil nanoemulsion exposed enhanced efficacy against *Bemisia tabaci* with sustained release over two weeks (Modafferi et al. 2024). Nanosystems distribute biopesticides precisely to target pests, reducing non-target impacts and minimizing exposure to the environment and beneficial insects. *Bacillus subtilis* was encapsulated in hydrogels, which decreased negative impacts on pollinators while preserving their ability to combat soil infections (Hashem et al. 2019). Nanodelivery technologies represent a big leap forward in biopesticide application, providing better performance, lower environmental effect, and alignment with sustainable agriculture goals.

7.3 Mechanisms and Interactions of Nanoparticles with Pests and Plants

Nanoparticles (NPs) are materials at the nanoscale that have drawn a lot of interest in agricultural applications because of their special characteristics and actions. As pests and diseases pose an extending threat to the worldwide agricultural industry, nanoparticles present creative approaches to crop management and protection. Optimizing their use and reducing potential dangers require an understanding of how nanoparticles, pests, and plants interact. This section explores the physiological reactions of plants, the effects of nanoparticles on pest organisms, and the environmental consequences of their use.

7.3.1 Physicochemical Properties and Behavior

Numerous physicochemical characteristics, such as size, shape, surface charge, and composition, affect how nanoparticles interact with pest species. Since they are usually between 1 and 100 nanometers in size, nanoparticles might behave differently from larger particles. Because of their high surface area to volume ratio, which raises their reactivity, nanoparticles are more bioavailable to pests (Altammar 2023). Insect pests' cuticles can be penetrated by titanium dioxide nanoparticles, enabling internal delivery and interaction with cellular components (Chandoliya et al. 2024). Furthermore, the shape of nanoparticles affects how they are absorbed and transported through biological systems; spherical and rod-shaped nanoparticles may have different absorption kinetics (Hoshyar et al. 2016).

7.3.2 Absorption Mechanisms

Numerous pathways, such as passive diffusion, endocytosis, and receptor-mediated uptake, allow nanoparticles to penetrate pest species. The characteristics of the nanoparticles' surface frequently determine the mechanism of entrance. By interacting electrostatically with negatively charged elements of the insect cuticle, positively charged nanoparticles can enhance absorption. Research has shown that pests may efficiently absorb metal nanoparticles including zinc oxide and silver nanoparticles, which can have systemic impacts (Sánchez-López et al. 2020).

7.3.3 Mode of Action

Nanoparticles have the ability to interfere with cellular functions inside pest species, which can have a variety of biological impacts. The production of reactive oxygen species (ROS), which can result in oxidative stress and harm cellular constituents such as lipids, proteins, and nucleic acids, is one of the main developments (Manke et al. 2013). It has been demonstrated that AgNPs cause ROS generation in a variety of insect pests, which leads to cell death and decreased reproductive success (Muhammad Jafir et al. 2023). Furthermore, nanoparticles have the ability to disrupt hormonal systems, which can impact development. According to studies, certain nanoparticles have the ability to imitate or interfere with endocrine signaling pathways, which can cause developmental defects in pest insects (Iavicoli et al. 2017).

7.3.4 Targeted Delivery of Pesticides

A possible method for increasing the efficacy of pest management is the combination of nanoparticles and conventional pesticides. Lower dosages and less environmental pollution are made possible by nanoscale expressions that increase the solubility and bioavailability of active substances. When environmental triggers like pest contact or particular temperature conditions occur, nanoencapsulation of pesticides can prevent degradation of the active components and allow for regulated release (Sajad et al. 2023). This focused strategy maximizes pest management effectiveness while lowering collateral harm to non-target creatures and ecosystems.

7.4 Plant-Response Mechanisms

Nanoparticles interact with plants in a variety of intricate ways that go beyond their pesticidal action because of their distinct physicochemical characteristics. These interactions have a major impact on the physiology, growth, and defense mechanisms of plants, which in turn shapes how they react to biotic and abiotic stimuli. The size, shape, chemical makeup, and surface charge of nanoparticles moderate their effects on plant systems by controlling their absorption, translocation, and ensuing biological consequences (Tortella et al. 2023).

7.4.1 Influence on Plant Physiology

Nanoparticles are mainly absorbed by plants through their roots, though they can also be absorbed through the leaves, when sprayed on. Nanoparticles can move via the xylem and phloem after internalization, reaching different tissues. By enhancing light uptake and chlorophyll production, nanoparticles like those created of zinc oxide, titanium dioxide, or silica have been demonstrated to increase photosynthetic activity (Wang et al. 2023a, b). This improvement is especially helpful in situations where plants are unable to maintain sufficient photosynthetic rates, such as low light levels or soils lacking in nutrients. Furthermore, certain nanoparticles function as micronutrient carriers, providing vital nutrients like magnesium, zinc, and iron more effectively than traditional fertilizers. These micronutrients are essential for energy production, enzymatic processes, and the synthesis of important macromolecules, all of which enhance the general health and productivity of plants (Ayaz et al. 2024).

7.4.2 Promotion of Growth and Biomass Accumulation

It has been reported that nanoparticles promote root and shoot growth, hence resulting in the increase of biomass accumulation. Low quantities of silver nanoparticles have been shown to encourage lateral root formation and root elongation, most likely by altering auxin distribution and signaling pathways (Abdou et al. 2022). The strength and fundamental integrity of cell walls are also improved by silicon nanoparticles, which offer mechanical support that can increase a plant's resistance to physical stresses and lodging. Furthermore, genes involved in growth control can have their expression changed by nanoparticles. Research shows that certain nanoparticle exposure increases genes linked to cell division, elongation, and differentiation, which speeds up development and shortens the time to maturity (Abdou et al. 2022).

7.4.3 Activation of Defense Mechanisms

In addition to encouraging growth, nanoparticles are essential for strengthening plant defenses against environmental stresses, pests, and diseases. Nanoparticles have the ability to function as elicitors in plant systems, causing reactive oxygen species (ROS) to be produced and systemic acquired resistance (SAR) pathways to be activated. It has been demonstrated that pathogenesis-related (PR) proteins, which are elemental elements of the plant immune system, are expressed more when copper nanoparticles are present (Tortella et al. 2023). Additionally, secondary metabolites like phenolics, flavonoids, and terpenoids—which are essential for warding off pests and preventing microbial infections, are delivered more readily by nanoparticles. These substances support the plant's long-term stress tolerance in addition to bolstering its immediate defense (Martínez-Chávez et al. 2024).

Additionally, by altering stress-responsive pathways, nanoparticles can decrease damage under abiotic stressors such drought, salt, and heavy metal toxicity. According to research, silicon and titanium nanoparticles enhance osmotic adjustment and water-use efficiency (WUE), allowing plants to sustain turgor and metabolic activity in water-limited environments (Daler et al. 2024; Grewal et al. 2024).

Similarly, at times of extreme stress, cerium oxide nanoparticles absorb excess ROS to prevent oxidative damage to biological modules (Sadidi et al. 2020).

7.4.4 Potential Trade-Offs and Risks

Although nanoparticles have provided several advantages for improving plant responses, there may be trade-offs in how they cooperate. Excessive nanoparticle concentrations can cause phytotoxicity, which is typified by oxidative damage, decreased root elongation, and poor germination. The overproduction of ROS causes DNA damage, protein denaturation, and lipid peroxidation, which in turn causes this toxicity. Comprehending these dose-dependent effects is essential to maximizing the use of nanoparticles in agriculture (Hasanuzzaman et al. 2021). The way that nanoparticles interact with plants highlights their potential as efficient instruments to improve resistance and growth. Nanoparticles present a possible path for sustainable agricultural developments by altering physiological processes, encouraging biomass buildup, and triggering defensive mechanisms. But weighing their advantages against any drawbacks is still essential, requiring more investigation into their long-term effects on ecosystems and plant health (Thabet and Alqudah 2024).

7.4.5 Stress Mitigation and Resilience

Apart from improving pest resistance, nanoparticles can assist plants in managing abiotic stresses including salinity and drought. Nanoparticles can increase a plant's resistance to unfavorable climatic conditions by enhancing water retention and encouraging root extension (El-Saadony et al. 2022). According to studies, plant tissues' osmotic potential can be increased by silver and gold nanoparticles, improving their resistance to drought. By guaranteeing crop viability in shifting climates, this ability to reduce abiotic stress promotes agricultural sustainability in addition to improving plant health (Thabet and Alqudah 2024).

7.5 Environmental Impacts and Nanotoxicity

The environmental effects and possible toxicity of nanoparticles raise serious concerns, despite the fact that they provide exciting prospects for agricultural innovation. The chemistry and biology of soil can be altered by the introduction of nanoparticles into agricultural ecosystems, potentially resulting in ecological imbalances. Research has indicated that nanoparticles have the ability to endure in the environment, impacting nutrient cycling mechanisms and soil microbial communities (Mittal et al. 2020). The physicochemical characteristics of nanoparticles distress their mobility in soil and water, which affects their bioavailability to soil organisms and plants. Determining the long-term impacts of nanoparticles on soil health and ecosystem functioning requires an understanding of their environmental destiny (Belal and El-Ramady 2016).

7.5.1 Nanotoxicity to Non-target Organisms

Research is still being done to determine whether nanoparticles are harmful to non-target animals, such as beneficial insects, soil fauna, and aquatic organisms. Certain nanoparticles can have cytotoxic effects on valuable insects as indicated by toxicological evaluations, resulting in decreased survival and poor reproductive success (Zoe et al. 2023). Reduced foraging behavior and higher mortality rates in honeybees have been associated with exposure to silver nanoparticles (Liu et al. 2021).

Additionally, fish and other aquatic species are at risk from the leaching of nanoparticles into aquatic habitats, which may cause bioaccumulation and biomagnification in food webs (Inês et al. 2019).

7.5.2 Regulatory Considerations and Risk Assessment

It is imperative to create thorough regulatory frameworks to control the use of nanoparticles in agriculture due to the possible environmental hazards they pose. There are currently few regulations that definitely address nanomaterials, which emphasizes the necessity of thorough risk evaluations that take into account their special characteristics and interactions in biological systems (Kumari et al. 2023). Standardized testing processes and safety evaluations must be created in order to guarantee the ethical application of nanotechnology in agriculture while preserving environmental integrity and public health.

Research on how nanoparticles interact with plants and pests is a complicated and multidimensional field with great capability to advance sustainable farming methods. We can maximize the potential of nanoparticles while addressing the safety and environmental problems related to their use by comprehending their mode of action within pest species, their routes of absorption, and their effects on plant physiology and immune responses (Ahmad et al. 2024). Developing responsible and efficient nanotechnology products that improve agricultural resilience and productivity requires ongoing research and innovation in this area (Abdul et al. 2024; Muhie 2022).

7.6 Advantages and Limitations of Nanotechnology in Crop Protection

7.6.1 Sustainability Benefits

The integration of nanotechnology into sustainable agricultural practices, including urban and rural farming, offers transformative benefits by reducing chemical inputs, enhancing resource efficiency, and minimizing the environmental footprint of food production. These innovations are pivotal for achieving global food security while safeguarding ecosystems (Gari et al. 2022). Nanotechnology significantly reduces the dependency on synthetic fertilizers and pesticides, which are prevalent in conventional farming. Excessive use of agrochemicals often results in soil degradation, water contamination, and harm to beneficial soil microbiota (Tudi et al. 2021). Nanofertilizers and nanopesticides enable controlled and precise delivery of nutrients and pest control agents, reducing chemical wastage and environmental contamination. For instance, nanoencapsulation technologies ensure the slow release of nutrients, enhancing plant uptake efficiency while minimizing leaching.

Nanotechnology advances precision agriculture by improving the targeted application of inputs. Nanosensors monitor real-time soil health, pest presence, and crop conditions, enabling the precise application of nano-formulated fertilizers and pesticides only where needed. These sensors integrate seamlessly with technologies like remote sensing and drone-assisted spraying, further reducing input wastage and environmental impact (Sanyaolu and Sadowski 2024). Nanotechnology drives efficient resource utilization. Nano-filtration systems enhance water use efficiency by improving irrigation methods like drip systems and rainwater harvesting, reducing water wastage by up to 60% compared to conventional systems. Additionally, the incorporation of nanoparticles in composting accelerates nutrient recycling, enriching soil organic matter and improving nutrient availability. These innovations align with sustainable practices by lowering the demand for external resources.

Crop diversification and intercropping, complemented by nanotechnology, optimize land productivity. The use of nano-biofertilizers in legume-based rotations enhances nitrogen fixation from the atmosphere, decreasing the reliance on synthetic nitrogen fertilizers. Nanomaterials also improves soil structure and microbial activity, further supporting sustainable land use. Nanotechnology significantly mitigates greenhouse gas emissions in agriculture. Nano-coatings on fertilizers minimize nitrogen volatilization, reducing emissions from synthetic fertilizer production. In rice farming, the use of nano-silica particles supports water management practices like alternate wetting and drying (AWD), and lowering methane emissions without compromising yields (Abdul et al. 2024). Additionally, conservation tillage practices integrated with nanomaterials enhance carbon sequestration, contributing to climate change mitigation. Agroecological applications of nanotechnology preserves biodiversity by minimizing chemical disturbances and maintaining habitat connectivity. Nanopesticides for instance target specific pests while protecting beneficial organisms, fostering ecosystem resilience (Tahat et al. 2020).

7.6.2 Challenges and Limitations

7.6.2.1 Economic Viability

The economic feasibility of adopting advanced technologies, such as nanotechnology in agriculture, is pivotal for ensuring widespread adoption, particularly among smallholder farmers. The costs of production and scalability of such innovations often dictate their accessibility, necessitating careful consideration of economic factors. The production of nanomaterials, including nanofertilizers and nanopesticides, involves high initial investments due to the specialized equipment and expertise required (Yadav et al. 2023a, b). Additionally, processes such as nanoparticle synthesis, functionalization, and testing contribute to the overall cost. Despite these challenges, economies of scale have shown potential to reduce per-unit costs as production volumes increase (Altammar 2023). Innovations in green synthesis methods, using natural precursors, also offer cost-effective alternatives to conventional methods.

Scaling up nanotechnology applications remains a significant hurdle due to technical and logistical challenges. While laboratory-level applications have demonstrated promise, the transition to field-scale deployment requires consistent performance across diverse agro-climatic conditions (Yadav et al. 2023a, b). Investment in pilot-scale facilities and collaborative public-private partnerships can bridge this gap, enabling the development of scalable and cost-effective production models. For smallholder farmers, access to nanotechnology-based solutions depends on affordability and awareness. Subsidies, government incentives, and extension programs are crucial to making these technologies accessible. Furthermore, tailored solutions, such as nanofertilizers designed for small-scale application, can address specific needs of resource-constrained farmers (Malik et al. 2023). Successful integration also requires robust distribution networks and training initiatives to build capacity among smallholders.

7.6.2.2 Safety Concerns

The application of nanotechnology in agriculture raises critical safety concerns, particularly regarding its environmental and ecological implications. Addressing these concerns requires a comprehensive understanding of potential toxic effects and the establishment of clear regulatory frameworks. Nanomaterials, due to their small size and high reactivity, can interact with soil, water, and biotic systems in unpredictable ways. Studies have shown that nanoparticles such as silver and zinc oxide can adversely affect soil microbial communities, potentially disrupting nutrient cycling and ecosystem services (Strekalovskaya et al. 2024). Similarly, concerns have been raised about bioaccumulation of nanomaterials in food chains, leading to toxic effects on higher trophic levels.

Nanotechnology-based pesticides, while reducing chemical load, may exhibit prolonged persistence in the environment, leading to unintended consequences. To mitigate these risks, researchers advocate for lifecycle assessments and ecotoxicological studies to evaluate the long-term impacts of nanomaterial use in agriculture. The lack of standardized regulations for nanotechnology in agriculture presents a significant barrier to its adoption. Regulatory frameworks must address issues such as nanoparticle characterization, safety thresholds, and risk assessment protocols (Iavicoli et al. 2017; Abdul et al. 2024). International cooperation and harmonization of standards are essential to streamline approval processes and facilitate global trade of nano-enabled agricultural products.

7.6.2.3 Public Perception and Ethical Concerns

Public acceptance of nanotechnology in food production hinges on addressing concerns related to safety, ethics, and transparency. Building trust among consumers and stakeholders is vital for the successful integration of these technologies. The introduction of nanotechnology in food systems has faced skepticism due to perceived risks and a lack of awareness. Public hesitancy often stems from limited knowledge about the benefits and safety of nanotechnology (van Dijk et al. 2015). Transparent communication, highlighting the advantages of nanotechnology, such as improved nutrient use efficiency and reduced pesticide residue, can alleviate concerns. Moreover, involving stakeholders in decision-making processes and conducting public consultations can foster acceptance (Mittal et al. 2020). Educational campaigns and labeling initiatives can further bridge the knowledge gap, ensuring consumers are informed about nano-enabled products. The ethical implications of using nanotechnology in agriculture also warrant attention. Issues such as equity in access, potential monopolization by large corporations, and unintended consequences on vulnerable populations must be addressed (Schulte and Salamanca-Buentello 2007). Ethical frameworks should prioritize inclusivity and sustainability, ensuring that the benefits of nanotechnology reach marginalized groups and do not exacerbate existing inequalities.

7.7 Current Trends and Research in Nanotechnology for Crop Protection

Nanotechnology has emerged as a transformative tool in modern agriculture, offering innovative solutions to longstanding challenges in crop protection. The global demand for sustainable agricultural practices is intensifying as climate change, pests, and diminishing natural resources exert unprecedented pressure on food systems. Nanotechnology, with its potential to improve resource efficiency, reduce chemical inputs, and enhance the precision of pest and disease management, is poised to play a pivotal role in addressing these challenges (Shang et al. 2019). Nanomaterials, including nanopesticides, nanofertilizers, and nanosensors, are the cornerstone of nanotechnology applications in agriculture. Their unique physicochemical properties, such as high surface area-to-volume ratios, tunable surface functionalities, and enhanced reactivity, enable their use in targeted and efficient crop protection strategies. For instance, nanoformulations of pesticides can increase the bioavailability and stability of active ingredients, reducing the amount required and minimizing off-target effects (Gari et al. 2022). Similarly, nanoscale carriers can deliver fertilizers or plant nutrients in a controlled manner, ensuring optimal uptake while minimizing leaching and environmental pollution (Shang et al. 2019; Tang et al. 2024).

One of the key trends in nanotechnology research for crop protection is the development of environmentally benign and biodegradable nanomaterials. Green synthesis methods, using plant extracts or microbial systems, are gaining attention for producing nanoparticles with reduced toxicity and enhanced biocompatibility (Alsaiari et al. 2023). These approaches not only address safety concerns but also aligns with the principles of sustainable agriculture. For instance, biosynthesized nanoparticles have shown efficacy against a range of pathogens and pests, including *Magnaporthe oryzae* and *Helicoverpa armigera*, without causing significant harm to beneficial organisms (Mittal et al. 2020). Nanosensors are another area of significant advancement, offering real-time monitoring of crop health, pest infestations, and environmental conditions. These sensors, often embedded in precision agriculture systems, enable early detection of stressors and allow for timely interventions, thereby reducing crop losses (Ahmad et al. 2024). Carbon nanotube-based sensors have been developed to detect volatile organic compounds emitted by plants under pest attack, providing an early warning system for farmers (Yadav et al. 2023a, b).

The use of nanotechnology in managing pesticide resistance is also a growing research focus. Conventional pesticide application often leads to the development of resistance in pest populations, necessitating the use of higher doses or alternative chemicals. Nanopesticides, with their ability to penetrate pest integuments more effectively and deliver active ingredients in a sustained-release manner, offer a solution to this problem (Islam et al. 2020). Moreover, combining nanotechnology with genetic and molecular tools could further enhance its efficacy in tackling resistance mechanisms in pests and pathogens. Despite these advancements, the adoption of nanotechnology in crop protection is not without challenges. Regulatory frameworks for nano-enabled agricultural products remain underdeveloped, particularly in addressing safety, environmental impact, and standardization of nanomaterials (Iavicoli et al. 2017; Ahmad et al.2024). Furthermore, public perception and ethical concerns about the use of nanotechnology in food production present additional hurdles to widespread adoption. Transparent communication and rigorous safety assessments are essential to build trust among stakeholders (Iavicoli et al. 2017).

Future research is increasingly focusing on the integration of nanotechnology with other cutting-edge fields, such as synthetic biology, artificial intelligence, and big data analytics, to create holistic and adaptive agricultural systems. This convergence has the potential to revolutionize crop protection by enabling smarter, more sustainable, and highly efficient farming practices (Tripathy et al. 2024). In conclusion, nanotechnology is reshaping the landscape of crop protection by providing innovative tools to enhance agricultural productivity and sustainability. While promising strides have been made, continued interdisciplinary research and policy development are crucial to ensure the safe and equitable deployment of these technologies. The integration of nanotechnology with sustainable agricultural practices holds the promise of securing food systems against the growing challenges of the twenty-first century.

7.7.1 Emerging Technologies

The integration of nanotechnology with advanced biotechnological tools has opened new frontiers in crop improvement and protection. Cutting-edge research in this domain focuses on nanoscale genome editing, DNA-targeted nanoparticles, and bio-nanotechnology, each offering innovative solutions to enhance agricultural productivity and sustainability. These advancements hold the potential to revolutionize crop management practices, address global food security challenges, and mitigate the impacts of climate change on agriculture.

Nanoscale Genome Editing

Nanoscale genome editing is a transformative approach that combines nanotechnology with CRISPR-Cas systems to enable precise genetic modifications in plants. By using nanoparticles as delivery vehicles, researchers have enhanced the efficiency and specificity of gene-editing tools. Lipid-based nanoparticles have been used to deliver CRISPR-Cas9 components directly into plant cells, bypassing traditional methods like *Agrobacterium*-mediated transformation or gene guns, which often have lower efficiency and higher off-target effects (Cunningham et al. 2018). Gold nanoparticles and mesoporous silica nanoparticles (MSNs) are particularly promising due to their biocompatibility and ability to encapsulate and protect CRISPR components from degradation. Studies have shown that nanoparticle-mediated delivery of CRISPR machinery can edit genes related to stress tolerance, pest resistance, and nutrient use efficiency with unprecedented precision (Vats et al. 2022). Targeted modifications in rice (*Oryza sativa*) genes involved in drought resistance have been achieved using nanoscale genome-editing platforms, highlighting their potential in climate-resilient agriculture (Ahmar et al. 2021; Vats et al. 2022).

DNA-Targeted Nanoparticles
DNA-targeted nanoparticles represent another frontier in agricultural nanotechnology, offering a means to precisely alter gene expression or silence harmful genes. These nanoparticles are designed to bind specifically to target DNA sequences, delivering genetic material or inhibitors such as RNA interference (RNAi) molecules. Polymeric nanoparticles, for instance, have been employed to transport RNAi molecules into insect pests, effectively silencing essential genes and reducing pest populations without the environmental hazards associated with chemical pesticides (Chen et al. 2023). In plants, DNA-targeted nanoparticles have been utilized to enhance resistance to pathogens by modulating gene expression in response to infection. For instance, chitosan-based nanoparticles loaded with pathogen-responsive promoters have shown potential in inducing systemic acquired resistance (SAR) in crops. This approach not only improves crop resilience but also reduces the need for chemical inputs, aligning with the principles of sustainable agriculture (Noman et al. 2023).

Bio-Nanotechnology
Bio-nanotechnology, the intersection of nanotechnology and biological systems, is driving innovations in crop protection and enhancement. This field focuses on the design and application of biologically derived nanoparticles, such as those synthesized using plant extracts, bacteria, or fungi. These "green" nanoparticles are biodegradable, environmentally friendly, and often exhibit unique biological activities that can be harnessed for agricultural purposes (Gari et al. 2022). A notable application of bio-nanotechnology is the development of antimicrobial nanoparticles to combat plant pathogens. Silver nanoparticles synthesized using *Bacillus subtilis*, for example, have demonstrated potent activity against fungal pathogens like *Magnaporthe oryzae*, which causes rice blast disease (Ibrahim et al. 2020). Similarly, nanoparticles derived from plant extracts have been shown to enhance seed germination and root development while protecting crops from microbial infections (Sajad et al. 2023). Bio-nanotechnology is also advancing in the area of smart delivery systems. Bio-nanocapsules made from plant proteins have been engineered to deliver agrochemicals in a controlled manner, ensuring that active ingredients are released only under specific environmental conditions, such as changes in pH or temperature (Gari et al. 2022). These systems not only improve the efficiency of

agrochemical use but also reduce environmental contamination. The integration of nanoscale genome editing, DNA-targeted nanoparticles, and bio-nanotechnology represents a paradigm shift in agricultural research. Future studies aim to refine these technologies by enhancing their efficiency, reducing costs, and addressing safety concerns. Collaborative efforts between researchers, policymakers, and industry stakeholders will be critical to translating these cutting-edge innovations into scalable and sustainable agricultural solutions.

7.7.2 Case Studies of Successful Applications

Nanotechnology has shown considerable promise in enhancing crop protection, with several successful applications across a range of crops. These advancements have not only improved pest and disease management but also offered sustainable solutions by reducing the reliance on traditional chemical pesticides. Below, we highlight some specific examples where nanotechnology has significantly contributed to crop protection, providing a glimpse into its potential for the future.

7.7.2.1 Nanoparticles for Fungal Disease Management in Rice

A notable example of nanotechnology in crop protection is the use of silver nanoparticles to combat fungal pathogens in rice. Rice is highly susceptible to various fungal diseases, including *Magnaporthe oryzae*, the causal agent of rice blast disease. Recent studies have demonstrated that AgNPs exhibit strong antifungal properties against *M. oryzae* by disrupting cell wall integrity and inhibiting spore germination (Ibrahim et al. 2020). Furthermore, AgNPs have been shown to enhance plant defense mechanisms by stimulating the production of reactive oxygen species (ROS) and inducing systemic acquired resistance (SAR), which improves the overall disease resistance of rice plants (Ahmad et al. 2024). This application not only offers an effective alternative to chemical fungicides but also promotes environmentally friendly farming practices by reducing pesticide residues in rice products.

7.7.2.2 Nanoencapsulation for Controlled Release of Pesticides in Cotton

Another successful application of nanotechnology in crop protection is the use of nanoencapsulation systems for controlled release of pesticides. In cotton farming, pests such as *Helicoverpa armigera* and *Spodoptera litura* pose significant challenges to crop yields. A study by Manna et al. (2023) demonstrated that the encapsulation of the insecticide chlorpyrifos in biodegradable polymer nanoparticles significantly improved its efficacy and reduced its environmental impact. The controlled release of the pesticide over an extended period ensures continuous protection against pests while minimizing the risk of over-application, thus reducing the overall pesticide load in the environment. Moreover, nanoencapsulation has been found to improve the stability of pesticides under adverse environmental conditions, such as high temperatures and UV radiation, which often degrade conventional formulations (Zainab et al. 2024).

7.7.2.3 Nanoseed Coatings for Improved Germination and Pest Protection in Maize

Nanotechnology has also been utilized for enhancing seed quality and pest protection through the application of nanocoatings. In maize, nanocoatings that incorporate biocidal agents and growth-promoting substances have been found to improve seed germination and protect against early-season pests, such as *Diabrotica virgifera* (Tarigan et al. 2024). Nanocoatings containing silica and titanium dioxide nanoparticles not only provide a physical barrier to pests but also release bioactive compounds that stimulate plant growth, resulting in higher germination rates and stronger seedlings. This technology has the potential to reduce the need for chemical pesticides in the early stages of crop growth, leading to more sustainable farming practices (Tortella et al. 2023).

7.7.2.4 Nanoparticle-Mediated Delivery of Biocontrol Agents in Tomatoes

The use of nanotechnology to enhance the delivery of biocontrol agents represents an innovative strategy for integrated pest management (IPM). In tomatoes, researchers have developed nanocarriers that deliver biocontrol agents such as *Bacillus thuringiensis* (Bt) and *Trichoderma* spp. directly to the site of infection, improving their efficacy against insect pests and plant pathogens (Tyagi et al. 2024). Nanoparticles, such as chitosan and poly(lactic-co-glycolic acid) (PLGA), have been used to encapsulate these biocontrol agents, protecting them from environmental degradation and enhancing their persistence in the field (Yang et al. 2022). This approach allows for the precise and controlled release of biocontrol agents, ensuring that they remain active for longer periods and are more effective in suppressing pest populations.

7.8 Research Gaps and Future Directions

While the successful applications of nanotechnology in crop protection are promising, several research gaps remain that must be addressed to ensure widespread adoption and long-term sustainability of these technologies.

7.8.1 Ecotoxicity Studies

The ecotoxicological responses of nanoparticles has gradually become an emerging issue and an additional research area. Furthermore, more studies need to focus on the effects of nanoparticle in controlling diseases and pests, as well as their effects on non-target beneficial animals in both agriculture and aquatic ecosystems. Several studies have also suggested that essential nanoparticles for soil health and nutrient cycling like silver and zinc can also cause possible harm to soil microbes (Yousef et al. 2023). On the other hand to reduce these problem, the bioaccumulation evaluations and long-term ecological research on those nanoparticles must be suggested for their beneficial impact on the larger ecosystem in agriculture field.

7.8.2 Impact on Beneficial Insects

While nanoparticles may offer targeted protection against pests, their non-selective toxicity to beneficial species could disrupt ecosystem services and harm biodiversity. This issue that the use of nanoparticles affects beneficial insects like pollinators and natural predators of crop pests has also gained much attention and is a source of current investigation (Chaud et al. 2021; Ahmad et al. 2024). More studies on ecologically safe formulations of nanoparticles and its selectivity to dangerous pests and sparing beneficial organisms is of utmost importance. Similarly, research should also focus on the effectiveness of nanoparticles on insect behavior, reproduction, and survival rates, particularly on important pollinators like bees (Yousef et al. 2023).

7.8.3 Long-Term Effects on Soil Health

The possible effects of nanotechnology on soil health remain unknown. Some nanoparticles are introduced into soil which helps to change the physical and chemical nature of the soil. Some of the nanoparticles are known for the role they play in improving soil nutrient content and microbial activity (Munazza et al. 2023). Prolonged use of silver nanoparticles can alter the structure of soil, which disrupts the microbial population (Sharma et al. 2023). Studies have shown that nanotechnology has become a long-lasting, effective, and focused solution due to its ability to improve crop security. However, more studies are needed to understand the long-term interactions between nanoparticles and soil ecosystems, as well as about their effectiveness on soil fertility and plant growth over time (Sajad et al. 2023). Several studies have also suggested about the potential of nanotechnologies in sustainable agriculture, which helps fill the research gaps along with a guaranteed both efficient and ecologically conscious crop protection techniques (Shang et al. 2019).

7.9 Policy, Regulations, and Adoption Challenges in Nanotechnology for Agriculture

7.9.1 Regulatory Landscape

Despite its beneficial properties, nanotechnology faces several challenges due to the unique characteristics of nanoparticles. Regional differences in regulatory frameworks reflect different priorities in risk evaluations for the environment and human health. The European Union prioritizes preventive measures with strict regulatory restrictions, whereas other countries like the USA use a case-by-case approach based on current chemical safety legislation (Kumari et al. 2023). Lack of an internationally recognized definition of nanotechnology makes it difficult to develop uniform standards to reduce the risk as well to get its approval. Global trade and exchange of scientific knowledge regarding agricultural nanotechnology applications are hampered by this lack (Iavicoli et al. 2017). Furthermore, standardized tools for evaluating the risks of nanoparticle are provided by frameworks such as those from the Organisation for Economic Co-operation and Development (OECD). In developing countries there is a lack of capacity to evaluate the impacts of nanoparticle on human health which leads the uneven adaptation and more environmental risks. Additionally, administrative burden and costs associated with compliance discourage smaller enterprises from entering the market (Ma et al. 2024; Stander and Theodore 2011).

7.9.2 Adoption Challenges

In the field of agriculture, adoption of nanotechnology is complicated due to socio-economic disparities and gaps in technological accessibility. Smaller firms and under developed regions find it difficult to adopt technologies that are expensive and require a large investment (OECD 2010). The limited availability of cost-effective solutions in nanotechnology, together with an inadequate system for dissemination, further compounds this issue. Side by side these issues form a critical barriers towards the adoption of this technology and also public perception. Policymakers are influenced to proceed cautiously by consumer and advocacy groups' due to mistrust regarding the safety of nanoparticles in food systems and their effects on the environment. To increase public trust and guarantee safe and fair application of nanotechnology in the agricultural sector, collaboration between scientists, regulators, and stakeholders is crucial (Amini et al. 2024).

Recommendations

To advance safe and sustainable application of nanotechnology in agriculture, efforts must focus on:

1. Developing internationally harmonized definitions and regulatory standards.
2. Increasing investments in R&D for long-term toxicity studies of nanoparticles.
3. Enhancing access to affordable, scalable nanotechnology solutions, particularly for developing nations and small enterprises.

7.10 Barriers to Adoption of Nanopesticides and Nanofertilizers

Adoption of nanopesticides and nanofertilizers in agricultural practices is challenging, which can be broadly categorized into three key areas: lack of technical know-how, high costs, and regulatory bottlenecks.

7.10.1 Lack of Technical Know-how

The limitation of handling technical expertise, like nanotechnology-based products, for farmers and agricultural stakeholders is one of the main obstacles to adoption. Lack of awareness among farmers about the specific advantages of nanopesticides and nanofertilizers in the field of agriculture, such as improved bioavailability, targeted delivery, and reduced environmental, also has a huge impact on adopting these technologies (Tang et al. 2024; Ahmad et al. 2024). To overcome these problems, developing countries should mitigate awareness programs to educate farmers about the proper handling, use, and possible hazards of nanoproducts (Yadav et al. 2023a, b). This knowledge gap hinders the effective use and dissemination of these advanced technologies in the agricultural field.

7.10.2 High Costs

Another major problem in adopting this nanotechnology or the nanoproduct is their high costs. Compared to conventional fertilizers or pesticide, the production and commercialization of nano-based products are resource intensive and expensive. For example, the production of nano-based pesticides requires advanced manufacturing techniques and specialized equipment, which adds to their cost (Chaud et al. 2021). So, farmers under strict financial restraints often are not able to invest in these expensive nano-based technologies, creating a gap between large-scale commercial farmers and smaller stakeholders (Dhillon and Moncur 2023).

7.10.3 Regulatory Bottlenecks

The regulatory landscape of nanopesticides and nanofertilizers are complicated and frequently differ from one place to another. The rules and regulations for using nanotechnologies in the agriculture field are still being developed to assess, evaluate, and standardize their risk. Due to the lack of precise rules and regulations for registration, labeling, and marketing of nanoproducts, both producers and consumers have little knowledge on such concepts and are left behind (Kumari et al. 2023). Further, the startups and smaller businesses are also discouraged from investing in nanoscale developments due to lengthy and expensive approval procedures (Tang et al. 2024). Moreover, from the concerns about the toxicological impact and the environmental persistence of nanomaterials, it is recommended that rigorous testing protocols and long-term studies be conducted on these materials before regulatory permission can be given. The broad use of nano-agricultural products is delayed for this cautious approach which is required for overall consumer and stakeholder safety (Kah 2015).

Recommendations
To overcome these barriers, multi-stakeholder collaborations involving policymakers, researchers, and industry leaders are essential. Efforts should focus on:

1. Increasing funding for farmer education and outreach programs to bridge the technical knowledge gap.
2. Establishing financial support mechanisms, such as subsidies or incentives, to reduce the cost burden on small-scale farmers.
3. Streamlining regulatory frameworks to provide clear, science-based guidelines for the approval and use of nanopesticides and nanofertilizers while ensuring safety and sustainability.

7.11 Guidelines for Safe and Responsible Use of Nanomaterials

In any operating situation, but especially in those involving new technologies, dangerous chemicals, or high-stakes decisions, it is imperative to apply safe and responsible practices. Based on current research, this part provides important guidelines for guaranteeing security, efficient risk management, and responsible handling of nanomaterial in agriculture. Developing strong safety procedures is essential to using nanomaterials responsibly. Clear standard operating procedures (SOPs) for handling, storing, and disposing of nanomaterials at every stage should be part of these protocols. To guarantee that staff members are knowledgeable about safety procedures and danger management, regular training sessions must be held. In order to properly handle calamities, companies also require comprehensive contingency plans. To reduce operational dangers, a methodical approach to safety management that includes real-time monitoring and risk assessment tools is essential (Buitrago et al. 2021).

An essential part of overseeing the safe application of nanomaterials is risk assessment. Frameworks for quantitative risk assessment (QRA), such as Hazard and Operability Studies (HAZOP) and Failure Mode and Effects Analysis (FMEA), are especially useful for detecting deficiencies and reducing hazards. In addition to mitigating recognized hazards, a proactive strategy is crucial. Using predictive analytics to foresee new risks is part of this (Kuempel et al. 2012). Such progressive approaches guarantee the systematic management of possible risks, including those that are not immediately obvious. In order to avoid mishaps and guarantee operational safety when using nanomaterials, proper handling procedures are essential.

Maintaining adherence to safety regulations requires fostering an organizational culture that prioritizes safety. To enable employees to report safety issues or near-miss situations without fear, open lines of communication should be set up. Positive habits can be reinforced by praising and rewarding safe practices. In this process, leadership is essential because it establishes safety priorities and gives workers the freedom to put safety first at work. To make sure safety precautions are applicable and efficient in a profession that is changing quickly, periodic evaluation and continual improvement are crucial. Frequent audits assist in finding flaws in present processes, and feedback systems enable the integration of useful observations from stakeholders and employees. Safety recommendations must be updated to reflect new laws and scientific findings as nanotechnology develops. Organizations can minimize risks to human health and the environment while promoting innovation and operational efficiency by putting these suggestions into practice and ensuring the safe and responsible use of nanomaterials (Kuempel et al. 2012; Tarek et al. 2015).

7.12 Future Prospects and Conclusion

7.12.1 Vision for the Future

Combination of cutting-edge technologies and nanotechnology has a revolutionary possibility for sustainable agriculture. The future of nanotechnology with the integration of Internet of Things (IoT), and artificial intelligence (AI) may become a significant way towards precision farming. Furthermore, nanosensors can be utilized to gather and evaluate real-time data on environmental condition, nutrient levels, pest presence, and soil health, which will allow for timely interventions such as targeted pesticide application or better irrigation schedules, thereby minimizing resource waste and greatly reducing environmental damage (Sharma et al. 2023; Yadav et al. 2023a, b).

Nanoparticles like stimuli-responsive nanoparticles shows potential efficacy in nutrition delivery as well as crop protection via controlled release systems. Their maximum efficiency depends on environmental cues such as pH, temperature, or moisture levels. Moreover, biodegradable nanomaterials are becoming a promising approach, in line with the ideas of circular economy to overcome environmental buildup and nanotoxicity. Beside these, urban farming methods like hydroponic and vertical farming can also benefit from the use of nanoparticles. So, ongoing research should focus on nanomaterials that help in light collection for photosynthesis and on nanocoatings for energy-efficient greenhouses. The combination of these developments with IoT-based monitoring can supply the expanding urban food demand (Tortella et al. 2023; Sharma et al. 2023).

To achieve accurate field conditions, monitoring the nanopesticides and nanofertilizers offer a more effective substitute for traditional agrochemicals. Furthermore, antibacterial nanocoatings and nanotechnology may help reduce post-harvest losses in terms of better packaging and storage. However, there are a few issues that were identified such as, the effect of nanomaterials on the environment and human health, along with the requirement for cost-effective scaling up and broad acceptance of this approach by stakeholders and regulators. Stakeholder participation, strong regulatory frameworks, and interdisciplinary research will be needed to identify all problems concerning nanomaterials (Ahmad et al. 2024).

The contribution of nanotechnologies in the field of agriculture brings an era of change with environmental friendly, efficient, and sustainable products. Besides, these technologies also greatly contribute to environmental preservation and global food security. Researchers, legislators, and industry stakeholders can work together to make a more resilient and environmentally friendly agricultural future a reality.

References

Abbas M, Buntinx M, Deferme W, Peeters R (2019) (Bio)polymer/ZnO nanocomposites for packaging applications: a review of gas barrier and mechanical properties. Nano 9(10):1494

Abdul W, Muhammad M, Ullah S, Abdi G, Shah GM, Zaman W, Ayaz A (2024) Agriculture and environmental management through nanotechnology: eco-friendly nanomaterial synthesis for soil-plant systems, food safety, and sustainability. Sci Total Environ 926:171862

Abdou MA, Ibrahim H, Mohamed AA (2022) Impact of silver nanoparticles mixture with NAA and IBA on rooting potential of Psidium guajava L. stem cuttings. Egyptian J Chem 65(132):1119–1128

Abhiram G (2023) Contributions of nano-nitrogen fertilizers to sustainable development goals: a comprehensive review. Nitrogen 4(4):397–415

Adil M, Bashir S, Bashir S, Aslam Z, Ahmad N, Younas T, Asghar RMA, Alkahtani J, Dwiningsih Y, Elshikh MS (2022) Zinc oxide nanoparticles improved chlorophyll contents, physical parameters, and wheat yield under salt stress. Front Plant Sci 13:932861

Ahmad Z, Niyazi S, Firdoos A, Wang C, Manzoor MA, Ramakrishnan M, Upadhyay A, Ding Y (2024) Enhancing plant resilience: nanotech solutions for sustainable agriculture. Heliyon 10(23):e40735

Ahmar S, Mahmood T, Fiaz S, Mora-Poblete F, Shafique MS, Chattha MS, Jung K-H (2021) Advantage of nanotechnology-based genome editing system and its application in crop improvement. Front Plant Sci 12:663849

Aktar MW, Sengupta D, Chowdhury A (2009) Impact of pesticides use in agriculture: their benefits and hazards. Interdiscip Toxicol 2(1):1–12

Alghuthaymi MA, Abd-Elsalam KA, Abo Dalam HM et al (2002) Trichoderma: an eco-friendly source of nanomaterials for sustainable agroecosystems. J Fungi (Basel) 8(4):367

Ali AM, Bijay-Singh (2024) Silicon: a crucial element for enhancing plant resilience in challenging environments. J Plant Nutr 1:1–36

Ali S, Ahmad N, Dar MA et al (2023) Nano-agrochemicals as substitutes for pesticides: prospects and risks. Plants (Basel) 13(1):109

Alsaiari NS, Alzahrani FM, Amari A et al (2023) Plant and microbial approaches as Green methods for the synthesis of nanomaterials: synthesis, applications, and future perspectives. Molecules 28(1):463

Altammar KA (2023) A review on nanoparticles: characteristics, synthesis, applications, and challenges. Front Microbiol 14:1155622

Amini SM, Gilaki M, Karchani M (2024) Safety of nanotechnology in food industries. Electron Physician 6(4):962–968

Anjali CH, Sharma Y, Mukherjee A, Chandrasekaran N (2012) Neem oil (Azadirachta indica) nanoemulsion—a potent larvicidal agent against Culex quinquefasciatus. Pest Manag Sci 68(2):158–163

Anurag Y, Yadav K, Abd-Elsalam KA (2023) Exploring the potential of nanofertilizers for a sustainable agriculture. Plant Nano Biology 5:100044

Arora PK, Tripathi S, Omar RA et al (2024) Next-generation fertilizers: the impact of bionanofertilizers on sustainable agriculture. Microb Cell Factories 23:254

Ayaz A, Zaman W, Radák Z, Gu Y (2024) Green strength: the role of micronutrients in plant-based diets for athletic performance enhancement. Heliyon 10(12):e32803

Bai Y, Xu T, Zhang X (2020) Graphene-based biosensors for detection of biomarkers. Micromachines (Basel) 11(1):60

Belal ES, El-Ramady H (2016) In: Ranjan S, Dasgupta N, Lichtfouse E (eds) Nanoparticles in water, soils and agriculture. Nanoscience in Food and Agriculture 2. Sustainable Agriculture Reviews Volume 21

Buitrago E, Novello AM, Fink A, Riediker M, Rothen-Rutishauser B, Meyer T (2021) Nano safe III: a user friendly safety management system for nanomaterials in laboratories and small facilities. Nano 11(10):2768

Campos E, Oliveira J, da Silva C et al (2015) Polymeric and solid lipid nanoparticles for sustained release of Carbendazim and Tebuconazole in agricultural applications. Sci Rep 5:13809

Chandoliya R, Sharma S, Sharma V, Joshi R, Sivanesan I (2024) Titanium dioxide nanoparticle: a comprehensive review on synthesis. Applications and Toxicity Plants 13(21):2964

Chaud M, Souto EB, Zielinska A et al (2021) Nanopesticides in agriculture: benefits and challenge in agricultural productivity, toxicological risks to human health and environment. Toxics 9(6):131

Chen A, Halilovic L, Shay JH, Koch A, Mitter N, Jin H (2023) Improving RNA-based crop protection through nanotechnology and insights from cross-kingdom RNA trafficking. Curr Opin Plant Biol 76:102441

Cunningham FJ, Goh NS, Demirer GS, Matos JL, Landry MP (2018) Nanoparticle-mediated delivery towards advancing plant genetic engineering. Trends Biotechnol 36(9):882–897

Daler S, Kaya O, Korkmaz N, Kılıç T, Karadağ A, Hatterman-Valenti H (2024) Titanium nanoparticles (TiO2-NPs) as catalysts for enhancing drought tolerance in grapevine saplings. Horticulturae 10(10):1103

Dhillon R, Moncur Q (2023) Small-scale farming: a review of challenges and potential opportunities offered by technological advancements. Sustainability 15(21):15478

Easwaran C, Christopher SR, Moorthy G, Mohan P, Marimuthu R, Koothan V, Nallusamy S (2024) Nano hybrid fertilizers: a review on the state of the art in sustainable agriculture. Sci Total Environ 929:172533

Elbasuney S, El-Sayyad GS, Attia MS et al (2022) Ferric oxide colloid: towards Green nano-fertilizer for tomato plant with enhanced vegetative growth and immune response against Fusarium wilt disease. J Inorg Organomet Polym 32:4270–4283

El-Saadony MT, Saad AM, Soliman SM, Salem HM, Desoky E-SM, Babalghith AO, El-Tahan AM, Ibrahim OM, Ebrahim AAM, Abd El-Mageed TA, Elrys AS, Elbadawi AA, El-Tarabily KA, Abu Qamar SF (2022) Role of nanoparticles in enhancing crop tolerance to abiotic stress: a comprehensive review. Front Plant Sci 13:946717

Elsayed AA, El-Gohary A, Taha Z, Farag HM, Hussein MS, Abou AK (2022) Hydroxyapatite nanoparticles as novel nano-fertilizer for production of rosemary plants. Sci Hortic 295:110851

Feng J, Sun Y, Zhang K, Zhao Y, Ren Y, Chen Y, Zhuang H, Chen S (2022) Autonomous detection of Spodoptera frugiperda by feeding symptoms directly from UAV RGB imagery. Appl Sci 12(5):2592

Ferrier DC, Honeychurch KC (2021) Carbon nanotube (CNT)-based biosensors. Biosensors 11(12):486

Filyushin MA, Shagdarova BT, Shchennikova AV, Il'ina AV, Kochieva EZ, Varlamov VP (2022) Pretreatment with chitosan prevents *Fusarium* infection and induces the expression of Chitinases and β-1,3-Glucanases in garlic (*Allium sativum* L.). Horticulturae 8(5):383

Fu Z, Chen K, Li L, Zhao F, Wang Y, Wang M, Shen Y, Cui H, Liu D, Guo X (2018) Spherical and spindle-like abamectin-loaded nanoparticles by flash nanoprecipitation for southern root-knot nematode control: preparation and characterization. Nano 8(6):449

Ganachari SV, Shilar FA, Patil VB, Khan TMY, Saleel CA, Ali MA (2024) Optimizing ammonia detection with a polyaniline–magnesia nano composite. Polymers 16(20):2892

Gangwar J, Kadanthottu Sebastian J, Puthukulangara Jaison J, Kurian JT (2023) Nano-technological interventions in crop production—a review. Physiol Mol Biol Plants 29(1):93–107

Gari SGN, Thattantavide A, Kumar A (2022) Nanotechnology—big impact: how nanotechnology is changing the future of agriculture? J Agric Food Res 10:100457

Grewal S, Boora R, Kumari S, Thakur R, Goel S (2024) Fascinating aspects of nanosilicon enabled plant stress tolerance—a comprehensive review. Plant Nano Biol 8:100077

Gujar GT, Kalia V, Kumari A et al (2007) Helicoverpa armigera baseline susceptibility to Bacillus thuringiensis cry toxins and resistance management for Bt cotton in India. J Invertebr Pathol 95(3):214–219
Hasanuzzaman M, Raihan MRH, Masud AAC et al (2021) Regulation of reactive oxygen species and antioxidant defense in plants under salinity. Int J Mol Sci 22(17):9326
Hashem A, Tabassum B, Fathi Abd Allah E (2019) Bacillus subtilis: a plant-growth promoting rhizobacterium that also impacts biotic stress. Saudi J Biol Sci 26(6):1291–1297
Hoshyar N, Gray S, Han H, Bao G (2016) The effect of nanoparticle size on in vivo pharmacokinetics and cellular interaction. Nanomedicine (Lond) 11(6):673–692
Huang B, Chen F, Shen Y, Qian K, Wang Y, Sun C, Zhao X, Cui B, Gao F, Zeng Z et al (2018) Advances in targeted pesticides with environmentally responsive controlled release by nanotechnology. Nano 8(2):102
Hua Y, Younes A, Ki-Hyun K (2022) Molecularly imprinted polymers for sensing gaseous volatile organic compounds: opportunities and challenges. Environ Pollut 311:119931
Iavicoli I, Leso V, Beezhold DH, Shvedova AA (2017) Nanotechnology in agriculture: opportunities, toxicological implications, and occupational risks. Toxicol Appl Pharmacol 329:96–111
Ibrahim E, Luo J, Ahmed T, Wu W, Yan C, Li B (2020) Biosynthesis of silver nanoparticles using onion Endophytic bacterium and its antifungal activity against rice pathogen *Magnaporthe oryzae*. J Fungi (Basel) 6(4):294
Inês F, Venâncio C, Lopes I, Oliveira M (2019) Nanoplastics and marine organisms: what has been studied? Environ Toxicol Pharmacol 67:1–7
Islam W, Tayyab M, Khalil F, Hua Z, Huang Z, Han YH (2020) Chen, silicon-mediated plant defense against pathogens and insect pests. Pestic Biochem Physiol 168:104641
Ismail T, Keratum A, El-Hetawy L (2022) Formulation of abamectin and plant oil-based nanoemulsions with efficacy against the two-spotted spider mite Tetranychus urticae (Acari: Tetranychidae) under laboratory and field conditions. Appl Biol Chem 65:61
Jafir M, Irfan M, Zia-ur-Rehman M, Hafeez F, Ahmad JN, Sabir MA, Zulfiqar U, Iqbal R, Zulfiqar F, Moosa A (2023) The global trend of nanomaterial usage to control the important agricultural arthropod pests: a comprehensive review. Plant Stress 10:100208
Jia X, Shi N, Tang W, Su Z, Chen H, Tang Y, Sun B, Zhao L (2022) Nano-hydroxyapatite increased soil quality and boosted beneficial soil microbes. Plant Nano Biology 1:100002
John H, Johnson L, Worthley C, Losic D (2017) 5—nanopesticides: a review of current research and perspectives, editor(s): Alexandru Mihai Grumezescu. New Pesticides and Soil Sensors, Academic, pp 193–225
Kah M (2015) Nanopesticides and nanofertilizers: emerging contaminants or opportunities for risk mitigation? Front Chem 3:64
Kah M, Hofmann T (2014) Nanopesticide research: current trends and future priorities. Environ Int 63:224–235
Kalogiannidis S, Kalfas D, Chatzitheodoridis F, Papaevangelou O (2022) Role of crop-protection technologies in sustainable agricultural productivity and management. Land 11(10):1680
Karlsson Green K, Stenberg JA, Lankinen Å (2020) Making sense of integrated pest management (IPM) in the light of evolution. Evol Appl 13(8):1791–1805
Karrat A, Amine A (2020) Recent advances in chitosan-based electrochemical sensors and biosensors. Arab J Chem Environ Res 7(2):66–93
Karthik P, Saravanaraj A, Vijayalakshmi V, Ragavan KV, Vaidyanathan VK (2024) Chapter 15—nanobiopesticides: significance, preparation technologies, safety aspects, and commercialization for sustainable agriculture, editor(s): Kamel a. Abd-Elsalam. In: Nanobiotechnology for plant protection. Nanohybrid Fungicides, Elsevier, pp 367–393
Kashyap B, Kumar R (2021) Sensing methodologies in agriculture for monitoring biotic stress in plants due to pathogens and pests. Inventions 6(2):29
Kilani-Morakchi S, Morakchi-Goudjil H, Sifi K (2021) Azadirachtin-based insecticide: overview, risk assessments, and future directions. Front Agron 3:676208
Kuempel ED, Geraci CL, Schulte PA (2012) Risk assessment and risk management of nanomaterials in the workplace: translating research to practice. Ann Occup Hyg 56(5):491–505

Kumalasari MR, Alfanaar R, Andreani AS (2024) Gold nanoparticles (AuNPs): a versatile material for biosensor application. Talanta Open 9:100327
Kumari R, Suman K, Karmakar S, Mishra V, Lakra SG, Saurav GK, Mahto BK (2023) Regulation and safety measures for nanotechnology-based agri-products. Front Genome Ed 5:1200987
Li G, Liu Z, Gao W, Tang B (2022) Recent advancement in graphene quantum dots based fluorescent sensor: design, construction and bio-medical applications. Coord Chem Rev 478:214966
Liu X, Jiang H, Xu H, Shang S, Wang D, Bai Y, Zeng F (2024) Volatile organic compounds as early detection indicators of wheat infected by *Sitophilus oryzae*. Foods 13(21):3390
Liu Y-J, Jing Z, Bai X-T, Diao Q-Y, Wang J, Wu Y-Y, Zhao Q, Xia T, Xing B, Holden PA, Ge Y (2021) Nano-La2O3 induces honeybee (Apis mellifera) death and enriches for pathogens in honeybee gut bacterial communities. Front Microbiol 12:780943
Ma X, Tian Y, Yang R et al (2024) Nanotechnology in healthcare, and its safety and environmental risks. J Nanobiotechnol 22:715
Makadia HK, Siegel SJ (2011) Poly lactic-co-glycolic acid (PLGA) as biodegradable controlled drug delivery carrier. Polymers (Basel) 3(3):1377–1397
Malik S (2024) Data-driven decision-making: leveraging the IoT for real-time sustainability in organizational behavior. Sustainability 16(15):6302
Malik S, Muhammad K, Waheed Y (2023) Nanotechnology: a revolution in modern industry. Molecules 28(2):661
Manke A, Wang L, Rojanasakul Y (2013) Mechanisms of nanoparticle-induced oxidative stress and toxicity. Biomed Res Int 2013:942916
Manna S, Roy S, Dolai A, Ravula AR, Perumal V, Das A (2023) Current and future prospects of "all-organic" nanoinsecticides for agricultural insect pest management. Front Nanotechnol 4:1082128
Martínez-Chávez LA, Hernández-Ramírez MY, Feregrino-Pérez AA, Esquivel EK (2024) Cutting-edge strategies to enhance bioactive compound production in plants: potential value of integration of elicitation, metabolic engineering, and green nanotechnology. Agronomy 14(12):2822
Melanie M, Miranti M, Kasmara H et al (2022) Nanotechnology-based bioactive antifeedant for plant protection. Nanomaterials (Basel) 12(4):630
Miguel-Rojas C, Pérez-de-Luque A (2023) Nanobiosensors and nanoformulations in agriculture: new advances and challenges for sustainable agriculture. Emerg Top Life Sci 7(2):229–238
Mittal D, Kaur G, Singh P, Yadav K, Ali SA (2020) Nanoparticle-based sustainable agriculture and food science: recent advances and future outlook. Front. Nanotechnol. 2:579954
Modafferi, Antonino, Urbaneja, Alberto, Aure, Cristina M, Laudani, Francesca, Palmeri, Vincenzo, Giunti G, Campolo, Orlando, Pérez-Hedo Meritxell (2024) Green pest control strategies: essential oil-based nano-emulsions for Delottococcus Aberiae De Lotto. (Hemiptera: Pseudococcidae) Management
Muhie SH (2022) Novel approaches and practices to sustainable agriculture. J Agric Food Res 10:100446
Munazza I, Khan F, Ahmed T, Noman M, Zulfiqar F, Rizwan M, Chen J, Siddique KHM, Li B (2023) Nanobiotechnology to advance stress resilience in plants: current opportunities and challenges. Materials Today Bio 22:100759
Mustafa IF, Hussein MZ (2020) Synthesis and technology of nanoemulsion-based pesticide formulation. Nanomaterials (Basel) 10(8):1608
Noman M, Ahmed T, Wang J et al (2023) Nano-enabled crop resilience against pathogens: potential, mechanisms and strategies. Crop Health 1(15)
Nongbet A, Mishra AK, Mohanta YK et al (2022) Nanofertilizers: a smart and sustainable attribute to modern agriculture. Plants (Basel) 11(19):2587
OECD (2010) The impacts of nanotechnology on companies: policy insights from case studies. OECD Publishing, Paris
Pretty J, Bharucha ZP (2015) Integrated pest management for sustainable intensification of agriculture in Asia and Africa. Insects 6(1):152–182
Rathore A, Hasan W, Ramesha NM, Pujar K, Singh R, Panotra N, Satapathy SN (2024) Nanotech for crop protection: utilizing nanoparticles for targeted pesticide delivery. Uttar Pradesh J Zool 45(6):46–41

Ray PC, Yu H, Fu PP (2009) Toxicity and environmental risks of nanomaterials: challenges and future needs. J Environ Sci Health C Environ Carcinog Ecotoxicol Rev 27(1):1–35

Rodkate N, Suwanwong S, Srisawat K, Timsorn K (2023) Graphene quantum dots based fluorescent biosensor for nitrogen detection of urea fertilizer. J Phys Conf Ser 2653(1):012071

Rodríguez-Varillas S, Fontanil T, Obaya ÁJ, Fernández-González A, Murru C, Badía-Laíño R (2022) Biocompatibility and antioxidant capabilities of carbon dots obtained from tomato (Solanum lycopersicum). Appl Sci 12(2):773

Rosli MAF, Syed Jaafar SN, Azizan KA, Yaakop S, Aizat WM (2024) Omics approaches to unravel insecticide resistance mechanism in *Bemisia tabaci* (Gennadius) (Hemiptera: Aleyrodidae). Peer J 12:e17843

Sajad K, Zahoor M, Sher Khan R, Ikram M, Islam NU (2023) The impact of silver nanoparticles on the growth of plants: the agriculture applications. Heliyon 9(6):e16928

Sadidi H, Hooshmand S, Ahmadabadi A et al (2020) Cerium oxide nanoparticles (nanoceria): hopes in soft tissue engineering. Mol 25(19):4559

Sánchez-López E, Gomes D, Esteruelas G et al (2020) Metal-based nanoparticles as antimicrobial agents: an overview. Nanomaterials (Basel) 10(2):292

Santos TS, de Souza Varize C, Sanchez-Lopez E, Jain SA, Souto EB, Severino P, Mendonça MC (2022) Entomopathogenic fungi-mediated AgNPs: synthesis and insecticidal effect against *Plutella xylostella* (Lepidoptera: *Plutellidae*). Materials 15(21):7596

Sanyaolu M, Sadowski A (2024) The role of precision agriculture technologies in enhancing sustainable agriculture. Sustainability 16(15):6668

Saurabh K, Prakash V, Dubey AK et al (2024) Enhancing sustainability in agriculture with nanofertilizers. Discov Appl Sci 6:559

Schulte PA, Salamanca-Buentello F (2007) Ethical and scientific issues of nanotechnology in the workplace. Environ Health Perspect 115(1):5–12

Shang Y, Hasan MK, Ahammed GJ, Li M, Yin H, Zhou J (2019) Applications of nanotechnology in plant growth and crop protection: a review. Molecules 24(14):2558

Sharma A, Tselykh A et al (2023) Artificial intelligence and internet of things oriented sustainable precision farming: towards modern agriculture. Open Life Sci 18(1):20220713

Sireesha M, Babu VJ, Kiran ASK, Ramakrishna S (2018) A review on carbon nanotubes in biosensor devices and their applications in medicine. Nano 4(2):36–57

Stander L, Theodore L (2011) Environmental implications of nanotechnology—an update. Int J Environ Res Public Health 8(2):470–479

Strekalovskaya EI, Perfileva AI, Krutovsky KV (2024) Zinc oxide nanoparticles in the "soil–bacterial community–plant" system: impact on the stability of soil ecosystems. Agronomy 14(7):1588

Tahat M, Alananbeh KM, Othman YA, Leskovar DI (2020) Soil health and sustainable agriculture. Sustainability 12(12):4859

Tang Y, Zhao W, Zhu G, Tan Z, Huang L, Zhang P, Gao L, Rui Y (2024) Nano-pesticides and fertilizers: solutions for global food security. Nano 14(1):90

Tarek RM, Steevens JA, Thomas TA, Linkov I (2015) The challenges of nanotechnology risk management. Nano Today 10(1):6–10

Tarigan SI, Kiss J, György T, Doan NPY, Toepfer S (2024) Effects of microbial biostimulants on maize and its Pest, the Western corn rootworm. Diabrotica virgifera virgifera Agronomy 14(10):2239

Tavallali V, Karimi S (2023) Developing nano silicon-salicylic acid complex for inducing root-knot nematode resistance in tomato. SILICON 15:7335–7343

Thabet SG, Alqudah AM (2024) Unraveling the role of nanoparticles in improving plant resilience under environmental stress condition. Plant Soil 503:313–330

Tortella G, Rubilar O, Pieretti JC et al (2023) Nanoparticles as a promising strategy to mitigate biotic stress in agriculture. Antibiotics (Basel) 12(2):338

Tripathy A, Patne AY, Mohapatra S, Mohapatra SS (2024) Convergence of nanotechnology and machine learning: the state of the art, challenges, and perspectives. Int J Mol Sci 25(22):12368

Tudi M, Daniel Ruan H, Wang L et al (2021) Agriculture development, pesticide application and its impact on the environment. Int J Environ Res Public Health 18(3):1112

Tyagi A, Lama Tamang T, Kashtoh H, Mir RA, Mir ZA, Manzoor S, Manzar N, Gani G, Vishwakarma SK, Almalki MA et al (2024) A review on biocontrol agents as sustainable approach for crop disease management: applications, production, and future perspectives. Horticulturae 10(8):805

Udomkun P, Chandi K, Boonupara T, Kaewlom P (2024) Innovative approaches: exploring nano-biochar technology's impact on soil properties, Alachlor retention, and microbial populations. Environ Technol Innov 35:103659

Upadhyay PK, Dey A, Singh VK et al (2023) Conjoint application of nano-urea with conventional fertilizers: an energy efficient and environmentally robust approach for sustainable crop production. PLoS One 18(7):e0284009

van Dijk H, Fischer ARH, Marvin HJP, van Trijp HCM (2015) Determinants of stakeholders' attitudes towards a new technology: nanotechnology applications for food, water, energy and medicine. J Risk Res 20(2):277–298

Vats S, Kumawat S, Brar J, Kaur S, Yadav K, Magar SG, Jadhav PV, Salvi P, Sonah H, Sharma S, Deshmukh R (2022) Opportunity and challenges for nanotechnology application for genome editing in plants. Plant Nano Biol 1:100001

Vrignaud S, Benoit JP, Saulnier P (2011) Strategies for the nanoencapsulation of hydrophilic molecules in polymer-based nanoparticles. Biomaterials 32(33):8593–8604

Wang D, Jia M, Wang L, Song S, Feng J, Zhang X (2017) Chitosan and β-cyclodextrin-epichlorohydrin polymer composite film as a plant healthcare material for Carbendazim-controlled release to protect rape against *Sclerotinia sclerotiorum* (Lib.) de Bary. Materials 10(4):343

Wang D, Saleh NB, Byro A et al (2022) Nano-enabled pesticides for sustainable agriculture and global food security. Nat Nanotechnol 17(4):347–360

Wang X, Xie H, Wang P, Yin H (2023a) Nanoparticles in plants: uptake, transport and physiological activity in leaf and root. Materials (Basel) 16(8):3097

Wang Z et al (2023b) Sex pheromone receptor-derived peptide biosensor for efficient monitoring of the cotton bollworm Helicoverpa armigera. ACS Sensors 8:363

Wilson SK, Pretorius T, Naidoo S (2023) Mechanisms of systemic resistance to pathogen infection in plants and their potential application in forestry. BMC Plant Biol 23(1):404

Wu X, Yu Y, Baerson SR et al (2017) Interactions between nitrogen and silicon in rice planthopper Nilaparvata lugens. Front Plant Sci 8:28

Yadav A, Yadav K, Abd-Elsalam KA (2023a) Nanofertilizers: types, delivery and advantages in agricultural sustainability. Agrochemicals 2(2):296–336

Yadav A, Yadav K, Ahmad R, Abd-Elsalam KA (2023b) Emerging frontiers in nanotechnology for precision agriculture: advancements. Hurdles and Prospects Agrochemicals 2(2):220–256

Yang F, Cabe M, Nowak HA, Langert KA (2022) Chitosan/poly(lactic-co-glycolic)acid nanoparticle formulations with finely-tuned size distributions for enhanced Mucoadhesion. Pharmaceutics 14(1):95

Yin J, Su X, Yan S, Shen J (2023) Multifunctional nanoparticles and nanopesticides in agricultural application. Nanomaterials (Basel) 13(7):1255

Yousef HA, Fahmy HM, Arafa FN et al (2023) Nanotechnology in pest management: advantages, applications, and challenges. Int J Trop Insect Sci 43:1387–1399

Zainab R, Hasnain M, Ali F et al (2024) Prospects and challenges of nanopesticides in advancing pest management for sustainable agricultural and environmental service. Environ Res 261:119722

Zhang H, Zheng T, Wang Y, Li T, Chi Q (2024) Multifaceted impacts of nanoparticles on plant nutrient absorption and soil microbial communities. Front Plant Sci 15:1497006

Zoe LH, David SR, Rajabalaya R (2023) Chitosan nanoparticle toxicity: a comprehensive literature review of in vivo and in vitro assessments for medical applications. Toxicol Rep 11:83–106

Engineered Nano-Biochar from Lignocellulosic Residues: Innovative Approaches for Sustainable Agriculture and Environmental Remediation

8

Edgar Vázquez-Núñez, Hermes Pérez-Hernández, Alma Hortensia Serafin Muñoz, Norma L. Gutiérrez Ortega, Gabriela Medina-Perez, and Ileana Vera-Reyes

Abstract

This chapter aims to provide an in-depth examination of engineered nano-biochar (NBC) manufactured from lignocellulosic residues as an innovative option for improving agricultural sustainability and tackling environmental issues. This chapter will explore the methods for transforming accessible and renewable lignocellulosic residues into high-value nano-biochar with improved characteris-

E. Vázquez-Núñez (✉)
Grupo de Investigación sobre Aplicaciones Nano y Bio Tecnológicas para la Sostenibilidad (NanoBioTS), Departamento de Ingenierías Química, Electrónica y Biomédica, División de Ciencias e Ingenierías, Universidad de Guanajuato, Campus León, Loma del Bosque, León, Guanajuato, Mexico
e-mail: edgar.vazquez@ugto.mx

H. Pérez-Hernández
InstitutoNacional de InvestigacionesForestales, Agrícolas y Pecuarias (INIFAP). Campo Experimental Edzná, Campeche, C.P, Mexico

A. H. S. Muñoz · N. L. Gutiérrez Ortega
Departamento de IngenieríaCivil y Ambiental, División de Ingenierías, Campus Guanajuato, Universidad de Guanajuato, Guanajuato, Gto, Mexico

G. Medina-Perez
ICAP—Institute of Agricultural Sciences, Autonomous University of the State of Hidalgo, Tulancingo de Bravo, Hidalgo, Mexico

I. Vera-Reyes
Centro de Investigación en QuímicaAplicada (CIQA), Departamento de Biociencias y Agrotecnología, Saltillo, Coahuila, Mexico

S. S. Purewal et al. (eds.), *Agri-Nanotechnology: Innovations for Sustainable Agriculture and Environmental Restoration*,
https://doi.org/10.1007/978-981-96-9756-4_8

tics. The topic will address the selection of feedstocks, advanced synthesis methods, and the comprehensive evaluation of the resultant nano-biochar, emphasizing its remarkable potential for enhancing soil health, augmenting agricultural yields, and eliminating pollutants from water systems. This chapter will illustrate the diverse advantages of nano-biochar by including new research findings and case examples, therefore advocating for its significance in a circular bioeconomy and sustainable development.

Keywords

Feedstocks · Lignocellulosic residues · Nano-biochar · Soil health · Sustainability

8.1 Introduction

Biochar (BC) is the product of the thermochemical conversion of lignocellulosic biomass under limited oxygen conditions that is widely used as an effective soil amendment. This material contributes to environmental remediation due to its high adsorptive performance during removal of aqueous pollutants. Bulk biochar, generally ranging in the millimeter scale, plays a key role in improving water retention capacity, fertility of soils, releasing nutrients, and enhancing soil microbial activity (Arabzadeh Nosratabad et al. 2024; Gwenzi et al. 2015). Nano-biochar (NBC), a more refined form of biochar, possesses an increased surface area and enhanced reactivity, showing significant promise in addressing environmental contaminants and mitigating abiotic stresses in agriculture, particularly the low availability of nutrients and water scarcity (Sani et al. 2023a, b). NBC shows unique properties, such as its high adsorption capacity and ability to sequester carbon, making it an ideal material for environmental remediation and sustainable agriculture. A novel approach for improving NBC performance includes the integration of metal oxide nanoparticles into its structure, optimizing biochar's functionality for a wider range of applications from pollutant removal to soil enhancement (Bidi et al. 2021).

The selection of raw materials is crucial for defining the physicochemical properties of biochar. Several studies have variated the feedstock from wood, agricultural residues, or grass residues, obtaining distinct surface characteristics, content of specific nutrients, adsorptive performance and stability, offering versatile applications in environmental or agricultural contexts (Tomczyk et al. 2020). The aim of this chapter is to explore different synthesis methods of NBC, the impact of the feedstock selection on BC and NBC properties, and the most relevant applications of this material in agriculture and environmental remediation.

8.2 Synthesis Methods: From Biochar to Nano-Biochar

NBC offers sustainable solutions for removing soil contaminants and improving soil quality for agriculture. Due to its high surface area, presence of functional groups, porosity and capacity of sequester carbon, NBC removes efficiently heavy metals from polluted soil and water; meanwhile, in agricultural soils, these materials enhance soil fertility, nutrient release, water holding capacity and microbial activity. Production methods significantly influence its physicochemical properties, impacting its performance in environmental remediation (Sani et al. 2023a, b).

The main methods for producing biochar are mainly based on the general properties and physicochemical characteristics of the biochar to be obtained, which will depend directly on the operational factors of the reactor, the chemical composition of the precursor material, and the economic viability, to be considered as a sustainable solution for both the farmer and the environmental remediation (Gole et al. 2024).

The production of biochar typically occurs at 300–700 °C with little to no oxygen, though some processes require up to 1600 °C. These high energy costs pose a major challenge for large-scale production. Additional issues include limited control over material properties, reaction conditions, and atmospheric pollutant emissions (Arabzadeh Nosratabad et al. 2024).

Current research aims to improve production processes for sustainability and environmental acceptability. An innovative approach is nano-biochar, with structural sizes of 1–100 nm. Compared to macro-biochar, nano-biochar offers higher surface area, greater porosity, smaller hydrodynamic radius, stronger negative zeta potential, better oxygen-containing functional groups, and fewer carbon defects (Osman et al. 2023).

Nanometric sizes enhance the physicochemical, textural, structural, and morphological properties of nanocarbons. Production methods enable integrating nanomaterials like zerovalent iron, graphene oxide, or zeolites into the nano-biochar matrix, functionalizing it for applications in agriculture, environmental remediation, and various industries (Amdeha 2024).

The methods of synthesis of nano-biochar are generally classified into two: (i) indirect methods, i.e. the pristine biochar is first obtained from the raw material, and then they are taken to a nanometric size; and (ii) direct nano-biochar, where nano-biochar are generated directly from the raw material.

With respect to the integration of nanoparticles, two processes can also be considered: the first is the direct incorporation of these during the obtaining of the nano-biochar, or the second that would involve incorporating the nanoparticles once the nano-biochar is available (Heinrich et al. 2023; Zheng et al. 2023).

8.2.1 Indirect Methods for Obtaining Nano-Biochar

The synthesis of nano-biochar involves two main stages: first, obtaining biochar, and second, reducing it to the nanometric size. Biochar is commonly produced through pyrolysis, hydrothermal carbonization, and gasification.

(a) Pyrolysis: It is the thermochemical decomposition of biomass in the absence of oxygen, occurring between 200 and 600 °C (Tomczyk et al. 2020). This process produces biochar, biooil, and/or syngas, depending on the temperature used. Lower temperatures favor biochar production, while higher temperatures maximize biooil or syngas yields. The rate of heating also influences biochar properties. Pyrolysis can be classified as conventional, rapid, or ultra-fast. Conventional pyrolysis occurs between 300 °C and 500 °C with a slow heating rate (5–7 °C min^{-1}) and extended residence times, producing high-quality biochar. Rapid pyrolysis, conducted at temperatures between 400 °C and 650 °C with heating rates exceeding 100 °C/s and short residence times (0.5–2.0 s), focuses on biooil and gas production, with biochar as a by-product. Ultra-fast pyrolysis occurs at temperatures from 700 °C to 1000 °C, with heating rates exceeding 500 °C/s, producing primarily gases and biooils and minimal biochar (less than 15%) (Alvarez-Chavez et al. 2019).

Each pyrolysis method affects the quality and quantity of the products produced. The choice of method depends on the desired end-product, with conventional pyrolysis prioritizing biochar and rapid or ultra-fast pyrolysis optimizing liquid and gaseous outputs. Understanding the specific requirements of each method is crucial for tailoring production to meet environmental or agricultural goals (Hublikar et al. 2024). Figure 8.1 shows a generalized method for production of biochar via pyrolysis.

(b) Hydrothermal synthesis: Hydrothermal synthesis, or hydrothermal carbonization, treats biomass with hot compressed water at high pressure to form a carbon-rich material. This method operates at temperatures between 100 and 300 °C and pressures between 5 and 70 bar, using autoclaves and typically requiring longer reaction times. Unlike pyrolysis, hydrothermal synthesis can process wet feedstock without the need for drying, making it more energy-efficient. It also produces biochar enriched with sulfur functional groups and achieves high conversion efficiency at relatively low temperatures. The solid product, called hydrocarbon, exhibits improved combustion properties, such as higher ignition temperatures and broader combustion ranges. Additionally,

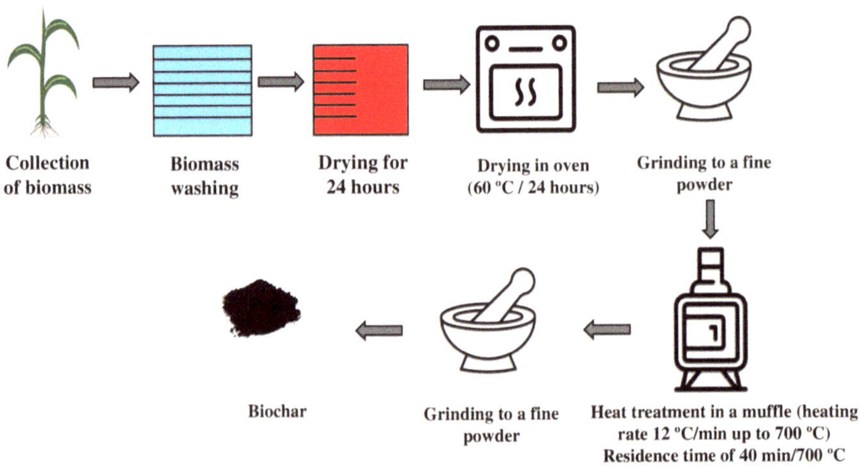

Fig. 8.1 Description of the synthesis method for biochar via pyrolysis

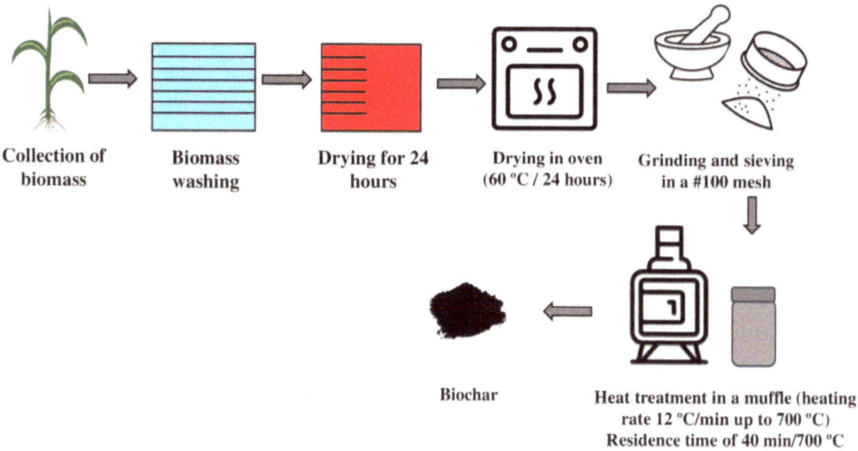

Fig. 8.2 Description of the synthesis method for biochar using hydrothermal process

hydrothermally produced biochar shows reduced metal concentrations compared to pyrolytic biochar, which often retains high levels of metals from the original biomass (Ighalo et al. 2023). Fig. 8.2. represents the hydrothermal process to produce biochar.

(c) Gasification: It is a thermochemical conversion method where biomass is partially oxidized with a small amount of oxygen at temperatures between 600 °C and 1400 °C, producing synthesis gas (syngas) and biochar. This process uses a gasifying agent, such as oxygen, air, or water vapor, to heat the biomass. It is cost-effective, simple, and can produce clean syngas with high energy capacity, suitable for small-scale applications. Biochar produced by gasification typically has high carbon content, a large surface area, and low ash content. The process occurs in four stages: first, the biomass is heated to 100 °C to dry; second, pyrolysis occurs between 300 °C and 500 °C, breaking biomass into smaller molecules; third, water vapor and CO_2 facilitate reduction; and fourth, the carbonaceous fraction reacts with the gasifying agent. The process yields approximately 50% syngas and 50% biochar (Ighalo et al. 2024).

8.2.1.1 Methods of Obtaining Nano-Biochar from Biochar

Biochar can later be converted into nano-sized biochar or nano-biochar, by various conventional physical or chemical methods, mainly using mechanical methods, which are low, and fulfill the function of reducing particle size.

(a) Grinding: This process involves using steel or ceramic balls in a mill to break biochar into smaller, nanometric structures. The particle size depends on the mill's rotation speed and grinding time. This process can be performed wet or dry, with dry grinding preferred for its simplicity. However, a key disadvantage is the production of heterogeneous nanoparticle sizes, with particles prone to reaggregation. Achieving sizes below 100 nm typically requires grinding times exceeding 24 h, influenced by factors such as the biochar-to-ball ratio and the balls' size, weight, and number. The number of balls can range from tens to thousands, with weights from milligrams to fractions of a kilogram and sizes from millimeters to centimeters. Double disc systems are an alternative for achieving a more homogeneous grind. Heat pretreatment of biochar at 80 °C for 24 h has been found to improve the grinding process, reducing both the time required and the resulting nanoparticle size. This method provides a practical approach to producing nanometric biochar particles for various applications, but the efficiency and consistency of the grinding process remain challenges that need further optimization.

(b) Hydrothermal Extraction: In this method, biochar is grinded, passed through a sieve, and added into a hydrothermal reactor, as it was reported by Guo et al. (2020). The mixture consisted in grinded biochar with concentrated strong acids (HNO_3 and H_2SO_4), this mixture reacts for 2 h at room temperature; this suspension is transferred and mixed. The solids are separated by vacuum filtration with a membrane. The membrane is dried in the oven, and the solid recovered is stored for further characterization and use.

(c) Sonication: It is an alternative technique to produce high-energy nano-biochar that uses high-energy ultrasonic radiation for the massive disintegration of biochar in suspension, which is generally suspended in water with a ratio of 1:40, with sonication times of 30 min. After sonication, it is suspended in an alkaline solvent (such as sodium hexametaphosphate) and separated by sieving and centrifugation. The nano-biochar obtained using this method develop unobstructed porosity.

(d) Pyrolysis by Microwave: Pyrolysis can also be performed using microwave energy, which involves dielectric heating. Microwaves penetrate biomass particles in suspension, vibrating at high frequencies to create alternating electric fields, friction, and collisions. This converts electromagnetic energy into thermal energy, resulting in a temperature gradient with higher core and lower surface temperatures. The process ensures even heating and influences biochar properties, such as achieving nanometric sizes. Wallace et al. reported producing nano-biochar through microwave pyrolysis in a batch reactor. Microwave energy levels were varied (2100, 2400, and 2700 W) while maintaining constant parameters. Biochar, acting as a carbon microwave absorber, was introduced at 10% by weight in an N2 atmosphere. The process yielded biooil and nano-biochar (Wallace et al. 2019).

8.2.1.2 Direct Methods for Obtaining Nano-Biochar

These methods would involve the generation of nano-sized biochar from biomass directly. Few reported direct methods succeed in generating nano-biochar (Chaubey et al. 2024).

(a) *Hydrothermal Synthesis:* Few studies report the synthesis of nano-biochar through acid-treated hydrothermal methods. Biomass, such as soybean straw and animal waste, is digested with acids in high-pressure hydrothermal reac-

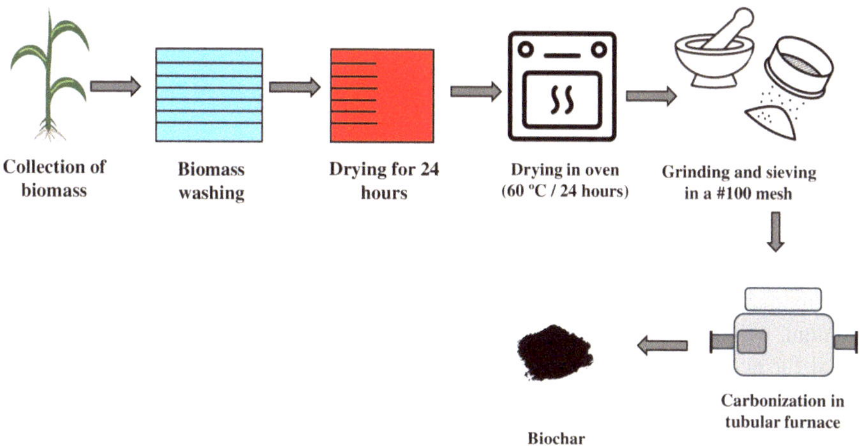

Fig. 8.3 Overview of a generalized method for nano-biochar production

tors. To obtain nanoparticles, multiple centrifugation cycles (1000–2500 rpm) are performed for 2–30 min. This process yields particles ranging from 3 to 6 nm. Acid-treated hydrothermal biochar is often called "biochar dots," "biochar nanodots," or "biochar nanoparticles" (Guo et al. 2020).

(b) *Conventional Grinding and Pyrolysis:* Biomass, dried at 90 °C for 3 hours, is ground into micro powder. The crushed biomass is then pyrolyzed at 600 °C for 2 hours in a sealed steel container with limited oxygen and cooled to room temperature. Macro/nano-biochar powders are obtained using a 400-mesh sieve (0.037 mm) (Ahuja et al. 2020). This method is schematized in Fig. 8.3 as follows.

8.2.1.3 Integration of Nanoparticles during Biochar Synthesis

Unmodified biochar has limited effectiveness, but activation techniques enhance surface functional groups and activity. Incorporating nanoparticles, such as metal oxides, graphene, or nanotubes, into biochar creates nano-biochar or nanohybrid carbon. These processes often include biomass pretreatment to modify its physicochemical properties for biochar generation. Energy-efficient methods involve simultaneous pyrolysis of biomass and nanoparticle precursors. Techniques like coprecipitation and impregnation followed by thermochemical treatments, such as hydrothermal processes, can be used to produce nano-biochar (Omiri et al. 2022).

Coprecipitation/impregnation: This involves combining a metal ion or salt precursor (e.g., Zn, Mg, Al, Fe, Co, Ni) with a reagent that induces precipitation. Biochar is dispersed in a solution containing metal precursors to incorporate metal oxides or nanoparticles. This process operates at low temperatures and does not require additional reactors but can be time-consuming due to steps like washing and drying. For instance, sugarcane bagasse-Fe_3O_4 is prepared by coprecipitating bagasse residue with $FeSO_4{\cdot}7H_2O$ and $FeCl_3{\cdot}6H_2O$ in ammonia, followed by calcination. Similarly, metal nitrates ($Fe(NO_3)_3{\cdot}9H_2O$, $Zn(NO_3)_2{\cdot}6H_2O$, $Ni(NO_3)_2{\cdot}6H_2O$)

with calcined *Acacia nilotica* seed shell ash produce a sol-gel of $Ni_{0.5}Zn_{0.5}Fe_2O_4$ magnetic nanoparticles, calcined to create nanohybrid biochars (Chausali et al. 2021).

8.3 Selection of Lignocellulosic Feedstocks, Variations, and Their Influence

The type of biomass used significantly impacts the structural properties of biochar and its derivates. For instance, wood biomass is known for low moisture and ash content, woody biochar typically has a high BET specific surface area and porosity, ideal for heavy metal adsorption, meanwhile agricultural residues, like rice husks and sunflower husks, offer a cost-effective and sustainable alternative, such as in a study by El-Gamal et al. (2023), biochar composites from sunflower seed husks (SHB) and rice straw (RSB) combined with a bacterial consortium (*Proteus mirabilis* and *Raoultella planticola*) were evaluated for remediation. Kinetic experiments at 30 °C and 150 rpm achieved complete removal of 700 mg L^{-1} lead (70 mg g^{-1}) in 48–66 hours and phenol (69.90 mg g^{-1}) in 42–48 hours. This performance was due to the synergy between the biochar's adsorption and the bacteria's metabolic activity. The study highlighted bacterial consortium-biochar composites as efficient, cost-effective eco-sorbents for detoxifying organic and inorganic pollutants.

The biochar obtained from agricultural residues generally exhibit high ash content and unique mineral compositions that influence their effectiveness in contaminant immobilization (Li et al. 2017). Campos and De la Rosa (2020) evaluated the immobilization potential of biochar from rice husk (RHB), olive pit (OPB), and wood chips (CWB) for Cu^{2+} and Pb^{2+} in aqueous solutions and acidic soils. RHB and CWB effectively removed Cu^{2+} and Pb^{2+} from solutions, with maximum adsorption capacities of 30.77 mg g^{-1} and 58.82 mg g^{-1} for Cu^{2+}, and 19.34 mg g^{-1} and 77.52 mg g^{-1} for Pb^{2+}. The adsorption followed the Langmuir isotherm model. Applying 5% RHB and CWB to contaminated soils improved physicochemical properties, reducing trace element bioavailability and supporting *Brassica rapa pekinensis* growth. These results confirm that RHB and CWB are efficient for heavy metals immobilization in polluted soils.

Wood-Based Feedstocks An important characteristic of wood-based biochar is that they are limited in nutrient concentration, which makes them more suitable for soil conditioning instead of being applied as direct fertilizers (Bolan et al. 2022; Mou 2024).

Most common wood-based residues are described as follows:

(a) *Agricultural Residues:* There are diverse type of agricultural residues used for biochar production, among them rice husks, wheat straw, and corn stalks; the biochar obtained from these feedstocks contain higher nutrient levels, especially potassium and phosphorus, which make them adequate for enhancing soil productivity. This type of biochar, however, shows lower specific surface areas than wood-based biochar, impacting in reduced contaminant adsorption capacity (Enaime and Lübken 2021). The production of BC using agricultural residues promotes the reduction of greenhouse gas emissions since this residual biomass is converted instead of being burnt, promoting carbon sequestration (Patel and Panwar 2023).

(b) *Grasses and Herbaceous Materials:* This feedstock produces biochar with moderate-to-low content of carbon and high concentration of minerals, such as potassium and calcium (Jouiad et al. 2015).

Often underutilized, grass biomass can yield carbonaceous materials with optimal properties for agricultural applications, supporting European agriculture and ecology. Sadaka et al. (2014) studied biochar from switchgrass carbonization, showing that as temperatures increased from 300 °C to 400 °C and residence times extended from 1 to 3 hours, biochar yield dropped from 82.6% to 35.2%, and volatile solids content decreased from 72.1% to 43.9%. Simultaneously, heating value (21.9 MJ kg^{-1}) and fixed carbon content (44.9%) increased, demonstrating efficient biomass energy use for high-quality solid fuel and carbon sequestration. Late-harvest grass is also being explored for farm-level biochar production, utilizing up to 35% of biomass energy for heating while contributing to carbon sequestration. Economic analysis highlights cost benefits, particularly for agronomic applications. However, regulatory challenges in Europe, especially in Germany, limit adoption. A SWOT analysis in Brandenburg identified opportunities like local carbon sink development but highlighted obstacles, including inadequate infrastructure and regulatory issues (Heinrich et al. 2023).

These findings highlight the dual benefits of using grasslands and agricultural residues for biochar, addressing underutilization while contributing to negative emissions through carbon sequestration and reduced greenhouse gas emissions.

(c) *Manure and Biosolids:* Although not strictly lignocellulosic, manures and biosolids are often used to produce biochar due to their high nutrient content, especially nitrogen, phosphorus, and sulfur. This biochar is typically less stable but highly effective as fertilizers and soil amendments (Piash et al. 2022). Banik et al. (2021) found that using biochar as a stabilizer in swine manure improved soil organic matter without affecting pH or total nitrogen levels. Biochar-treated manure significantly increased soil organic matter compared to conventional manure. However, no significant changes were observed in soil ammonium levels or plant biomass nutrient concentrations of nitrogen (N), phosphorus (P), and potassium (K). Manure-biochar mixtures reduced Mehlich3 extractable soil phosphorus (P), KCl extractable nitrate (NO^{3-}), and diluted Mehlich3 extractable copper (Cu) and zinc (Zn). These findings suggest biochar can reduce nutrient loss in early plant growth stages, positioning manure-biochar mixtures as a more effective soil amendment than conventional manure.

8.3.1 Pretreatment Methods for Optimizing Feedstock Properties

Pretreatment enhances feedstock properties, improving biochar yield and quality by addressing heterogeneity, high ash content, and limited functionality.

8.3.1.1 Physical and Mechanical Pretreatments

Physical processes like milling and sieving reduce biomass particle size, creating a uniform feedstock for pyrolysis. Uniformity ensures consistent heating, enhancing biochar's surface area and porosity. Smaller particles also facilitate the impregnation of chemical agents during pretreatments, especially in fibrous materials like rice straw. Mechanical pretreatment, such as milling, significantly improves biochar properties, increasing its effectiveness in environmental applications and agronomic benefits. For example, biochar from corncob subjected to high-energy milling showed reduced particle sizes (947.9 nm to 799.2 nm) and increased surface area (10.115 $m^2 g^{-1}$ to 18.303 $m^2 g^{-1}$) as milling speed varied from 300 to 700 rpm. This increase allowed a better interaction of biochar with soils, increasing the soil nutrient content and water retention, improving crucial characteristics for enhanced productivity in places where ecosystems face water scarcity and limitations in organic matter content (Subagio et al. 2021).

Mechanical pretreatment alters biochar chemical characteristics. In a study published by Lyu et al. (2018), it was found that this pretreatment improved biochar ability to remove nickel (Ni(II)) from aqueous solutions. Milling increased biochar's surface area and chemically modified the material by introducing oxygen-containing functional groups (carboxyl, lactonic, hydroxyl), enhancing electrostatic attraction and surface complexation with metal ions. The kinetics of treated and untreated biochar showed important differences, where ball-milled biochar performed fastest adsorption kinetics, demonstrating the potential of mechanical pretreatment to optimize biochar performance in environmental applications.

In another study, Manikandan et al. (2013) studied the effect of sieving and grinding processes on biochar performance when added to agricultural soil. The impact of two particle size ranges (0.06–0.5 mm and 2–4 mm) was investigated, revealing that smaller particles had a greater influence on soil properties. Sieved biochar, particularly smaller particles, exhibited a larger liming effect, raising soil pH by 0.3 units more than larger particles. Smaller particles also enhanced soil water retention, with sieved biochar retaining 91%–258% more water than ground biochar. Plant responses to particle size varied; ryegrass performed better with larger particles, while velvetleaf showed the highest growth response to smaller, sieved biochar.

These findings highlight the importance of particle size and mechanical pretreatment in tailoring biochar for specific applications. Milling and sieving improve biochar's physical and chemical properties, making it more effective for soil quality improvement, nutrient retention, and mitigating environmental contaminants in agricultural systems.

8.3.1.2 Thermal Pretreatments

Heat pretreatments like torrefaction partially pyrolyze biomass at 200–300 °C under inert conditions, reducing hydrophilicity, increasing energy density, and enhancing thermal stability. Torrefied biomass, with higher calorific value and lower moisture content, is ideal for biochar as a solid fuel. Wang et al. (2019) found that thermal pretreatment significantly reduces PM10 emissions during combustion. Biomass subjected to torrefaction at 300 °C (T-300), slow pyrolysis at 500 °C (S-500), or hydrothermal carbonization at 240 °C (H-240) and combusted at 1400 °C showed lower PM10 emissions, especially PM0.3 ($\leq$0.3 μm). This reduction was influenced by Cl release during pretreatment, forming gaseous KCl during combustion, and increased ash content affecting PM1–10 emissions. When pre-treated biochar was co-combusted with Ping Ding Shan bituminous coal, PM10 emissions, especially PM0.3, were further reduced due to interactions between biomass-derived gaseous species and aluminosilicates in the coal, like kaolinite. These interactions facilitated the capture of gaseous species and coalescence of sticky minerals, effectively lowering PM0.3 and PM1–10 emissions. The study emphasized thermal pretreatment's role in improving biochar's environmental performance, particularly in co-combustion scenarios.

Particularly, regarding the role of thermal pretreatment, such as torrefaction, Kwoczynski and Čmelík (2021) investigated the role of thermal pretreatment, such as torrefaction, in biochar production, analyzing twelve biomass wastes from the Czech Republic, including vegetable oil, wine, beer, sugar, flour, and wood processing. Key parameters assessed were nutrient and toxic element content, organic carbon, and thermal stability. To determine crystalline phases, thermogravimetric analysis (TGA) and X-ray diffraction (XRD) were used. Extracted rapeseed scrap resulted as the most suitable feedstock due to its high nutrient content and thermal stability. The torrefaction resulted in a process with high potential to convert biomass wastes into biochar for diverse applications. This technology produced torrefied biomass with improved energy density and reduction of PM_{10} emissions during combustion. It is highlighted that torrefaction pretreatment enhances biochar properties, reducing particle emissions and enabling the use of biomass for addressing global challenges.

8.3.1.3 Chemical Pretreatments

Chemical pretreatments, like acidic, alkaline, or salt solutions, enhance biochar's adsorption capacity and nutrient retention. Several studies have reported the benefits of using chemical compounds to improve the characteristics of biochar. For instance, Alvarez-Chavez et al. (2019) reported a reduction of ash content of the biochar prevenient of rice straw when the feedstock is chemically treated with diammonium phosphate and washed with HCl.

In another study, Zhou et al. (2019) proved that chemical pretreatment enhanced rice straw biochar's ability to adsorb hexavalent chromate [Cr(VI)]. Biochar was treated with acid, basic, or organic solutions (0.5 mol L^{-1} HCl, 0.5 mol L^{-1} NaOH, deionized water, or ethanol), and it was observed that NaOH-treated biochar showed the highest point zero charge (PZC) and Cr(VI) adsorption capacity under slightly acidic conditions due to enriched aromatic CO– and phenolic –OH groups. HCl treatment reduced ash content, protonated functional groups, and enhanced Cr adsorption, with acid-pretreated biochar removing over 90% of total Cr at pH 2. Ethanol and water treatments had minimal impact. X-ray Absorption Near-Edge Spectroscopy (XANES) revealed Cr(VI) adsorption occurred via reduction to Cr(III) and anionic adsorption. These findings highlight the potential of chemical pretreatment to improve rice straw biochar for Cr-polluted water remediation.

Regarding soil improvement, Griffin et al. (2022) reported that chemical pretreatment of rice straw (RS) with ammonium sulfate (AS) and diammonium phosphate (DAP) significantly enhanced biochar properties for soil applications. Pyrolysis experiments with pre-treated RS at varying concentrations (0.1, 0.2, and 0.5 mol L − 1), dwell times (30 and 60 minutes), and temperatures (300–600 °C) showed improved characteristics. Pretreatment reduced decomposition rates, increased char yields, and lowered ash content compared to untreated RS. Optimized conditions—400 °C, 30 min dwell time, and 0.2 mol L^{-1} DAP—produced a 52.2% biochar yield, 19.27 MJ kg^{-1} heating value, and 35.19% ash content. These results highlight chemical pretreatment's ability to improve biochar's thermal properties and nutrient composition, enhancing its effectiveness as a soil amendment for sustainable agriculture.

In another study, Rizwan et al. (2020) evaluated biochar from agricultural waste for environmental remediation, enhancing adsorptive properties through NaOH pretreatment. Rice and cotton straws were treated with dilute NaOH to open their lignocellulosic structure, followed by impregnation with Fe-Co nitrate or H_3PO_4. Biochar produced via slow pyrolysis exhibited specific surface areas (SSAs) from 12.26 to 581.13 m^2/g, total pore volumes (TPVs) from 0.033 to 0.3736 cm^3/g, and average pore sizes (APSs) from 2.57 to 10.76 nm.

These pretreatments significantly improved physicochemical properties, achieving high anion exchange capacity (251.78–810.13 mmol/kg at pH 3.5) and cation exchange capacity (108.22–464.67 mmol/kg at pH 7.0). Pb2+ and Cd2+ adsorption capacities increased by 23.07–82.74% and 16.90–556.33%, respectively, compared to pristine biochar. The Fe-Co-composite biochar showed superior performance for divalent metal adsorption, making it an effective soil amendment and alternative adsorbent for remediation. Finally, the role of chemical pretreatment improves the biochar performance in agriculture and remediation processes.

8.3.1.4 Biological Pretreatments

Enzymatic or microbial pretreatments break down lignocellulosic structures, mainly polymeric complexes, which modify the surface of biochar and improve nutrient release. This pretreatment includes the use of microorganisms or enzymes to degrade lignin and hemicellulose, facilitating the access to cellulose in further processes like pyrolysis or hydrothermal treatments (Wang et al. 2024a, b). When this method is combined with hydrothermal or chemical pretreatments, other benefits are observed, for instance, the recovery of fermentable sugars. This reason makes biological pretreatments a strategy technically and economically attractive.

Regarding the performance of enzymes, it has been identified that dosage, temperature, and processing time are crucial conditions for obtaining high-quality biochar with high lignin content and improved adsorption performance. In a study published by Li et al. (2024), biomass retreated enzymatically to yield a material with a reduction in dissolved solids (TDS) content (96–99%) and increased adsorption capacity by 52.29% compared to untreated biochar. These results contribute to the understanding of parameters that influence the stability of biochar from biomass biologically treated.

8.4 Physical and Chemical Properties of Nano- and Biochar

Thermal methods, like torrefaction and hydrothermal carbonization (HTC), enhance carbon stability, hydrophobicity, and energy density, meanwhile mechanical methods influence the particle size, increasing surface area and pore structure. Chemical treatments using acids or alkaline solutions modify surface chemistry affecting the ion exchange capacity of the materials. Finally, the use of enzymatic or microbial processes has shown a strong influence on the incorporation of certain functional groups on the biochar surface improving the nutrient release.

8.4.1 Surface Area, Porosity, and Particle Size Distribution

Nanobiochar's surface area, ranging from 10.8 to 1736 m^2/g, is a critical factor that impact on contaminant adsorption, nutrient retention, and catalytic performance (Leng et al. 2021). As it has been stated, the feedstock type, processing conditions, and post-treatment methods influence on the carbonaceous material characteristics. For instance, higher pyrolysis temperatures (>450 °C) affect the porosity due to the removal of volatile compounds, forming microporous structures (Tomczyk et al. 2020). When NBC is dispersed in water, particle size plays a crucial role; the range from 10 to 100 nm allows a better biochar dispersion, increasing the contact with aqueous pollutants and promoting mass transference and boosting adsorptive processes (Wang et al. 2024a, b).

Several analytical techniques have been used to visualize surface morphology, particle size, and pore structures. Scanning electron microscopy (SEM) and transmission electron microscopy (TEM) allow identify in different scale ranges the outer and inner structures of materials. When it is needed to evaluate the adsorption-desorption isotherms, Brunauer–Emmett–Teller (BET) analysis is frequently used since it offers information such as surface area, pore volume, and pore size distribution, which are critical metrics for assessing the adsorption capacity of NBC (Zheng et al. 2020). It is proved that higher micropore volumes enhance the material's ability to adsorb small molecules, while mesopores contribute to increased nutrient retention and catalytic efficiency (Leng et al. 2021).

8.4.2 Functional and Chemical Stability

The presence of functional groups in NBC determines its chemical properties (Liu et al. 2018). For instance, these groups determine cation exchange capacity (CEC) in soilor selectivity for adsorbing pollutants; particularly hydroxyl (-OH), carboxyl (-COOH), and carbonyl (C=O) groups are introduced during the pyrolysis process and can be further enhanced through chemical treatments like acid washing (e.g., HCl) or oxidation (e.g., H_2O_2) (Tomczyk et al. 2023).

The techniques used to characterize functional groups are Fourier-transform infrared spectroscopy (FTIR), which identifies specific chemical bonds, and X-ray photoelectron spectroscopy (XPS), which provides detailed surface chemistry information.

There exist certain considerations that allow us to determine the presence and duration of functional groups on NBC surface, for instance the presence of aromatic and graphitic carbon structures. Authors have proved that high-temperature pyrolysis promotes high content of aromatic structures, indicated as low H/C and O/C aromatic ratios (Das et al. 2021; Xu et al. 2021). This information is useful for estimating the resistance to degradation and longevity in an NBC environment.

The resistance to decomposition can be measured by thermogravimetric analysis (TGA), which gives information about thermal and chemical stability of NBC. In addition to this, the zeta potential of NBC surface charge is essential to determine the dispersibility of NBC in aqueous solutions. When a high zeta potential is obtained, it can be inferred that the material will show better colloidal stability, that is important for water treatment applications; this property is determined by dynamic light scattering (DLS) technique, which measures particle size and zeta potential.

8.4.3 Analytical Techniques for Characterizing Engineered Nano-Biochar

8.4.3.1 Morphological and Structural Analysis

The following techniques provide information about the surface, size, and morphology of NBC:

- *Scanning Electron Microscopy (SEM):* SEM provides high-resolution images of the surface morphology, particle size, and pore distribution of NBC.
- *Transmission Electron Microscopy (TEM):* This is used when it is required to know particle size, crystallinity, and pore morphology. TEM offers a deeper detail of NBC structure.
- *Atomic Force Microscopy (AFM):* AFM is used when the understanding of the material's topography is required to understand its mechanical behavior.
- *Field Emission Scanning Electron Microscopy (FESEM):* This technique provides enhanced resolution of the material, allowing a detailed examination of morphology and surface.

8.4.3.2 Surface Area and Porosity Analysis

Surface area and porosity are fundamental characteristics that influence the performance of materials in applications such as adsorption and catalysis. A higher surface area provides more active sites for chemical interactions, while well-defined porosity ensures efficient transport of molecules within the material. These parameters determine the efficiency of NBC in capturing contaminants, hosting catalytic reactions, or facilitating nutrient retention in environmental and industrial applications. Quantitative analyses of surface area and pore structure are essential to optimize NBC for these advanced functional uses. For this purpose, the following techniques are used:

- *Brunauer–Emmett–Teller (BET) Analysis:* BET quantifies the specific surface area, pore size, and volume through nitrogen adsorption-desorption isotherms, crucial for understanding the adsorption capacity of NBC.
- *Nitrogen Adsorption-Desorption Isotherms:* These isotherms provide pore size distribution, surface area, and total pore volume.
- *Small-Angle X-ray Scattering (SAXS):* SAXS analyzes nanoscale structures, including pore sizes and aggregation, providing insight into the hierarchical organization of NBC.

8.4.3.3 Chemical Composition and Functional Group Analysis

Understanding the chemical composition and surface functional groups of NBC is essential for evaluating its interactions with environmental contaminants. The presence and nature of specific functional groups dictate NBC's adsorption capacity, reactivity, and selectivity for pollutants. The chemical composition provides insight into its stability, catalytic potential, and compatibility with applications, enabling precise tailoring of its properties for environmental remediation and resource management. For this purpose, the following techniques are used:

- *Fourier Transform Infrared Spectroscopy (FTIR):* Identifies functional groups and surface bonding.
- *X-ray Photoelectron Spectroscopy (XPS):* Determines elemental composition and specific functional groups at the surface.
- *Raman Spectroscopy:* Assesses crystallinity, defects, and functional groups by analyzing vibrational modes.
- *Electron Paramagnetic Resonance (EPR) Spectroscopy:* Detects unpaired electrons and paramagnetic centers, revealing electrochemical properties.

8.4.3.4 Thermal and Oxidative Stability Analysis

Evaluating the stability of NBC under varying conditions is essential for understanding its durability and performance in practical applications. Stability analyses provide insights into how the material responds to thermal, chemical, and environmental stresses, ensuring its reliability and effectiveness in long-term usage scenarios such as environmental remediation, catalysis, and soil amendment, for instance:

- *Thermogravimetric Analysis (TGA):* Measures weight loss against temperature, providing insights into thermal stability.
- In Situ *FTIR Spectroscopy:* Tracks chemical changes during thermal treatment, offering real-time insights into functional group evolution.

8.4.3.5 Crystallinity and Structural Order

The crystallinity and structural order of NBC are pivotal factors influencing its performance in adsorption, catalysis, and electrical applications. A higher graphitic content enhances thermal stability, conductivity, and chemical durability, while structural defects and amorphous regions contribute to reactivity and active site availability.

- *X-ray Diffraction (XRD):* Provides insights into the graphitic content, crystalline phases, and overall structural order of N-BC, offering a detailed understanding of its phase composition.
- *High-Resolution Transmission Electron Microscopy (HR-TEM):* Enables visualization at the atomic level, revealing structural arrangements, defects, and the degree of graphitization critical to optimizing N-BC for specific applications.

8.4.3.6 Dispersibility and Stability in Solution

Dispersibility and stability of NBC determine its performance during environmental remediation. These properties define the distribution and interaction of NBC with aqueous pollutants:

- *Dynamic Light Scattering (DLS):* This technique provides details of particle size distribution and hydrodynamic radius, explaining the behavior of suspended NBC.
- *Zeta Potential Measurements:* This technique gives information about the surface charges and colloidal stability, which are crucial for ensuring long-term dispersion of NBC during its application for contaminant removal.

8.4.3.7 Other Novel Techniques

Among the advanced techniques for comprehending the physical and chemical structure of NBC are found:

- *X-ray Absorption Spectroscopy (XAS):* This technique is essential for providing information about oxidation states that are useful for describing the interaction mechanisms of NBC and how these factors control the removal of pollutants.
- *Gas Chromatography-Mass Spectrometry (GC-MS):* The amount of volatile organic compounds (VOCs) released by NBC can be identified by using this technique, which can be useful for determining the environmental impact and chemical stability of NBC during its application.

8.4.4 Interaction of Biochar with Soil and Water Systems

Biochar and NBC modify soil structure by enhancing porosity, aggregation, and bulk density, improving water infiltration, aeration, and root penetration (Hussain et al. 2020). The porous nature of these carbonaceous materials increases water retention in soils with high concentration of sand, while in soils with high amount of clay, it reduces waterlogging (Feng et al. 2022; Schubert and Qadir 2024). The addition of NBC alters soil properties, for instance, Jia et al. (2024) published a study where biochar was applied to saline clay loam soils at varying rates (7.5, 15, and 30 t ha^{-1}) observing an improvement of hydrological functions and structural properties of soils. Soil aggregate stability increased by 15–30% with increasing biochar application rates, with the highest connectivity index (2.36) and fractal dimension (2.56) observed at 30 t ha^{-1}. Biochar reduced small pores (<50 μm) and increased large pores (>300 μm), enhancing soil pore connectivity. This resulted in reductions in soil penetration resistance and increases in soil saturated hydraulic conductivity (46%), plant available water (27%), and least limiting water range (40%) compared to the control. These improvements were positively correlated with large pore size and connectivity, highlighting the critical role of biochar-induced pore structure alterations in improving soil hydrological functionalities for sustainable agriculture in saline soils. Based on bulk-biochar behavior, it can be hypothesized that nano-biochar may provide similar results regarding improvements in soil structure, hydrological functions, and nutrient retention. Additionally, due to its enhanced surface area, reactivity, and functional group availability, nano-biochar may further amplify benefits such as increased soil aggregate stability, improved pore connectivity, and greater mitigation of abiotic stresses, making it a promising tool for sustainable agriculture and environmental remediation. Certainly, the incorporation of biochar and NBC into soil enhances fertility by acting as a reservoir for essential nutrients, such as nitrogen (N), phosphorus (P), and potassium (K). Its high cation exchange capacity (CEC) facilitates nutrient retention, reducing leaching and increasing nutrient bioavailability (Ullah et al. 2024). Addition of biochar and nano-biochar modifies soil pH that enhances nutrients uptake by plants and affects biological processes by increasing soil microbial population, enzyme activity and diversity of microorganisms.

8.4.4.1 NBC in the Rhizosphere: Plant-Soil Interactions

Addition of biochar benefits rhizosphere by improving aggregate stability, nutrient cycling, and porosity. Particularly addition of aged biochar has demonstrated to improve microbial colonization and plant growth. NBC promotes the presence of nitrogen-fixing bacteria and mycorrhizal fungi boosting plant resilience. NBC stimulates carbon cycling and organic matter decomposition reducing erosion risks (Palansooriya et al. 2019). BC and NBC promote beneficial microbial populations in polluted soils, for instance, Bandara et al. (2022) reported that the addition of BC derived from poultry litter to polluted acidic soils reduced cadmium (Cd) bioaccumulation and restored bacterial ecology, improving soil health over time. Furthermore, BC protects soil organic matter, preventing its decomposition and stabilizing it within the soil matrix, creating a long-term carbon reservoir and supporting microbial colonization (Zheng et al. 2022). It can also be observed that NBC affects soil structure, surface chemistry, and microporosity (de Jesus Duarte et al. 2019; Yaashikaa et al. 2020). It is important to highlight that addition of NBC can suppress pathogens, enhancing plant health and soil resilience (Rajput et al. 2022). However, it is crucial to determine its concentration before its addition to soil, since it affects its toxicity toward soil microbes, influenced by root exudates (Sani et al. 2023a, b).

8.4.4.2 Enhancing Plant Growth and Stress Resilience

Including biochar and nano-biochar into soil supports plant growth by improving water retention, nutrient availability, and root development. These materials can mitigate abiotic stress i.e., water scarcity or salinity. The mechanism is based on the immobilization of toxic ions and promoting microbial detoxification. One example of what was mentioned earlier is provided by Tourajzadeh et al. (2024b) who conducted an experiment in greenhouse conditions to assess the effects of nano-biochar (NBC) and reduced irrigation on mitigating salinity and drought stress in quinoa. The factorial experimental design included three levels of salinity (1, 4, and 7 dS m^{-1}), irrigation depths (60%, 80%, and 100%), and NBC application rates (0%, 2%, and 4%). Results showed that 2% NBC significantly alleviated stress, increasing seed yield by 40.75% under salinity and 26.55% under drought conditions. Improvements were also observed in spike weight (SW), weight of 1000 seeds (WOS), and leaf area index (LAI), with increases of 13.88% and 10.89%, respectively. The highest water productivity (WP) of 12.51 kg m^{-3} was achieved with 60% irrigation depth, 2% NB, and 1 dS m^{-1} salinity. Physiological improvements included a 50.16% reduction in proline and a 22.62% reduction in carbohydrate content. Seed protein content increased under moderate salinity and reduced irrigation conditions. The study highlights the potential of NBC to mitigate salinity and drought stress, improve quinoa growth, and enhance water use efficiency in arid and semi-arid regions. Another study published by Abd El-Moaty et al. (2024) investigated the use of nano-biochar derived from peanut shells as a soil amendment to enhance faba bean growth under drought stress. The NBC, synthesized via mechanical grinding, had a bulk density of 0.307 g cm^{-3}, cation exchange capacity of 9.6 cmol kg − 1, zeta potential of −15.5 mV, and an average particle diameter of 21.86 nm. It significantly increased soil water holding capacity and organic matter by 66% and 220%, respectively. Nano-silica or nano-iron oxides were incorporated to improve nutrient retention and stress mitigation. The amendment enhanced growth parameters like stem height, leaf area index, pod number, and chlorophyll content. It also reduced oxidative damage, as indicated by lower lipid peroxidation and electrolyte leakage. These results showed a strong correlation between soil parameters and the plant response, highlighting the potential of the combined use of nanoparticles and biochar derived of peanut shell for improving soil health and enhancing resistance in water scarcity regions.

8.4.4.3 Applications in Water Remediation

There exists strong evidence of prevalent interactions between biochar and NBC and aqueous pollutants; these interactions respond to high porosity, surface area, and functional groups like hydroxyl, carboxyl, and phenolic moieties of these materials. Nanobiochar, with amplified reactivity and its numerous functional sites, is effective in addressing complex contaminants.

Among the mechanisms that facilitate contaminant binding to surfaces, we can identify those reported by Pathak et al. (2024):

- *Electrostatic Interactions:* The negatively charged surface of biochar adsorbs positively charged ions, such as Pb^{2+} and Cd^{2+}, particularly under acidic conditions.
- *Hydrophobic Interactions:* Nonpolar regions of biochar capture organic pollutants, including petroleum hydrocarbons.
- *Redox Reactions:* Functional groups, such as quinones, mediate the transformation of toxic species, such as Cr(VI), into less harmful forms.

Nano-biochar introduces advanced mechanisms, including cation-π interactions, which enhance the sequestration of heavy metals such as Hg^{2+}.

Biochar and NBC are recognized for its ability to remove heavy metals and organic pollutants from water, achieving over 90% removal efficiency for contaminants like Cu(II) and Cr(VI) under optimized conditions (Neeli et al. 2020). Nano-biochar, with its enhanced reactivity and surface area, interestingly extends these applications to emerging contaminants such as antibiotic resistance genes (ARGs) and microplastics, degrading or immobilizing them through advanced chemical interactions (KE et al. 2024; Rajput et al. 2024). The performance of biochar is influenced by several factors, including pH sensitivity, aging effects, and feedstock selection (Khare 2021).

Pollutant adsorption is highly pH-dependent, with acidic conditions favoring metal ion adsorption and alkaline environments supporting anion retention. One example has been reported by (Ramanayaka et al. 2020a) where NBC prepared via the top-down mechanized grinding of dendro-biochar demonstrated significant potential for removing emerging contaminants such as oxytetracycline (OTC), glyphosate (GL), hexavalent chromium (Cr(VI)), and cadmium (Cd(II)) from aqueous media. The study performed adsorption edge and isotherm experiments under varying pH conditions (3 to 8) and an NBC dosage of 0.5 gL^{-1}. It was observed that pH strongly influenced the adsorption process, with the optimal pH for OTC and Cd(II) being 9, while GL and Cr(VI) achieved their best removal efficiencies at pH 7 and 4, respectively.

Characterization revealed the flakey, graphitic nature of NBC with a surface area of 28 m^2/g and functional groups, such as single bond-OH, double bond C-O, single bond-NH, and CH^3, playing a critical role in adsorption. The isotherm analysis suggested a positive cooperative adsorption mechanism, best described by the Hills equation, and demonstrated high partition coefficients for the removal of all four contaminants compared to other adsorbents. These results highlight the effectiveness of NBC, produced through mechanical grinding, as a promising material for the remediation of various pollutants in aqueous environments, with pH and surface functionalization being key factors in optimizing its performance.

Aging enhances biochar's surface functionalization but may reduce structural integrity over time. An interesting study published by Cheng et al. (2024) reported that field-aged and fresh biochar significantly influence the behavior of herbicides and the composition of soil microbial communities. Using 14C-labeled technology, the study evaluated soil amended with field-aged biochar, fresh biochar, and a control without biochar. The results showed that aged biochar enhanced simazine leaching and mineralization by 4.8% and 1.66%, respectively, compared to fresh biochar ($P < 0.05$). The authors concluded that the biochar treatments efficiently adsorbed pesticides.

The addition of BC altered microbial community structure; for instance, the abundance of arbuscular mycorrhizal fungi (AMF) increased by 1.03 and 1.16-fold in aged and fresh biochar treatments, respectively ($P < 0.05$). It was observed a contrary effect related to fungal populations, observing that it was increased in aged biochar compared to fresh biochar treatments. Additionally, eukaryotic populations were reduced in aged and aged biochar-amended soils (1.02 and 1.14-fold respectively) ($P < 0.05$), confirming that field aging processes alter the dynamics of microbial communities in soils polluted with pesticides.

The relevance of these findings is based in the description of the biochar effect on soil microbial populations shifts, rather than on the herbicide adsorption. The observed increases in arbuscular mycorrhizal fungi and changes in fungal and eukaryotic populations demonstrate that biochar, particularly when aged, actively modifies the soil's biological environment. These dual effects underscore the potential of biochar not only as a tool for contaminant management but also to enhance soil health and ecosystem resilience.

8.5 Environmental Applications

8.5.1 Water Remediation: Case Studies on Water Treatment Systems Integrating Engineered NPs-Biochar

The challenge for the agricultural sector in the twenty-first century is to increase agricultural production to feed the world's growing population while maintaining the health of the required natural resources and the integrity of the agroecosystem (Sani et al. 2023b). The application of biochar is widespread as an effective means to boost sustainable agriculture while having a negligible influence on ecosystems and the environment. Currently, several studies have focused on evaluating the effectiveness of using biochar nanoparticles derived from agricultural residues in wastewater treatment to make it suitable for irrigation (Amdeha 2024; Mazarji et al. 2023).

Nanobiochar, unlike traditional biochar, has improved specific surface area, adsorption capacity, and mobility properties in soil, enabling it to promote soil properties, crop growth, and environmental remediation. Biochar-based nanocomposites can be synthesized to develop unique composites that combine the advantages of biochar with nanoparticles. In most cases, the composites possess greatly improved functional group composition, pore properties, surface active sites, catalytic degradability, and separability. Biochar nanocomposites are an excellent way to combine the advantages of nano-BC and other nanomaterials, as well as to improve the physical and chemical characteristics of biochar itself. More recently, several nanocomposite materials based on a wide range of substrates have been produced to purify wastewater, including nuclear effluents.

For instance, in one study, biochar chemically modified using sodium hydroxide and hydrochloric acid was used to remove lead from wastewater. The removal efficiency varied in the different cycles, generally showing a slight decrease in efficiency in successive cycles. This method ensures that biochar maintains its adsorption activity, making it a sustainable option for wastewater treatment (Alsawy et al. 2022).

Another study used three different sources of wastewater from agricultural drainage water, industrial effluents, and municipal wastewater (Soliman and Shawky 2024). The aim of this study was to evaluate the heavy metals removal capacity of biochar produced from three different agricultural residues, i.e., rice straw (RS), palm leaves (PF), and sugarcane residues (SCR). The sample analysis provided information that allowed the proposal that the treated wastewater suitable for irrigation after biochar removal treatment. The wastewater samples contained a variety of heavy metals, including aluminum, mercury, silver, barium, calcium, cadmium, chromium, copper, iron, magnesium, manganese, lead, potassium, strontium, zinc, arsenic, and bismuth. Thus, nanobiochar materials derived from rice straw, palm leaves, and sugarcane residues showed high adsorption and removal capacity of pollutants from wastewater. In particular, the nanobiochar derived from sugarcane residues showed the highest reduction in parameter values and the highest percentage of heavy metal removal, 83.33%. These results underline the potential of nanobiochar-based treatments to improve wastewater quality and promote sustainable water management practices.

One of the techniques to generate biochar-MNPs and improve the capacity of nanocarbons in water pollutant removal treatment is by pretreatment of biomass, which involves altering the feedstock before converting it into biochar by processes such as pyrolysis and calcination. For example, biomass can be impregnated with various metal salts (such as $FeCl_3$, $Fe(NO_3)_3$, $AlCl_3$, $MgCl_2$, $MnCl_2$, $CaCl_2$, and $ZnCl_2$) to perform biomass pretreatment prior to pyrolysis. Under this context, in a study, $Ni_{0.5}Zn_{0.5}Fe_2O_4$ magnetic nanoparticles supported on *Acacia Nilotica* seed husk ash (ANSA) were synthesized and applied for effective adsorption of Pb(II) ions from water (Omidvar-Hosseini and Moeinpour 2016). An aqueous mixture of $Fe(NO_3)_3{\cdot}9H_2O$, $Zn(NO_3)_2{\cdot}6H_2O$, $Ni(NO_3)_2{\cdot}6H_2O$, and ANSA metal nitrates was dissolved in water, to which extracted egg white solution was added. The Langmuir model described the adsorption process, and the kinetics were fitted to the pseudo-second-order rate equation. The nanoparticles could be easily separated from water using an external magnet, making them a practical and environmentally friendly solution to treat water pollution caused by Pb(II) ions. In this work, biochar was produced from raw corn stover at four different temperatures: 500, 600, 700, and 800 °C. This modified biochar, uniformly coated with nano ZnO/ZnS, showed superior porosity (with a BET surface area of 397.4 m^2/g and a total pore volume of 0.43 cm^3/g) compared to unmodified biochar (with a BET surface area of 102.9 m^2/g and a total pore volume of 0.20 cm^3/g). Batch adsorption tests indicate that the obtained nano ZnO/ZnS-modified biochar showed a high capacity to adsorb Pb(II), Cu(II), and Cr(VI), with maximum adsorption capacities significantly higher than those of ordinary biochar.

Another study prepared a nanocomposite of Fe_2O_3 and biochar by pyrolyzing paper and paper sludge (PPS) impregnated with $FeCl_3$ at 750 °C (Chaukura et al. 2017). The PPS biomass was immersed in $FeCl_3$ solution for 2 h and then air-dried at 80 °C for another 2 h. The $FeCl_3$-impregnated PPS, with a mass/volume ratio of 1:3, was pyrolyzed at 750 °C for 2 h in an N_2 atmosphere, producing Fe_2O_3-biochar. This nanocomposite was used for the effective removal of methyl orange (MO) from contaminated wastewater, and the results showed that its MO adsorption capacity was 52.79% higher than that of biochar.

8.5.1.1 Role of Nano-BC in the Sustainable Treatment of Polluted Water Under the Framework of the Bioeconomy

The circular bioeconomy is a methodical strategy that helps the economy while protecting the environment. Obtaining nano-BC from biomass residues and using them to treat pollutants in irrigation water and wastewater can help conserve water reserves and generate agroecosystem sustainability (Sani et al. 2023b). It therefore contributes to fostering a circular bioeconomy, and in turn helps to mitigate climate change.

Currently, several researchers have focused on the use of waste biomass as feedstock or even the utilization of biomass from plants that generate eutrophication in aqueous ecosystems. One study is novel; it obtained nanobiochar from water hyacinth (WH) and black tea waste (TW) as a promising approach for water treatment due to their unique properties that play an effective role in the adsorption of metals (Elbehiry et al. 2022). In this study, monoadsorption and multiadsorption systems of cadmium (Cd), chromium (Cr), and nickel (Ni) on biochar and nanobiochar from water hyacinth (BWH and NBWH) and black tea waste (BTW and NBTW) were investigated as potential low-cost and environmentally friendly sorbents for the removal of the above heavy metals (HM) from aqueous solutions. WH and TW were collected locally, prepared, and preserved until use in the experiment. Nanobiochar was prepared by grinding, characterized, and stored in airtight containers until use. A batch experiment was designed in mono and competitive systems to study the adsorption equilibrium behavior of HMs on biochar and nanobiochar. The Freundlich and Langmuir isotherm models fit the monoadsorption and competitive adsorption equilibrium results. The Freundlich isotherm model provided a better fit. Furthermore, it was observed that NBWH and NBWT effectively removed Cd in the mono system by ≥99.8, especially at the smallest concentration, whereas NBWT and BTW removed ≥99.8% and 99.7% in the competitive system, respectively. In the mono and competitive systems, NBTW nanobiochar removed more than 98.8

Cr. The sorbents were less effective in removing Ni compared to Cd and Cr. However, their efficiency was also very high. The results revealed that Cd was the most removed metal by sorbents, nanobiochar were better than biochar in removing HMs, and the results also indicated that the co-occurrence of multi-metals could fully occupy the adsorption sites on biochar and nanobiochar.

Nanobiochar adsorption is a great method for cleaning contaminants due to its large surface area and versatility. It is an inexpensive and environmentally friendly solution, often made from waste materials. Mohan (2024) published a study where corn straw nanobiochar (MNBC) was used to adsorb acrylamide. MNBC was chemically treated by alkaline treatment followed by acid treatment. MNBC resulted as a very effective material to adsorb acrylamide due to its higher surface area and pores. It was found an increase in the acrylamide removal capacity from 10 to 120 min of contact time, reaching equilibrium at 100 min. A higher MNBC dose of 30 mg/L shows a higher removal efficiency (92.0%) compared to the lower MNBC dose (10 mgL^{-1}), which has a removal efficiency of 75.2%. It has been observed that MNBC has superior acrylamide adsorption capacity at pH 7.0. The maximum amount of acrylamide adsorbed by MNBC is 80 mgg^{-1}, and its adsorption behavior agrees with Freundlich and Langmuir isotherms. The MNBC from the experiment can be reused for acrylamide removal and remains effective for up to six cycles. The primary adsorption process involves MNBC pores, effective binding regions, and interactions with acrylamide functional groups. We performed experiments using a fixed-bed configuration, adjusting the height of the MNBC bed to generate breakthrough curves. Regarding the cost-effective analysis performance, it was confirmed that MNBC can be successfully used to treat water. MNBC can be used as an eco-friendly material that can be applied as an effective adsorbent, potentially reducing acrylamide pollution in aquatic ecosystems. This work may inspire future research on nanobiochar-based adsorbents for various pollutants and the development of efficient water treatment systems, benefiting both industrial and municipal wastewater treatment.

Another study explored the potential of nanobiochar synthesized by ultrasound and centrifugation from rice husk for the removal of dyes from water (Aziz et al. 2023). Nanobiochar was synthesized from rice husk biochar for the removal of safranin, malachite green, and a mixture of both in aqueous water. The biochar was synthesized by pyrolysis at 600 °C for 2 h. To convert it into nanobiochar, sonication and centrifugation techniques were applied. The dye adsorption kinetics followed the pseudo-second order model, suggesting dye removal by a chemical sorption process. High adsorption was found at higher nanobiochar concentration, high temperature, and neutral pH. The maximum removal percentages of safranin, malachite green, and a mixture of dyes were 91.7%, 87.5%, and 85%, respectively. Therefore, nanobiochar could be a solution for the removal of dyes in the treatment of water containing dyes.

In general, the integration of NBC in water treatment within the field of bioeconomy helps in carbon sequestration and the reduction of methane and nitrous oxide emissions from agriculture can be achieved with NBC applications, contributing to climate change mitigation (Sani et al. 2023b). In parallel, due to its cost-effectiveness, sustainability and environmental friendliness, waste-derived nano-BC can become the most viable alternative to conventional waste management strategies, contributing to the circular bioeconomy and the broader goal of achieving the Sustainable Development Goals (SDGs).

8.5.2 Removal of Heavy Metals, Organic Pollutants, and Nutrients Using Engineered NPs-Biochar Composites

The rapid advancement of industrial and agricultural activities has resulted in the widespread release of hazardous materials, including heavy metals, organic pollutants, and excess nutrients, into the environment. These contaminants pose significant risks to ecosystems and human health due to their persistence, toxicity, and potential to bioaccumulate. Among the innovative solutions developed to tackle this global challenge, engineered NBC, particularly biochar supported with metal nanoparticles (MNPs@BC), has emerged as a highly promising material for environmental remediation.

As it has been mentioned, BC and NBC exhibit excellent stability, adsorption capacity, and versatility in applications ranging from water purification to soil amendment. However, the integration of nanosized metal particles onto biochar surfaces significantly amplifies its functional capabilities. This hybrid material combines the high reactivity, large specific surface area, and catalytic properties of nanoparticles with the stability, ion-exchange capacity, and structural support of biochar. As a result, N-BC materials demonstrate enhanced performance in removing heavy metals (e.g., Pb(II), Cr(VI), and Cd(II)), organic contaminants (e.g., pesticides, dyes, and pharmaceuticals), and nutrients (e.g., phosphate and nitrate) from polluted water and soil. For instance, Zhou et al. (2022) evaluated the aqueous Pb and Cd removal using composites at different ratios of biochar (BC) to nanohydroxyapatite (nHAP), finding that Pb exhibited higher sorption capacity and immobilization efficiency compared to Cd. The authors identified that the sorption process was governed by dissolution-precipitation, cation exchange, complexation, and cation-π interactions, highlighting that dissolution-precipitation primarily affects Pb and cation exchange dominating the immobilization of Cd. A remarkable finding was that the sorption capacity, immobilization of pollutants and the proportion of nHAP increased, suggesting that composite material ratios can be selected for specific pollutant concentrations.

It is well documented that exist synergistic effects between biochar and metal nanoparticles, such as adsorption, reduction, precipitation, and ion exchange, improving remediation efficiency. In a study published by Fan et al. (2018) it was highlighted that the aging process incorporated oxygen groups on the biochar surface, influencing its adsorption behavior. In this study, aged wheat straw biochar was chemically modified via chemical oxidation with HNO_3-H_2SO_4 and NaOH-H_2O_2, observing a carbon loss and oxygen incorporation. The oxidation increased more developed mesopores, increasing specific surface area—126% for biochar treated with NaOH-H_2O_2 and 226% for biochar treated with 40% HNO_3-H_2SO_4. As a result of these changes, Cd-adsorption increased by 21.2%, attributed to both the increase in functional groups and the roughened surface from oxidation.

In a similar study, but focused on nitrogen-containing groups, Luo et al. (2023) demonstrated that grafting biochar with nitrogen functional groups, such as -NO_2, significantly increased its adsorption performance. Bagasse-based biochar was modified with various functional groups (-NO_2 > NH_2 > -CN > -SO_3H > -OH), observing that adsorption efficiency increased due to the electron-withdrawing ability of the incorporated groups, particularly -NO_2.

The novel approach of integrating biochar with metal nanoparticles has gained relevance due to its wide versatility in environmental remediation. This hybrid material has shown several mechanisms that can improve the removal of diverse contaminants. The modification of biochar surface can be performed during its synthesis by incorporating functional groups that enhance the interaction with pollutants and stability in aqueous media.

8.5.3 Soil Enhancement and Carbon Sequestration

8.5.3.1 Improvement of Soil Properties and Crop Yield

The use of NBC as fertilizer is obvious due to its high content of carbon and micro- and macronutrients. As it was mentioned, this material also serves as microbial communities support, enhancing soil physical and chemical properties (Abed Hussein et al. 2022; Khan et al. 2021; Ramadan and Abd-Elsalam 2020; Shafiq et al. 2023). This material also boosts plant stress tolerance by maintaining growth and activating systemic acquired resistance against biotic and abiotic stresses (Abd El-Moaty et al. 2024; Sammar Raza et al. 2023; Yousaf et al. 2024).

In this regard, it is proved that microbial communities play an essential role in plant-soil ecosystems, influencing soil health, fertility, nutrient cycling, carbon dynamics, and crop production (Herrera Paredes and Lebeis 2016; Ke and Miki 2015). Several studies have evaluated the effect of the addition of NBC on these microbial populations (Chausali et al. 2021; Sarma et al. 2024). The shift in soil pH and redox potential influences microbial activity and community composition (Khaliq et al. 2023; Nishshankage et al. 2024; Zhang and Shen 2022); however, some other studies have found opposite results, observing a negative effect on microbial ecology indexes, such as abundance or composition. It is under scrutiny if the addition of NBC affects specific microbial groups in rhizosphere, promoting the availability of certain nutrients and stimulants for plants, and as a consequence, enhancing their growth (Sani et al. 2023a, b). For example, Liao et al. (2019) reported that when adding BC to soil, changes in the densities of Firmicutes and Bacteroidetes were found, particularly for members as *Clostridium, Bacillus, Sporomusa, Desulfosporosinus*, and *Alicyclobacillus* while decreasing the amounts of *Proteobacteria* members utilizing plant-derived carbon, notably *Methylobacterium* and *Sphingomonas*.

Since biochar and nanobiochar enhance the release of dissolved organic matter, plants can easily uptake it as organic acids, humic compounds, and colloidal nutrients (Feng et al. 2021; Hernandez-Soriano et al. 2016); as it was observed by Liu et al. (2021) that when applied 0.1% of biochar to soil, increased shoot nitrogen, phosphorus, and potassium content by more than 1.5 times. Similar results were observed when dilute biochar was added to soil obtaining a higher leaf growth and photosynthetic activity (Kumar et al. 2021; Liu et al. 2022; Rizwan et al. 2019). Exogenous applications of nano-BC have been found to enhance seed germination and plant growth (Shafiq et al. 2023; Yue et al. 2019). Carbon nanoparticles derived from bulk biochar also improve early plant growth traits (Li et al. 2023; Yue et al. 2019); In Table 8.1 represents some experiments where are analyzed the effect of

Table 8.1 Nano-BC applications in plant trials improving crop yield

Nanobiochar or biochar-based nanomaterials	Application doses	Plant species	Effects at soil–plant interface	References
Nano-BC	1%, 3% and 5%	*Daucus carota* L.	Vegetative growth, photosynthetic pigments	Khaliq et al. (2023)
Nano-BC	0.7–1.0 mgL^{-1}	*Zea mais* L.	Increased yield of maize and grain weight	Yang et al. (2020)
Nano-BC	100 mgL^{-1} 400 $mg L^{-1}$	*Oryza sativa* L.	Higher dry biomass, plant height; increased root vitality and antioxidant enzyme activities; alleviated cd toxicity Application of nano-BC can promote rice seedling growth, enhance biomass, increase the root length chlorophyll concentration, reduce oxidative stress and lipid peroxidation	Yue et al. (2019) Shen et al. (2020)
Nano-BC	250 mgL^{-1}	*Allium ascalonicum*	Increased plant height and tuber weight	Purbalisa et al. (2021)
Nano-BC	50, 100, 250 $mg kg^{-1}$ rates in pots	*Brassica napus*	Improving yield	Mahmoud (2022)
Water-soluble carbon nanoparticles	50 mg/L	*Triticum aestivum subsp. aestivum*	Increased plant growth	Saxena et al. (2014)
Nanobiochar modified with ACC deaminase rhizobacteria	0.5% (5gkg^{-1})	*Mentha piperita L.*	Plant growth and minimizes Pb stress	Zafar-ul-Hye et al. (2021)
MgO biochar	1.0% wt	*Lolium multiflorum* Lam.	Increased phosphate uptake and enhanced yield	Manolikaki et al. (2016)

NBC on crop yield improvements as a result of nanobiochar interactions with organic compounds and plant root hairs, modifying surface chemistry and facilitating nutrient absorption (Brar et al. 2024; Gu et al. 2022; Li et al. 2023).

8.5.3.2 Potential for Long-Term Carbon Sequestration and Climate Change Mitigation

Soil carbon sequestration involves capturing atmospheric CO_2 and storing it in the soil through plant processes (Lal et al. 2015), also enhances food and nutritional security, water quality, promotes biodiversity, and supports elemental recycling. Recent results suggest that the optimal concentration of soil organic carbon (SOC) in the root zone ranges between 1.5% and 2.0%, however, it is important to note that SOC levels are influenced by diverse factors, i.e., land use, soil management, and farming systems (Souza et al. 2023; Franzluebbers 2024). More than 50% of the total soil carbon pool is between 0.3 and 1 m deep (Bhattacharya et al. 2016). However, soils in agroecosystems are often severely depleted of SOC and degraded. Restoring soil quality requires increasing SOC levels through best management practices, such as conservation agriculture, which promotes a positive carbon budget. This strategy aims to mitigate climate change while advancing global food security.

SOC naturally decomposes over time, releasing carbon dioxide into the atmosphere (Abbas et al. 2020a). Several studies indicate that incorporating nano-biochar into soil significantly boosts SOC levels. For instance, adding 8% nano-biochar can increase SOC by approximately 41%, with no notable reduction even after 210 days. This suggests that nano-biochar enhances SOC stability and reduces its mineralization through mechanisms such as (1) physical stabilization via aggregation, (2) binding with clay and silt particles, and (3) biochemical stabilization through the formation of resistant SOC compounds (Liu et al. 2015) see Fig. 8.4. Integrating SOC into biochar is likely the primary factor in improving stability. Research consistently shows that SOC decomposition slows with biochar application.

Regarding the changes in CO_2 emissions from soil caused bynano-biochar, the authors report that NBC can increase carbon dioxide emissions, but prolonged use tends to reduce them by enhancing soil moisture retention and limiting organic carbon decomposition (Abbas et al. 2020b; Hachem et al. 2022). This reduction could be linked to the microporous nature of C-based materials and its impact on soil microbial dynamics, contributing to long-term carbon sequestration.

Greenhouse gases (GHGs) significantly contribute to global warming and climate change, with agricultural activities during crop cultivation being a major source of GHGs like carbon dioxide (CO_2), methane (CH_4), and nitrous oxide (N_2O) (Malhi et al. 2021; Bhatti et al. 2024). Applying nano-biochar (nano-BC) has shown potential by influencing the mechanisms behind GHG generation and emissions.

Biochar can mitigate the release of GHGs such as nitrous oxide(N_2O) and methane (CH_4) (Shoudho et al. 2024). NBC reduces ammonia (NH_3) volatilization by enhancing soil properties such as temperature, pH, aeration, and organic matter mineralization (Sani et al. 2023b). Its high surface area adsorbs NH_3 and ammonium (NH_4^+) while enhancing nitrification, converting nitrogen compounds into less harmful forms, and decreasing N_2O accumulation (Sarma et al. 2024a). Furthermore,

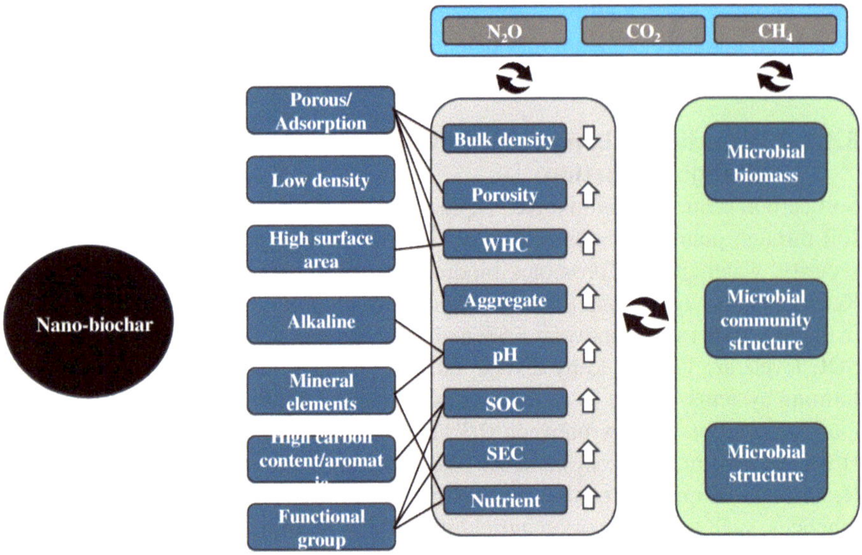

Fig. 8.4 Effect of nanobiochar on soil organic carbon content

NBC activates enzymatic and microbial activity, which consumes NH_4^+ and reduces its presence in the soil. It also increases soil porosity and aeration, reducing anaerobic respiration and N_2O emissions (Goswami et al. 2022; Saini 2024). Combined with organic or inorganic fertilizers or applied independently, nano-biochar can mitigate GHG emissions like CH_4, N_2O, and CO_2 without negatively impacting crop production. However, completely replacing inorganic fertilizers with nano-biochar can significantly reduce yields, making their combination practical. Studies have shown that modified biochar, such as that infused with nano-zero-valent iron (nZVI), can further improve its efficiency in reducing nitrate, N_2O, and CO_2 emissions by influencing microbial activity, denitrification genes, and methanogenesis processes (Koul and Taak 2018; Arabzadeh Nosratabad et al. 2024). Nevertheless, the impact of nano-biochar on GHG emissions varies depending on its composition and preparation method. For instance, phosphorus-modified biochar reduces CO_2 and N_2O emissions, while nZVI-modified biochar may increase them. These differences highlight the need for further research to understand the role of nano-BC in diverse soil types and agricultural conditions and optimize its application for sustainable farming.

8.5.4 Combined Agricultural and Environmental Benefits

High surface area, porosity, and surface chemistry make NBC an ideal prospect for use in various agricultural applications. In this context, the production of nano-biochar and its practical application in small orchards and commercial farms of vegetables, fruit trees, and industrial crops could be an approach that can particularly increase crop yields, increase carbon stocks, and decrease GHG emissions (Zheng et al. 2020). Studies suggest that nano-biochar is used as a fertilizer because of its high carbon and mineral content. However, although research studying the effect of NBC on plants has shown it to be effective for crop health and productivity, its supply is recommended in low and moderate concentrations between 10 and 400 ppm or mg kg^{-1} of dry soil (Pérez-Hernández et al. 2021).

8.5.4.1 Agricultural Benefits

Reports indicate that there is little knowledge available on the impact of nano-biochar on different morphological and biochemical characteristics of plants. Nevertheless, it is recognized for its nutrient uptake efficiency, water uptake efficiency, and water retention capacity. In general, nano-biochar applied to soil helps the plant resist biotic and abiotic stress. The high production of crops from biochar is in many cases synonymous with crop health. In this line, researchers demonstrated that Quinoa plants (*Chenopodium quinoa* Willd.) subjected to salinity and drought stress were mitigated with appropriate concentrations of 2% NBC (w/w), thus contributing to yield and improvement of nutraceutical quality. On the contrary, at concentrations of 4% (w/w), it affected the production of seeds, carbohydrates, and photosynthetic pigments (Tourajzadeh et al. 2024a). Other studies reported that low concentrations have a positive influence on wheat. Exposing plants to 1% (w/w) concentrations under drought conditions resulted in increased yield (Sammar Raza et al. 2023). To evaluate the synergistic effect of NBC on two wheat varieties, researchers demonstrated that soil or foliar applications resulted in plant improvements at concentrations of 1, 3, and 5%. At the sowing stage (soil NBC) and 30 days after germination (foliar NBC), plant height, fresh and dry biomass of stems, leaves, roots, and photosynthetic pigments increased (Muhammad Mehmood et al. 2024).

Although organic residues have been efficiently used to promote nutrient cycling and uptake, the effectiveness and changes in the soil are generally short-lived, and the amendments are mainly transformed into carbon dioxide. Therefore, it is here where NBC contributes effectively to soil productivity, improves pH soil, and cation exchange capacity, increases water retention, and increases beneficial soil microbial biomass, thus indirectly improving soil fertility. Using coconut waste and generating biochar and derivates solves two problems. On the one hand, the waste is used to produce biochar, and when added to the soil as a nanometric product, it improves the absorption of nutrients by plants (Ramadan et al. 2020). According to the authors, coconut bagasse husk with a rate of 30 t ha^{-1} improved the absorption of nitrogen and phosphorus, and increased the biomass in the corn crop (Gonzaga et al. 2018). In rice crops, interesting results have been shown after the application of biochar. When researchers treated the soil with biochar at concentrations of 0.25% (g g^{-1} soil), they observed increases in N and P uptake in rice plants by 40% and 46%, respectively, compared to inorganic fertilizer. In addition, the results revealed significant changes in microbial activity, which may have contributed to nutrient uptake in plant tissue (Chew et al. 2020). Other significant effects have been observed in the soil quality of vegetable crops. The authors demonstrated that the combination of foliar and soil at a concentration of 0.3% and 5% of NBC contributed to higher accumulations of N P K in carrot plants. Moreover, the effect indirectly impacted the physiology of the plant by increasing the content of chlorophyll, carbohydrates, phenols, and flavonoids (Khaliq et al. 2023). On the other hand, the joint application of amendments and NBC may also result in potential effects on maize production. Research has shown that compared to the control treatment (0 t ha^{-1} NBC) a concentration of 5 t ha^{-1} poultry manure +5 t ha^{-1} NBC contributed to high N uptake rates and consequently to grain yield (Das et al. 2024). However, studies suggest that the biochar derived from poultry manure and sewage sludge is better in terms of mineral contribution compared to plant remains since it has been shown that NBC, in addition to being rich in NPK, also contains considerable amounts of sulfur (S), magnesium (Mg), zinc (Zn), iron (Fe), manganese (Mn) and copper (Cu) (Amalina et al. 2022).

In other cases, NBC not only improves nutrient uptake but also soil health by increasing beneficial microorganisms and decreasing unwanted microorganisms in the soil. Reports suggest that while nano charcoal can have positive effects, adverse effects have also been reported. For example, the application of nano charcoal at rates of 3–10 μg g^{-1} substantially changes soil microbial communities, while also improving soil fertility and improving soil health from contaminants (Qian et al. 2018). In this regard, it was reported that the addition of NBC increased the microbial community even when the soil was contaminated with phenanthrene. The main reason is that NBC serves as food for bacteria, and carbon sources can absorb minerals in the soil, making the soil an enriched habitat for minerals and possibly other organic compounds (Song et al. 2019). Similarly, when the combined action of vermicompost + NBC added to soil containing corn plants was investigated, a reduction in ammonia emissions was confirmed, a higher absorption of N, P, and K, in addition to an increase in microbial biomass and improved growth parameters of the corn crop, compared to the vermicompost treatment and conventional fertilization alone. The authors highlighted that by reducing the supply of NPK fertilization, nitrous oxide emissions, a gas that contributes to global warming, are reduced, yield is increased, and soil health is maintained (Rashid et al. 2023). Therefore, the NBC, in addition to absorbing nutrients, also immobilizes hazardous organic and inorganic materials that can affect the soil and cause toxic effects on plants.

Also, NBC positively impacts arbuscular mycorrhizal fungi, and plants improve resistance to pathogens. The explanation for this is that the porous structure and high surface area of NBC, in addition to the changes in pH, salinity, and ion modifications generated in the soil, create suitable environments for microbial colonization and growth and, consequently, higher enzymatic activity (Brar et al. 2024). In this sense, from the perspective of soil and crop health, the authors suggest that with the gradual management or incorporation of NBC in agricultural systems, producers can reduce the use of fungicides and chemical compounds in general (Bolan et al. 2023). For further information, Brar et al. (2024) report more details on the positive implications of nano-biochar as fertilizers and disease control.

8.5.4.2 Environmental Benefits

According to the latest update on global pesticide consumption, China is the country that uses the most pesticides, followed by the United States, Brazil, and Argentina, with a consumption of 53%, 11%, and 6%, respectively. The problem is alarming since the sum of chemicals amounts to more than 1000 active ingredients, whose destination is the soil, water, and air (Brar et al. 2024). As a result of this problem, NBC is an efficient product for degrading organic compounds in agricultural soil and irrigation water. In addition, laboratory and field studies have shown that in addition to being efficient in soil decontamination, production costs are reduced due to the lower use of pesticides. The efficiency is due to its porous structure, high surface area, high functional groups on the surface, high colloidal stability, high zeta potential, high volatile matter content, and high agglomeration capacity, which are the characteristics that enable the removal or degradation of heavy metals and organic and inorganic compounds. In this same sense, the combination of biochar (BC) and NBC contributed significantly to the increase of the microbial community and soil decontamination in a bean experiment. The authors reported that after three weeks of exposure, the 1% concentration (BC/NBC) increased the populations of phosphorus-solubilizing and nitrogen-fixing bacteria and a significant reduction of the concentration of methomyl (MET) and the hydrophobic lambda-cyhalothrin (LCT) (Chandi et al. 2024).

Regarding the removal of heavy metals in water, it was demonstrated that manganese sulfide-functionalized NBC made from millet bran was able to reduce the Cd(II) concentration (Fan et al. 2020). The formation of complexes (MnS/MNOx) and the oxidation/reduction mechanisms in the surface of NBC were the predominant processes that facilitated the removal of cadmium. In a similar study, it was demonstrated that for chromium elimination, the reduction of bonds of the hydroxyl (-OH), carboxyl (-COOH), and carbonyl (-CO) groups present on the NBC surface facilitated the removal process (Ramanayaka et al. 2020).

It has been also explored and demonstrated that NBC can mitigate greenhouse gas emissions. However, the effectiveness of NBC depends on factors such as raw material, environmental factors, soil type, and application modes. According to the authors, NBC significantly retains soil organic carbon content by 70% to 80%, reducing the release of CO_2 into the atmosphere, as well as other gases such as nitrous oxide (N_2O) and methane (CH_4), major compounds that contribute to climate change (Chagas et al. 2022). Indeed, the authors demonstrated that NBC limits the movement of CH_4 into the atmosphere due to the porous structure of NBC and the pH conditions of the soil and feedstock. For example, authors proved that NBC in soils with a pH of 6.5–7.5 prevented CH_4 emissions, contrary to what occurred in acidic soils (higher increase in the gas). The explanation is that in soils with neutral and alkaline pH, the growth of Gram-negative bacteria (methanotrophs), the main microorganisms that oxidize methane, is promoted (Shakoor et al. 2021).

According to researchers, the acidic pH of the soil favors the emission of N_2O. However, it has been shown that pyrolysis temperatures above 500 °C favor the increase of denitrifying bacteria and consequently a reduction of N_2O to N_2. Additional factors such as the type of raw material, the application rate, the type of soil, soil moisture, pH, C/N ratio, nitrifying, and denitrifying bacteria, and various physicochemical properties of the soil contribute to the low emission of N_2O. For all the above, the application of NBC will be effective depending on the soil conditions and biological processes. Without a doubt, more knowledge generation is required to have a successful recommendation for the use of NBC in agricultural soils and the environment in general (Lyu et al. 2022).

There is extensive information on the effects that NBC can have on increased growth, yield, and crop protection, plant tolerance to biotic and abiotic stress, crop fertilization, development of microbial diversity, soil carbon sequestration, degradation of pollutants, mitigation of greenhouse gases and decontamination of irrigation water. A comprehensive summary of positive effects can be read in current research reports (Faizan et al. 2024; Hu et al. 2024; Pathak et al. 2024; Sani et al. 2023a, b; Sarma et al. 2024a, b; Sultan et al. 2024).

Synergistic Effects on Crop Health and Environmental Quality

From the perspective of the circular economy, reusing and recycling nutrients is an available option to increase soil fertility and health and protect the environment. All over the world, countries are concerned about reducing the pollution of agricultural and urban soils. In this line, the use of organic waste to produce NBC is becoming increasingly popular in agriculture. However, the new strategy with NBC is suggested to be supplied in low doses in co-fertilization with inorganic fertilizers. In this way, producers manage to increase crop yields and, at the same time, can maintain soil health and fertility by reducing external inputs (Rajput et al. 2022). Although the NBC effect has been studied for years, there are still gaps in our knowledge. However, it is expected that the technology will gradually be transferred to realistic field conditions. However, if nanotechnology were applied to small areas of local producers, it could be considered a giant step in promoting sustainability and the circular economy.

8.6 Challenges and Future Outlook

Scalability is arguably a crucial factor to consider when it comes to nano-biochar production. This involves the practicality of expanding operations from a small laboratory experiment to a full-scale production system. If there are limited possibilities for scalability, it could lead to concerns about obtaining raw materials and create logistical and financial obstacles for manufacturers. From a comprehensive standpoint, there are various factors that can impact the feasibility of expanding nano-biochar production. In this study, we categorize different scalability considerations into sustainability, regulatory, technology, and market factors. This section provides a detailed explanation of the various scalability considerations related to increasing the production of nano-biochar (Khare 2021).

The accessibility of raw materials from primary production plays a crucial role in the potential for large-scale synthesis, particularly in the context of sustainability. In the case of bottom-up production like nanobiochar, the availability of biochar is a significant obstacle for large-scale synthesis. While the agricultural sector generates residual biomass, this is hindered by the competition between land usage and the demand for food. Policies designed to mitigate the impact of climate change have caused concerns about unemployment and waste management. A study published by Shoudho et al. (2024) pointed out that this competition could lead to a scarcity of biochar/biomass, leading to increased production costs for bio-chemicals and bioenergy. Alternatively, using biochar dust from industrial sectors seems practical, but the synthesized nano-biochar from this type of biochar typically contains inorganic ash, potentially impacting cytotoxicity and leading to increased mortality of human and mammalian cells (Yaashikaa et al. 2020). Furthermore, biochar must be produced at a bench scale, requiring the construction of large kilns or biochar-making reactors. These challenges will impact the scalability of the technology, potentially rendering it unviable (Pierson et al. 2024).

Feedstock availability is also influenced by agricultural competition, with more producers in an area increasing the strain on resources. However, the role of demand-driven economics in supporting scalability should not be overlooked (Berazneva et al. 2021). For instance, a user might be willing to accept lower quality in nano-biochar processing if it significantly reduces costs. This approach effectively addresses efficiency and technical barriers through improved training and methods. Regulatory and certification bodies also play a critical role, especially during scaling. Compliance requires financial investment, and certification can shift production toward exporting to wealthier markets. To scale effectively, it's crucial for producers to engage locally, as the ethanol industry challenges illustrate (Igalavithana et al. 2022; Javaid et al. 2024).

Understanding the economic viability of nano-biochar production is the focus of this report. It is crucial to analyze the costs and benefits, as well as review financial reports, to make informed decisions about this business. While nanobiochar primarily offers significant environmental and social benefits, especially in developing countries, these are often harder to quantify than economic factors (Anupama and Khare 2021). Many factors such as production costs, product prices, demand and available financing affect the growth of NBC industry. For this reason, it is suggested a structured approach using a decision-support model. For instance, a study published by Ramadan et al. (2020) highlighted that the transition to biochar-based markets is influenced by local carbon prices and financing options, so that is needed to make critical strategic decisions. Consulting biochar producers and consumers may offer important insights into the carbon price market and the potential advantages of entering the biochar industry.

From a social perspective, NBC offers important potential benefits for smallholder agricultural communities in developing countries, enhancing food security and improving soil fertility by boosting physical and chemical soil properties (Kozłowski et al. 2023). The addition of NBC can be complemented with traditional soil practices, facilitating the transition toward more sustainable agricultural management. However, some ethical concerns must be considered, such as the potential risks associated with applying agricultural amendments and uncertainties about safety when these amendments are used in remote areas across different continents. This concerns have been present through the last years and are intrinsically related to the safe use of nanotechnology in agriculture, thus is essential to rigorously communicate the possible implications of these practices.

The successful commercial launch of nano-biochar products in the food, agriculture, and environmental sectors hinges on the active participation of all stakeholders. Your willingness to foster collaborations and build strong relationships is crucial. Post-conference priorities should focus on the following areas: (1) research, (2) infrastructure development, (3) regulation and standardization, and (4) business and commercialization. Implementing policies and incentives for large-scale production will enhance commercial nanobiochar products' feasibility and cost competitiveness. Increased scalability may lead to cost-effectiveness, achieved through optimized waste synergy at production facilities capable of high-output nanobiochar production. Establishing financing structures with local industry partners to efficiently develop cost-effective scaling methods for biochar production is crucial. Moreover, these industry partnerships will help clarify the use of technological process stream side streams and waste synergy strategies. If strategic partnerships are formed early, the commercial production of nano-processed biochar could become competitive within next years.

8.7 Conclusion

Nanobiochar made from lignocellulosic residues is a powerful example of how sustainability and innovation can come together to solve environmental and social challenges. By converting agricultural waste into a high-performing material, NBC helps tackle pollution, improve soil health, and support sustainable farming practices. Its versatility for being applied in environmental remediation and agriculture improvement have converted BC and NBC into an attractive option that naturally align with key Sustainable Development Goals (SDGs), such as ending hunger (SDG 2), ensuring responsible production (SDG 12), and combating climate change (SDG 13). Besides this, NBC has demonstrated to be a feasible tool for contributing to the reduction of greenhouse gas emission and promoting the adaptation to a changing climate.

NBC also impact societies. With this conversion technology, it is possible to transform agricultural residues into valuable resources, promoting a circular economy and incentivizing local markets. The use of these materials can increase the accessibility to water and restore degraded soils of vulnerable populations. Nevertheless, to maximize its impact, some challenges must be addressed, such as scaling production sustainability and adjusting the use of NBC to different ecosystems and conditions. NBC's potential can be unleashed by fostering multidisciplinary collaboration, integrating groups where scientist, policymakers, and local communities promote changes to more sustainable practices.

Acknowledgments E.V.-N. thanks to the Universidad de Guanajuato for its support through Grant No. 062/2024, titled "*Sustainable Valorization of Tequila Industry Residues: Hydrochar Production via Hydrothermal Carbonization of Bagasse/Vinasses and Its Impact on Soil Improvement and Controlled Nutrient Release.*" CIIC 2024-UGto.

References

Abbas F, Hammad HM, Ishaq W et al (2020a) A review of soil carbon dynamics resulting from agricultural practices. J Environ Manag 268:110319. https://doi.org/10.1016/j.jenvman.2020.110319

Abbas Q, Yousaf B, Ali MU et al (2020b) Transformation pathways and fate of engineered nanoparticles (ENPs) in distinct interactive environmental compartments: a review. Environ Int 138:105646

Abd El-Moaty HI, El-Dissouky A, Elhusseiny AF et al (2024) Low-cost nano biochar: a sustainable approach for drought stress mitigation in faba bean (Vicia faba L.). Frontiers. Plant Sci 15:1438893

Abed Hussein B, Mahdi AB, Emad Izzat S et al (2022) Production, structural properties nano biochar and effects nano biochar in soil: a review. Egyp J Chem 65:607–618. https://doi.org/10.21608/ejchem.2022.131162.5772

Ahuja N, Ansari A, Rajput R, Singh P (2020) Synthesis and characterization of zero Valent iron nanoparticles for textile wastewater treatment. Pollution 6:773–783. https://doi.org/10.22059/poll.2020.296735.740

Alsawy T, Rashad E, El-Qelish M, Mohammed RH (2022) A comprehensive review on the chemical regeneration of biochar adsorbent for sustainable wastewater treatment. NPJ Clean Water 5:29

Alvarez-Chavez BJ, Godbout S, Palacios-Rios JH et al (2019) Physical, chemical, thermal and biological pre-treatment technologies in fast pyrolysis to maximize bio-oil quality: a critical review. Biomass Bioenergy 128:105333

Amalina F, Syukor Abd Razak A, Krishnan S et al (2022) Advanced techniques in the production of biochar from lignocellulosic biomass and environmental applications. Cleaner Materials 6:100137. https://doi.org/10.1016/j.clema.2022.100137

Amdeha E (2024) Biochar-based nanocomposites for industrial wastewater treatment via adsorption and photocatalytic degradation and the parameters affecting these processes. Biomass Convers Biorefinery 14:23293–23318

Anupama KP (2021) A comprehensive evaluation of inherent properties and applications of nanobiochar prepared from different methods and feedstocks. J Clean Prod 320:128759. https://doi.org/10.1016/j.jclepro.2021.128759

Arabzadeh Nosratabad N, Yan Q, Cai Z, Wan C (2024) Exploring nanomaterial-modified biochar for environmental remediation applications. Heliyon 10:e37123. https://doi.org/10.1016/j.heliyon.2024.e37123

Aziz S, Uzair B, Ali MI et al (2023) Synthesis and characterization of nanobiochar from rice husk biochar for the removal of safranin and malachite green from water. Environ Res 238:116909

Banik C, Koziel JA, Bonds D et al (2021) Comparing biochar-swine manure mixture to conventional manure impact on soil nutrient availability and plant uptake—a greenhouse study. Land 10:372

Bandara T, Krohn C, Jin J et al (2022) The effects of biochar aging on rhizosphere microbial communities in cadmium-contaminated acid soil. Chemosphere 303:135153. https://doi.org/10.1016/j.chemosphere.2022.135153

Berazneva J, Woolf D, Lee DR (2021) Local lignocellulosic biofuel and biochar co-production in sub-Saharan Africa: the role of feedstock provision in economic viability. Energy Econ 93:105031. https://doi.org/10.1016/j.eneco.2020.105031

Bhattacharya SS, Kim K-H, Das S et al (2016) A review on the role of organic inputs in maintaining the soil carbon pool of the terrestrial ecosystem. J Environ Manag 167:214–227. https://doi.org/10.1016/j.jenvman.2015.09.042

Bhatti UA, Bhatti MA, Tang H et al (2024) Global production patterns: understanding the relationship between greenhouse gas emissions, agriculture greening and climate variability. Environ Res 245:118049. https://doi.org/10.1016/j.envres.2023.118049

Bidi H, Fallah H, Niknejad Y, Tari DB (2021) Iron oxide nanoparticles alleviate arsenic phytotoxicity in rice by improving iron uptake, oxidative stress tolerance and diminishing arsenic accumulation. Plant Physiol Biochem 163:348–357

Bolan N, Hoang SA, Beiyuan J et al (2022) Multifunctional applications of biochar beyond carbon storage. Int Mater Rev 67:150–200

Bolan S, Hou D, Wang L et al (2023) The potential of biochar as a microbial carrier for agricultural and environmental applications. Sci Total Environ 886:163968. https://doi.org/10.1016/j.scitotenv.2023.163968

Brar B, Saharan BS, Seth CS et al (2024) Nanobiochar: soil and plant interactions and their implications for sustainable agriculture. Biocatal Agric Biotechnol 57:103077. https://doi.org/10.1016/j.bcab.2024.103077

Campos P, De la Rosa JM (2020) Assessing the effects of biochar on the immobilization of trace elements and plant development in a naturally contaminated soil. Sustainability 12:6025

Chagas JKM, Figueiredo CCD, Ramos MLG (2022) Biochar increases soil carbon pools: evidence from a global meta-analysis. J Environ Manag 305:114403. https://doi.org/10.1016/j.jenvman.2021.114403

Chandi K, Udomkun P, Boonupara T, Kaewlom P (2024) Enhancing soil health, microbial count, and hydrophilic methomyl and hydrophobic lambda-cyhalothrin remediation with biochar and nano-biochar. Sci Rep 14:19551. https://doi.org/10.1038/s41598-024-70515-2

Chaubey AK, Pratap T, Preetiva B et al (2024) Definitive review of nanobiochar. ACS Omega acsomega.3c07804:12331. https://doi.org/10.1021/acsomega.3c07804

Chaukura N, Murimba EC, Gwenzi W (2017) Synthesis, characterisation and methyl orange adsorption capacity of ferric oxide–biochar nano-composites derived from pulp and paper sludge. Appl Water Sci 7:2175–2186

Chausali N, Saxena J, Prasad R (2021) Nanobiochar and biochar based nanocomposites: advances and applications. J Agric Food Res 5:100191

Cheng H, Xing D, Twagirayezu G et al (2024) Effects of field-aging on the impact of biochar on herbicide fate and microbial community structure in the soil environment. Chemosphere 348:140682

Chew J, Zhu L, Nielsen S et al (2020) Biochar-based fertilizer: supercharging root membrane potential and biomass yield of rice. Sci Total Environ 713:136431. https://doi.org/10.1016/j.scitotenv.2019.136431

Das SK, Ghosh GK, Avasthe R, Sinha K (2021) Compositional heterogeneity of different biochar: effect of pyrolysis temperature and feedstocks. J Environ Manag 278:111501

Das SK, Ghosh GKR, Choudhury BU et al (2024) Developing biochar and organic nutrient packages/technology as soil policy for enhancing yield and nutrient uptake in maize-black gram cropping system to maintain soil health. Biomass Conv Bioref 14:2515–2527. https://doi.org/10.1007/s13399-022-02300-y

de Jesus Duarte S, Glaser B, Paiva de Lima R, Pelegrino Cerri CE (2019) Chemical, physical, and hydraulic properties as affected by one year of Miscanthus biochar interaction with Sandy and Loamy tropical soils. Soil Systems 3:24

Elbehiry F, Darweesh M, Al-Anany FS et al (2022) Using biochar and nanobiochar of water hyacinth and black tea waste in metals removal from aqueous solutions. Sustainability 14:10118

El-Gamal EH, Rashad M, Saleh ME et al (2023) Potential bioremediation of lead and phenol by sunflower seed husk and rice straw-based biochar hybridized with bacterial consortium: a kinetic study. Sci Rep 13:21901

Enaime G, Lübken M (2021) Agricultural waste-based biochar for agronomic applications. Appl Sci 11:8914

Faizan M, Alam P, Kumari A et al (2024) Unraveling the nano-biochar mediated regulation of heavy metal stress tolerance for sustaining plant health. Plant Stress 14:100615. https://doi.org/10.1016/j.stress.2024.100615

Fan Q, Sun J, Chu L et al (2018) Effects of chemical oxidation on surface oxygen-containing functional groups and adsorption behavior of biochar. Chemosphere 207:33–40

Fan Z, Zhang Q, Li M et al (2020) Removal behavior and mechanisms of cd(II) by a novel MnS loaded functional biochar: influence of oxygenation. J Clean Prod 256:120672. https://doi.org/10.1016/j.jclepro.2020.120672
Feng Z, Fan Z, Song H et al (2021) Biochar induced changes of soil dissolved organic matter: the release and adsorption of dissolved organic matter by biochar and soil. Sci Total Environ 783: 147091. https://doi.org/10.1016/j.scitotenv.2021.147091
Feng X, Xia X, Chen S et al (2022) Amendment of crop residue in different forms shifted micropore system structure and potential functionality of macroaggregates while changed their mass proportion and carbon storage of paddy topsoil. Geoderma 409:115643
Franzluebbers A (2024) Root-zone enrichment of soil organic carbon and nitrogen under grazing and other land uses in a humid-temperate region. Grass Forage Sci 79:265–280. https://doi.org/10.1111/gfs.12665
Gole P, Raut K, Kandasubramanian B (2024) Polymer-based biochar materials for environmental remediation: a review. Hybrid Advances 6:100267. https://doi.org/10.1016/j.hybadv.2024.100267
Gonzaga MIS, Mackowiak C, De Almeida AQ et al (2018) Positive and negative effects of biochar from coconut husks, orange bagasse and pine wood chips on maize (Zea mays L.) growth and nutrition. Catena 162:414–420. https://doi.org/10.1016/j.catena.2017.10.018
Goswami L, Kushwaha A, Singh A et al (2022) Nano-biochar as a sustainable catalyst for anaerobic digestion: a synergetic closed-loop approach. Catalysts 12:186. https://doi.org/10.3390/catal12020186
Griffin G, Ward L, Madapusi S et al (2022) A study of chemical pre-treatment and pyrolysis operating conditions to enhance biochar production from rice straw. J Anal Appl Pyrolysis 163:105455
Gu YY, Zhang HY, Liang XY et al (2022) Impact of biochar and bioorganic fertilizer on rhizosphere bacteria in saline–alkali soil. Microorganisms 10:2310. https://doi.org/10.3390/microorganisms10122310
Guo F, Bao L, Wang H et al (2020) A simple method for the synthesis of biochar nanodots using hydrothermal reactor. MethodsX 7:101022. https://doi.org/10.1016/j.mex.2020.101022
Gwenzi W, Chaukura N, Mukome FN et al (2015) Biochar production and applications in sub-Saharan Africa: opportunities, constraints, risks and uncertainties. J Environ Manag 150:250–261
Hachem K, Ansari MJ, Saleh RO et al (2022) Methods of chemical synthesis in the synthesis of nanomaterial and nanoparticles by the chemical deposition method: a review. BioNanoScience 12:1032–1057
Heinrich T, Park H, Orozco R et al (2023) Biochar production from late-harvest grass–challenges and potential for farm-scale implementation. Sustain Prod Consum 37:256–267
Hernandez-Soriano MC, Kerré B, Kopittke PM et al (2016) Biochar affects carbon composition and stability in soil: a combined spectroscopy-microscopy study. Sci Rep 6:25127. https://doi.org/10.1038/srep25127
Herrera Paredes S, Lebeis SL (2016) Giving back to the community: microbial mechanisms of plant–soil interactions. Funct Ecol 30:1043–1052. https://doi.org/10.1111/1365-2435.12684
Hu Y, Cao Y, Ma C, Yan W (2024) Nano-biochar as a potential amendment for metal(loid) remediation: implications for soil quality improvement and stress alleviation. J Environ Manag 351:119658. https://doi.org/10.1016/j.jenvman.2023.119658
Hublikar LV, Shilar FA, Suliphuldevara Mathada B, Ganachari SV (2024) A comprehensive investigation of green solutions for sustainable wastewater remediation: a review. J Mol Liq 400:124532. https://doi.org/10.1016/j.molliq.2024.124532
Hussain R, Garg A, Ravi K (2020) Soil-biochar-plant interaction: differences from the perspective of engineered and agricultural soils. Bull Eng Geol Environ 79:4461–4481
Igalavithana AD, You S, Zhang L et al (2022) Progress, barriers, and prospects for achieving a "hydrogen society" and opportunities for biochar technology. ACS EST Eng 2:1987–2001. https://doi.org/10.1021/acsestengg.1c00510

Ighalo JO, Conradie J, Ohoro CR et al (2023) Biochar from coconut residues: An overview of production, properties, and applications. Ind Crop Prod 204:117300. https://doi.org/10.1016/j.indcrop.2023.117300
Ighalo JO, Ohoro CR, Ojukwu VE et al (2024) Biochar for ameliorating soil fertility and microbial diversity: from production to action of the black gold. iScience 28:111524. https://doi.org/10.1016/j.isci.2024.111524
Javaid SF, Dai M, Wu Y et al (2024) Production of biochar by slow and solar-biomass pyrolysis: focus on the output configuration assessment, adaptability, and barriers to market penetration. Arab J Sci Eng 49:7731–7750. https://doi.org/10.1007/s13369-023-08549-3
Jia A, Song X, Li S et al (2024) Biochar enhances soil hydrological function by improving the pore structure of saline soil. Agric Water Manag 306:109170
Jouiad M, Al-Nofeli N, Khalifa N et al (2015) Characteristics of slow pyrolysis biochars produced from Rhodes grass and fronds of edible date palm. J Anal Appl Pyrolysis 111:183–190
Ke PJ, Miki T (2015) Incorporating the soil environment and microbial community into plant competition theory. Front Microbiol 6:1066. https://doi.org/10.3389/fmicb.2015.01066
KE MR, Anil A, Maria HJ et al (2024) Recent advances in the application of engineered biochar for wastewater treatment. Catalytic applications of biochar for environmental remediation: a green approach towards environment restoration (Vol 1) 45–69
Khaliq H, Anwar S, Shafiq F et al (2023) Interactive effects of soil and foliar-applied Nanobiochar on growth, metabolites, and nutrient composition in Daucus carota. J Plant Growth Regul 42:3715–3729. https://doi.org/10.1007/s00344-022-10832-w
Khan HA, Naqvi SR, Mehran MT et al (2021) A performance evaluation study of nano-biochar as a potential slow-release nano-fertilizer from wheat straw residue for sustainable agriculture. Chemosphere 285:131382. https://doi.org/10.1016/j.chemosphere.2021.131382
Khare P (2021) A comprehensive evaluation of inherent properties and applications of nano-biochar prepared from different methods and feedstocks. J Clean Prod 320:128759
Koul B, Taak P (2018) Biotechnological Strategies for Effective Remediation of Polluted Soils. https://doi.org/10.1007/978-981-13-2420-8
Kozłowski M, Igwegbe CA, Tarczyńska A, Białowiec A (2023) Revealing the adverse impact of additive carbon material on microorganisms and its implications for biogas yields: a critical review. Materials 16:7250. https://doi.org/10.3390/ma16237250
Kumar A, Joseph S, Graber ER, Taherymoosavi S et al (2021) Fertilizing behavior of extract of organomineral-activated biochar: low-dose foliar application for promoting lettuce growth. Chem Biol Technol Agric 8:21. https://doi.org/10.1186/s40538-021-00222-x
Kwoczynski Z, Čmelík J (2021) Characterization of biomass wastes and its possibility of agriculture utilization due to biochar production by torrefaction process. J Clean Prod 280:124302
Lal R, Negassa W, Lorenz K (2015) Carbon sequestration in soil. Curr Opin Environ Sustain 15:79–86. https://doi.org/10.1016/j.cosust.2015.09.002
Leng L, Xiong Q, Yang L et al (2021) An overview on engineering the surface area and porosity of biochar. Sci Total Environ 763:144204
Li J, Liang N, Jin X et al (2017) The role of ash content on bisphenol A sorption to biochars derived from different agricultural wastes. Chemosphere 171:66–73
Li C, Wang Z, Xu Y et al (2023) Analysis of the effect of modified biochar on saline–alkali soil remediation and crop growth. Sustainability 15:5593. https://doi.org/10.3390/su15065593
Li B, Liu Y, Tong WK et al (2024) Effects of cellulase treatment on properties of lignocellulose-based biochar. Bioresour Technol 413:131452
Liao H, Li Y, Yao H (2019) Biochar amendment stimulates utilization of plant-derived carbon by soil bacteria in an intercropping system. Front Microbiol 10:1361. https://doi.org/10.3389/fmicb.2019.01361
Liu S, Zhang Y, Zong Y et al (2015) Response of soil carbon dioxide fluxes, soil organic carbon and microbial biomass carbon to biochar amendment: a meta-analysis. GCB Bioenergy 8:392–406. https://doi.org/10.1111/gcbb.12265
Liu G, Zheng H, Jiang Z et al (2018) Formation and physicochemical characteristics of nano biochar: insight into chemical and colloidal stability. Environ Sci Technol 52:10369–10379

Liu Z, Tian F, An X et al (2021) Microwave co-pyrolysis of biomass, phosphorus, and magnesium for the preparation of biochar-based fertilizer: fast synthesis, regulable structure, and graded-release. J Environ Chem Eng 9:106456
Liu X, Zhang J, Wang Q et al (2022) Improvement of photosynthesis by biochar and vermicompost to enhance tomato (*Solanum lycopersicum* L.) yield under greenhouse conditions. Plants 11:3214. https://doi.org/10.3390/plants11233214
Luo M, Jiang X, Liu Y et al (2023) Enhanced adsorption complexation of biochar by nitrogen-containing functional groups. J Environ Chem Eng 11:111194
Lyu H, Gao B, He F et al (2018) Effects of ball milling on the physicochemical and sorptive properties of biochar: experimental observations and governing mechanisms. Environ Pollut 233:54–63
Lyu H, Zhang H, Chu M et al (2022) Biochar affects greenhouse gas emissions in various environments: a critical review. Land Degrad Dev 33:3327–3342. https://doi.org/10.1002/ldr.4405
Malhi GS, Kaur M, Kaushik P (2021) Impact of climate change on agriculture and its mitigation strategies: a Review. Sustainability 13:1318. https://doi.org/10.3390/su13031318
Mahmoud E (2022) Effect of nanomaterials on soil quality and yield of canola (*Brassica napus*) grown in heavy clayey soils Asmaa El-shahawy a. Egypt J Soil Sci 62:361–371. https://doi.org/10.21608/EJSS.2022.152942.1522
Manikandan A, Subramanian K, Pandian K (2013) Effect of high energy ball milling on particle size and surface area of adsorbents for efficient loading of fertilizer. Magnesium 3:10–42
Manolikaki II, Mangolis A, Diamadopoulos E (2016) The impact of biochars prepared from agricultural residues on phosphorus release and availability in two fertile soils. J Environ Manage 181:536–543. https://doi.org/10.1016/j.jenvman.2016.07.012
Mazarji M, Bayero MT, Minkina T et al (2023) Nanomaterials in biochar: Review of their effectiveness in remediating heavy metal-contaminated soils. Sci Total Environ 880:163330
Mohan PR (2024) Effective acrylamide adsorption in aqueous environments using maize straw nanobiochar (MNBC). Nanotechnology for. Environ Eng 9:1–15
Mou M (2024) Innovation potential of wood-based biochar for soil improvement in forestry
Muhammad Mehmood H, Yasin Ashraf M, Iqra Almas H et al (2024) Synergistic effects of soil and foliar nano-biochar on growth, nitrogen metabolism and mineral uptake in wheat varieties. J King Saud Univ—Science 36:103392. https://doi.org/10.1016/j.jksus.2024.103392
Neeli ST, Ramsurn H, Ng CY et al (2020) Removal of Cr (VI), As (V), Cu (II), and Pb (II) using cellulose biochar supported iron nanoparticles: A kinetic and mechanistic study. J Environ Chem Eng 8:103886
Nishshankage K, Fernandez AB, Pallewatta S et al (2024) Current trends in antimicrobial activities of carbon nanostructures: potentiality and status of nanobiochar in comparison to carbon dots. Biochar 6:2. https://doi.org/10.1007/s42773-023-00282-2
Omidvar-Hosseini F, Moeinpour F (2016) Removal of Pb (II) from aqueous solutions using Acacia Nilotica seed shell ash supported Ni0. 5Zn0. 5Fe2O4 magnetic nanoparticles. J Water Reuse Desal 6:562–573
Omiri J, Snoussi Y, Bhakta A, et al (2022) Citric acid-assisted preparation of biochar loaded with copper/nickel bimetallic nanoparticles
Osman AI, Zhang Y, Lai ZY et al (2023) Machine learning and computational chemistry to improve biochar fertilizers: a review. Environ Chem Lett 21:3159–3244. https://doi.org/10.1007/s10311-023-01631-0
Palansooriya KN, Wong JTF, Hashimoto Y et al (2019) Response of microbial communities to biochar-amended soils: a critical review. Biochar 1:3–22
Patel MR, Panwar NL (2023) Biochar from agricultural crop residues: environmental, production, and life cycle assessment overview. Resources, Conservation & Recycling. Advances 200173
Pathak HK, Seth CS, Chauhan PK et al (2024) Recent advancement of nano-biochar for the remediation of heavy metals and emerging contaminants: mechanism, adsorption kinetic model, plant growth and development. Environ Res 255:119136. https://doi.org/10.1016/j.envres.2024.119136

Pérez-Hernández H, Pérez-Moreno A, Sarabia-Castillo CR et al (2021) Ecological drawbacks of nanomaterials produced on an industrial scale: collateral effect on human and environmental health. Water Air Soil Pollut 232:435. https://doi.org/10.1007/s11270-021-05370-2

Piash MI, Iwabuchi K, Itoh T (2022) Synthesizing biochar-based fertilizer with sustained phosphorus and potassium release: co-pyrolysis of nutrient-rich chicken manure and Ca-bentonite. Sci Total Environ 822:153509

Pierson D, Anderson N, Brewen J et al (2024) Beyond the basics: a perspective on barriers and opportunities for scaling up biochar production from forest slash. Biochar 6:1. https://doi.org/10.1007/s42773-023-00290-2

Purbalisa W, Zulaehahl, Paputri DMW (2021) Application of remediation treatment to reduce lead in the soil of shallot cropping. IOP Conf Ser: Earth Environ Sci 648(1):012063. IOP Publishing. https://doi.org/10.1088/1755-1315/648/1/012063

Qian H, Ke M, Qu Q et al (2018) Ecological effects of single-walled carbon nanotubes on soil microbial communities and soil fertility. Bull Environ Contam Toxicol 101:536–542. https://doi.org/10.1007/s00128-018-2437-y

Rajput VD, Minkina T, Ahmed B et al (2022) Nano-biochar: a novel solution for sustainable agriculture and environmental remediation. Environ Res 210:112891. https://doi.org/10.1016/j.envres.2022.112891

Rajput P, Kumar P, Priya A et al (2024) Nanomaterials and biochar mediated remediation of emerging contaminants. Sci Total Environ 170064:170064

Ramadan MM, Asran-Amal, Abd-Elsalam KA (2020) Micro/nano biochar for sustainable plant health: present status and future prospects. Carbon Nanomaterials for Agri-Food and Environmental Applications. Elsevier, In, pp 323–357. https://doi.org/10.1016/b978-0-12-819786-8.00016-5

Ramanayaka S, Tsang DC, Hou D et al (2020) Green synthesis of graphitic nanobiochar for the removal of emerging contaminants in aqueous media. Sci Total Environ 706:135725

Rashid MI, Shah GA, Iqbal Z et al (2023) Nanobiochar reduces ammonia emission, increases nutrient mineralization from vermicompost, and improves maize productivity. J Clean Prod 414:137694. https://doi.org/10.1016/j.jclepro.2023.137694

Rizwan M, Noureen S, Ali S et al (2019) Influence of biochar amendment and foliar application of iron oxide nanoparticles on growth, photosynthesis, and cadmium accumulation in rice biomass. J Soils Sediments 19:3749–3759. https://doi.org/10.1007/s11368-019-02327-1

Rizwan M, Lin Q, Chen X et al (2020) Synthesis, characterization and application of magnetic and acid modified biochars following alkaline pretreatment of rice and cotton straws. Sci Total Environ 714:136532

Sadaka S, Sharara MA, Ashworth A et al (2014) Characterization of biochar from switchgrass carbonization. Energies 7:548–567

Saini AK (2024) Effect of nitrogen loaded Nanobiochar and mineral fertilizer on soil properties and performance of basmati Rice

Sammar Raza MA, Shah AN, Shahid MA et al (2023) Nano-biochar enhances wheat crop productivity by vindicating the effects of drought: in relation to physiological and phenological stages. ACS Omega 8:37808–37819. https://doi.org/10.1021/acsomega.3c01629

Sani MNH, Amin M, Siddique AB et al (2023a) Waste-derived nanobiochar: a new avenue towards sustainable agriculture, environment, and circular bioeconomy. Sci Total Environ 905:166881. https://doi.org/10.1016/j.scitotenv.2023.166881

Sani MNH, Amin M, Siddique AB et al (2023b) Waste-derived nanobiochar: a new avenue towards sustainable agriculture, environment, and circular bioeconomy. Sci Total Environ 16:166881

Sarma HH, Borah SK, Das O et al (2024) Nano biochar: properties, preparation, key features and its impact on soil and crop ecosystem. J Adv Biol Biotechnol 27:66–78. https://doi.org/10.9734/jabb/2024/v27i7967

Sarma H, Shyam S, Zhang M, Guerriero G (2024a) Nano-biochar interactions within the rhizosphere and their implications for plant-soil dynamics. Soil Environ Health 10:100095

Sarma H, Shyam S, Zhang M, Guerriero G (2024b) Nano-biochar interactions with contaminants in the rhizosphere and their implications for plant-soil dynamics. Soil Environ Health 2:100095. https://doi.org/10.1016/j.seh.2024.100095

Saxena M, Maity S, Sarkar S (2014) Carbon nanoparticles in 'biochar' boost wheat (*Triticum aestivum*) plant growth. RSC Adv 4:39948–39954. https://doi.org/10.1039/c4ra06535b

Schubert S, Qadir M (2024) Amelioration and management strategies for salt-affected soils. In: Soil salinity and salt resistance of crop plants. Springer, pp 125–150

Shafiq F, Anwar S, Zhang L, Ashraf M (2023) Nano-biochar: properties and prospects for sustainable agriculture. Land Degrad Dev 34:2445–2463. https://doi.org/10.1002/ldr.4620

Shakoor A, Arif MS, Shahzad SM et al (2021) Does biochar accelerate the mitigation of greenhouse gaseous emissions from agricultural soil?—a global meta-analysis. Environ Res 202:111789. https://doi.org/10.1016/j.envres.2021.111789

Shen Y, Tang H, Wu W et al (2020) Role of nano-biochar in attenuating the allelopathic effect from Imperata cylindrica on rice seedlings. Environ Sci Nano 7:116–126. https://doi.org/10.1039/C9EN00828D

Shoudho KN, Khan TH, Ara UR et al (2024) Biochar in global carbon cycle: towards sustainable development goals. Curr Res Green Sustainable Chem 8:100409. https://doi.org/10.1016/j.crgsc.2024.100409

Soliman M, Shawky A (2024) Assessing the efficacy of agro-waste biochar-derived nanoparticles for purification of municipal wastewater, agricultural drainage water and industrial effluents. Egypt J Soil Sci 64

Song B, Chen M, Ye S et al (2019) Effects of multi-walled carbon nanotubes on metabolic function of the microbial community in riverine sediment contaminated with phenanthrene. Carbon 144:1–7. https://doi.org/10.1016/j.carbon.2018.12.016

Souza LFT, Hirmas DR, Sullivan PL et al (2023) Root distributions, precipitation, and soil structure converge to govern soil organic carbon depth distributions. Geoderma 437:116569. https://doi.org/10.1016/j.geoderma.2023.116569

Subagio A, Umiati N, Kholil M et al (2021) Effect of variations in milling speed with high energy milling treatment on surface area of biochar material. IOP Publishing 1943:012024

Sultan H, Li Y, Ahmed W et al (2024) Biochar and nano biochar: enhancing salt resilience in plants and soil while mitigating greenhouse gas emissions: a comprehensive review. J Environ Manag 355:120448. https://doi.org/10.1016/j.jenvman.2024.120448

Tomczyk A, Sokołowska Z, Boguta P (2020) Biochar physicochemical properties: pyrolysis temperature and feedstock kind effects. Rev Environ Sci Biotechnol 19:191–215

Tomczyk A, Kondracki B, Szewczuk-Karpisz K (2023) Chemical modification of biochars as a method to improve its surface properties and efficiency in removing xenobiotics from aqueous media. Chemosphere 312:137238

Tourajzadeh O, Piri H, Naserin A, Cahri MM (2024a) Effect of Nano biochar addition and deficit irrigation on growth, physiology and water productivity of quinoa plants under salinity conditions. Environ Exp Bot 217:105564. https://doi.org/10.1016/j.envexpbot.2023.105564

Tourajzadeh O, Piri H, Naserin A, Mahdi Cahri M (2024b) Effect of nano biochar addition and deficit irrigation on growth, physiology and water productivity of quinoa plants under salinity conditions. Environ Exp Bot 217:105564

Ullah MS, Malekian R, Randhawa GS et al (2024) The potential of biochar incorporation into agricultural soils to promote sustainable agriculture: insights from soil health, crop productivity, greenhouse gas emission mitigation and feasibility perspectives—a critical review. Rev Environ Sci Biotechnol 23:1–26

Wallace CA, Afzal MT, Saha GC (2019) Effect of feedstock and microwave pyrolysis temperature on physio-chemical and nano-scale mechanical properties of biochar. Bioresour Bioprocess 6:33. https://doi.org/10.1186/s40643-019-0268-2

Wang W, Wen C, Li C et al (2019) Emission reduction of particulate matter from the combustion of biochar via thermal pre-treatment of torrefaction, slow pyrolysis or hydrothermal carbonisation and its co-combustion with pulverized coal. Fuel 240:278–288

Wang A, Ma Y, Zhao D (2024a) Pore engineering of porous materials: effects and applications. ACS Nano 18:22829–22854

Wang L, Li G, Chen X et al (2024b) Extraction strategies for lignin, cellulose, and hemicellulose to obtain valuable products from biomass. Adv Compos Hybrid Mater 7:219

Xu S, Chen J, Peng H et al (2021) Effect of biomass type and pyrolysis temperature on nitrogen in biochar, and the comparison with hydrochar. Fuel 291:120128

Yaashikaa PR, Kumar PS, Varjani S, Saravanan A (2020) A critical review on the biochar production techniques, characterization, stability and applications for circular bioeconomy. Biotechnol Rep (Amst) 28:e00570. https://doi.org/10.1016/j.btre.2020.e00570

Yang Y, Zhou B, Hu Z, Lin H (2020) The effects of nano-biochar on maize growth in northern Shaanxi Province on the Loess Plateau. Appl Ecol Environ Res 18(2). https://doi.org/10.15666/aeer/1802_28632877

Yousaf W, Shah AA, Afzal MB et al (2024) Supplementation of nano-biochar improved growth and physiological attributes in wheat seedlings exposed to salt stress through enhanced activity of hydrolysing and nitrogen metabolic enzymes and regulation of crucial metabolites. S Afr J Bot 167:500–508. https://doi.org/10.1016/j.sajb.2024.02.039

Yue L, Lian F, Han Y et al (2019) The effect of biochar nanoparticles on rice plant growth and the uptake of heavy metals: implications for agronomic benefits and potential risk. Sci Total Environ 656:9–18. https://doi.org/10.1016/j.scitotenv.2018.11.364

Zafar-ul-Hye M, Tahzeeb-ul-Hassan M, Wahid A et al (2021) Compost mixed fruits and vegetable waste biochar with ACC deaminase rhizobacteria can minimize lead stress in mint plants. Sci Rep 11:6606. https://doi.org/10.1038/s41598-021-86082-9

Zhang J, Shen JL (2022) Effects of biochar on soil microbial diversity and community structure in clay soil. Ann Microbiol 72:35. https://doi.org/10.1186/s13213-022-01689-1

Zheng H, Zhang C, Liu B et al (2020) Biochar for water and soil remediation: production, characterization, and application. In: Jiang G, Li X (eds) A new paradigm for environmental chemistry and toxicology. Springer Singapore, Singapore, pp 153–196

Zheng X, Xu W, Dong J et al (2022) The effects of biochar and its applications in the microbial remediation of contaminated soil: a review. J Hazard Mater 438:129557

Zheng T, Ouyang S, Zhou Q (2023) Synthesis, characterization, safety design, and application of NPs@BC for contaminated soil remediation and sustainable agriculture. Biochar 5:5. https://doi.org/10.1007/s42773-022-00198-3

Zhou J, Chen H, Thring RW, Arocena JM (2019) Chemical pretreatment of rice straw biochar: effect on biochar properties and hexavalent chromium adsorption. Int J Environ Res 13:91–105

Zhou C, Song X, Wang Y et al (2022) The sorption and short-term immobilization of lead and cadmium by nano-hydroxyapatite/biochar in aqueous solution and soil. Chemosphere 286:131810

Nanotechnology and Agribusiness: Part-II

9

Eshani L. Lokuge and Gayan Priyadarshana

Abstract

Nanotechnology is a rapidly growing field that offers positive opportunities for enhancing agricultural productivity and sustainability within agribusiness. This technology gives significant improvement in nanoscale to create novel materials and devices that can be applied through the different agricultural domains. The usage of nanotechnology in agribusiness highlights several key areas, including plant protection, nutrient delivery, soil enhancement, plant breeding, food safety and quality, biopesticides, and precision agriculture. In plant protection, nanomaterials can improve the efficiency and environmental safety of pesticides by enabling targeted delivery systems and controlled release, thereby minimizing lots of disadvantages. Additionally, nanotechnology enhanced nutrient delivery systems that improve the bioavailability of fertilizers, while increasing the harvest of the fields and reduce the wastage. Moreover, the application of nanotechnology to soil management system, where nanoscale materials can improve the structure of soil, water retention capacity, nutrient availability and protect the health level of soil across microbial activities. In the context of food safety, nanotechnology can be used for the development of smart packaging materials where nano-sensors that can detect pesticides, pathogens in agri-products, ensuring food safety and quality. Despite its advantages, use of nanotechnology in agribusiness faces challenges, and therefore there is a need for further research and development to address potential environmental and health risks. Strategic

E. L. Lokuge · G. Priyadarshana (✉)
Department of Materials and Mechanical Technology, Faculty of Technology, University of Sri Jayewardenepura, Homagama, Sri Lanka
e-mail: gayanp@sjp.ac.lk

S. S. Purewal et al. (eds.), *Agri-Nanotechnology: Innovations for Sustainable Agriculture and Environmental Restoration*,
https://doi.org/10.1007/978-981-96-9756-4_9

implementation of nanotechnology in agribusiness includes integrating precision farming systems and developing eco-friendly alternatives for resource management, aligning with the industry's sustainability goals. Looking ahead, the future outlook for nanotechnology in agribusiness is promising, with potential breakthroughs in crop improvement, environmental protection, and addressing global food security challenges through sustainable farming innovations.

Keywords

Nanotechnology · Agribusiness · Sustainability · Precision farming

9.1 Introduction

Nanotechnology is a multidisciplinary area that involves the manipulation of matters at the nanoscale, normally defined as in the range of 1 to 100 nanometer (nm) in size. Within this range, materials show unique physical and chemical properties that differ significantly from their large-scale counterparts. This remarkable phenomenon is primarily due to the increased surface-to-volume ratio and the quantum effects that become more pronounced at the nanoscale. Nanotechnology as a cutting-edge technology has shown its potential in various fields like solar, electronics, optics, pharmaceutical, and agricultural and food industries (Setyowati Retno Djiwanti 2019). In developing countries, agriculture is a paramount sector in the national economy. Rising agricultural production rates play a crucial role in the improvement of a country's gross domestic product (GDP) (Chhipa 2019).

The use of nanotechnology makes farming more scientific and focused. Computers, remote sensing equipment, and global satellite positioning systems are used in precision farming to measure a variety of factors. Accurate real-time data enabled by nanotechnology is used for monitoring soil conditions, environmental changes, plant diseases, and overall plant health (Ahmad et al. 2020). Scientists are working to address challenges with nanotechnology in order to satisfy food production targets and other obstacles in agriculture, including food security for growing populations, decreased production on cultivable land, low efficiency of agricultural inputs, large areas of land that cannot be farmed, low shelf life of food production, and an increase in pest attacks and plant diseases (Bhagat et al. 2015). Hence, the improper use of chemicals is continuously reducing soil fertility. Plants only use a small percentage of these chemical agri-inputs, and the remaining chemicals are wasted and harm the ecosystem. Unwanted chemicals are caused by the leaching of chemical fertilizers and pesticides into the soil or by their escape into water bodies. By improving these instruments, nanotechnology in agriculture helps reduce the cost of pesticides and fertilizers. Methods based on nanotechnology can enhance the properties of agricultural inputs, such as controlled release, targeted delivery, improved solubility, and extended shelf life, due to which they are more effective and less likely to pollute the environment (McKee and Filser 2016; Mishra et al. 2014). In agribusiness, by applying these materials at a nanoscale, farmers can achieve more precise control over the release and effectiveness of fertilizers and pesticides, leading to better crop yields and reduced resource consumption. Furthermore, by lowering toxicity and boosting stability, green production of nanomaterials expanded the potential of metal and metal oxide nanoparticles in agriculture. Conventional agriculture has evolved into smart agricultural systems through the use of nanosensors for detecting pests and monitoring soil conditions (Chhipa 2019). According to the above study, nanotechnology plays a crucial role in soil health and management. Nanomaterials can be employed for soil remediation and for improving soil quality by increasing nutrient levels and promoting beneficial microbial activities. Their application can also lead to enhanced water retention, which is particularly important in dry regions facing climate change.

Moreover, the effects of nanotechnology go over agricultural yield to include food quality and safety. Advanced packing options that improve food preservation, lower spoilage, and traceability are made possible by nanotechnology. Food product safety and consumer confidence are increased by nanosensors' ability to identify pollutants and pathogens (Biswas et al. 2022; Singh et al. 2023).

Besides the positive impacts of nanotechnology on agribusiness, a few adverse effects have also been documented in terms of toxicity of plants and ecosystems. The use of different kinds of metal nanoparticles, metal-oxide nanoparticles, green-synthesized nanoparticles, and fertilizers or pesticides based on nanoformulations in agriculture, as well as their effects on plant growth and pathogen control, are discussed in this chapter. Additionally, nanoparticles are used as sensors for food safety, insect management, and soil nutrient detection. Agriculture is affected by nanotechnology in both positive and negative ways, but further study is needed to determine the long-term effects.

9.2 Basics of Nanotechnology

Nanomaterials are used in different industries such as catalyst, energy storage, medicine, and agricultural industries due to their unique properties (Bratovcic 2019; Gajanan and Tijare 2018; Khot et al. 2012). Surface effect and quantum effect of nanomaterials make it special rather than other materials (Roduner 2006). Therefore, nanomaterials play significant role in mechanical, thermal, magnetic, electronic, and optical sectors (Khot et al. 2012; Gade et al. 2010; Buzea et al. 2007). These nano-scale materials exhibit unique surface effects compared to micro- and bulk materials due to their high surface area and the large number of particles per unit mass. In nanomaterials, the percentage of atoms at the surface increases (Buzea et al. 2007; Roduner 2006). The physical and chemical characteristics of nanomaterials differ from those of their large-dimension counterpart due to these variations. According to the Gibbs-Thomson equation, this alteration has a direct impact on the nanomaterial's melting temperature (Joudeh and Linke 2022). Reactivity of nanomaterials increase with high surface and surface-to-volume ratios (Buzea et al. 2007). Also, the surface attributes have a major impact on their structure (Ikhmayies 2014). One important thing in surface impact is the nanomaterials' dispersity.

Strongly attractive interactions between particles can cause nanomaterials to aggregate and agglomerate, which affect negatively by reducing the surface area and other properties (Ashraf et al. 2018). This can be prevented by increasing the repulsive forces (Suttiponparnit et al. 2010). Nanomaterials show size-dependent properties in the nanoscale (1–100 nm) range, where quantum effects are involved (Ikhmayies 2014). As a result of this effect, some nanomaterials which don't have magnetic properties in bulk sizes including Pd, Au, and Pt become magnetic in nanoscale (Roduner 2006). Nanomaterials are mainly divided into four basic groups: zero-dimensional NMs, one-dimensional NMs, two-dimensional NMs, and three-dimensional NMs (Kolahalam et al. 2019).

In addition, nano-objects with external dimensions in the 1–100 nm range are considered nanoparticles. NPs have different sizes, shapes, and structures. They may be spherical, cylindrical, conical, tubular, hollow core, and spiral (Ealias and Saravanakumar 2017). NPs can be crystalline or amorphous in nature and can be either loose or agglomerated (Machado et al. 2015). Sometimes NPs can be uniform and consist of layers, including the surface layer, the shell layer, and the core layer (**Khan et al.** 2019). According to their composition, NPs can be divided into three main categories: organic, carbon-based, and inorganic (Ealias and Saravanakumar 2017).

Although there are various strategies for synthesizing nanomaterials, the two main ones are top-down and bottom-up. Lithography (Garg et al. 2020), mechanical or ball milling (Zhuang et al. 2016; Yadav et al. 2012), laser ablation (Zhang et al. 2017), sputtering (Ayyub et al. 2001), electron explosion arc discharge (Zhang et al. 2019), and thermal breakdown (Jain et al. 2020) are a few examples of top-down methods. Chemical vapor deposition (CVD) (Malandrino 2009), sol-gel (Danks et al. 2016), spinning, pyrolysis, and biological synthesis (Baig et al. 2021) are a few examples of bottom-up methods (Joudeh and Linke 2022; Ealias and Saravanakumar 2017). The study of the properties of matter at the nanoscale, with an emphasis on the unique, size-dependent properties of solid-state materials, is known as nanoscience (Ealias and Saravanakumar 2017). The production, engineering, and use of nanoparticles are all included in the topic of research known as nanotechnology. Size-dependent properties greatly affect the surface area, quantum effects, mechanical, thermal, magnetic, electronic, and optical properties, support for catalyst, and microbial activities (Joudeh and Linke 2022). Nanotechnology has garnered increasing attention year by year due to the innovative and promising applications of nanomaterials for the next industrial generation (Ealias and Saravanakumar 2017; Joudeh and Linke 2022).

Nowadays, nanomaterials has become crucial for human progress as a whole. For instance, employing green technology that solely uses nanomaterials is the only way to lower the risk of global warming and climate change in the first place. Furthermore, utilizing nanoparticles is more efficient than utilizing bulk materials. It is expected that nanotechnology, nanoscience, and nanomaterials will be crucial for global development in the future.

9.3 Application of Nanotechnology in Agribusiness

Nanotechnology plays a crucial role in agribusiness, including agriculture, food processing, soil features regulation, water system management, and pathogen detection.

9.3.1 Crop Improvement

Nanotechnology enhances materials at an incredibly small scale between 1 and 100 nm. And it shows significant improvements in agriculture, especially in crop improvement. This technology can improve the uptake of nutrients and pesticides through nano-encapsulation, allowing for controlled release and targeted delivery. It helps in seed coating to improve germination and plant resilience and also soil remediation to ensure the availability of suitable lands for crop plantation. Moreover, nanotechnology can control pests and diseases while reducing the need for chemical inputs, leading to more sustainable farming practices.

9.3.1.1 Nanofarming

Nanotechnology plays a significant role in the development of agriculture. Nanomaterials produced in various sizes and shapes are widely used in environmental science, agriculture, food processing, and medicine, which are always beneficial to agriculture. Nanotechnology can address serious challenges that agriculture faces, including decreased crop yields brought on by biotic and abiotic stresses, such as insufficient nutrients and environmental pollution. In recent years, farming has gained popularity due to the advancements of wireless networking and sensors used to track, evaluate, and regulate agricultural operations. Additionally, they are associated with site-specific crop management that covers a broad range of agricultural pre- and post-production activities, from the field crops to horticultural crops. A notable scientific achievement is the targeted delivery of CRISPR-associated protein, mRNA, and sgRNA for crop genetic modification using recent developments in tissue engineering and engineered nanomaterials. Nanosensors offer numerous opportunities for monitoring environmental stress and boosting plant' ability to fight off illnesses.

These specialized systems, including breeding programs, globalized supply chains, agrochemicals, and new machineries have increased the production of foods. Agro-nanotechnology has been introduced as a creative method for developing sustainable farming methods that can minimize several issues such as pollution, degradation, and climate change that currently farming faces.

9.3.1.2 Nanofertilizers

For crop development, nutrient supplements are important because they enhance crop productivity and soil fertility (Li et al. 2018). Nanofertilizers are a promising option for increasing global food production and promoting sustainable agriculture, as they offer an effective alternative to minimize the chronic problem of eutrophication (Shukla et al. 2019). Nanofertilizers are designed such as to control the nutrient flow according to the crop's need; therefore, nutrient losses can be minimized. 50–70% of nitrogen fertilizers, losses in the filed due to evaporation, leaching, and degrading, this decrease the efficiency of fertilizer and cause to increase the production cost (Yang et al. 2016; Miao et al. 2015). Nano-formulated nitrogenous fertilizers have the ability to release nutrients to the crop directly on time without any nutrient loss. Especially, this nano-formulated fertilizers avoid the contact of nutrients with other parts such as soils, water, air, and microbes (Dwivedi et al. 2016). For example, porous nanomaterials like zeolites, clay, or chitosan can control nitrogen losses while increasing the plant nutrient uptake process (Millán et al. 2008; Abdel-Aziz et al. 2016). Alternatively, pores can improve plant nutrients uptake, or they can give a passage to complex molecular transporters or root exudates by using ion channels, endocytosis, or the developing new pores 94.

Different studies state that NMs are applied as fertilizers in different forms: chitosan-based, nanocapsulate-based, nano-based slow-releasing fertilizers, urea coated with nano-bentonite, nanoengineered metal-based fertilizer, and nanogel-based and polyurethane-based fertilizers (El-Ramady et al. 2023; Yadav et al. 2023).

Agricultural productivity depends significantly on the fertilizers used. Green nanotechnology has brought notable advancements in global agriculture to help meet the growing food demand. Nanoaptamers are a recent innovation that can be used in fertilizer delivery systems. This can be used directly for targeted nutrient deliveries and for increasing the plant nutrient uptake from the soil (Rameshaiah et al. 2015; Kaushal and Wani 2017). Nanoaptamers can be used to deliver Au NPs and liposome NPs to plants by combining fertilizer components and the NPs. This system has high potential to safeguard the aptamer from degradation. The aptamer binds to a specific receptor on the plant cell and allows the release of the fertilizer to the cell (Naz et al. 2021). Carbon nanotubes, a type of nanofertilizer, stimulate crop growth by directly interacting with plant roots and supporting hormone synthesis. CNTs are tiny rolled-up structures of carbon atoms that can enhance soil quality by improving its structure and water-holding capacity, which positively affects overall plant growth. Even low amounts of CNTs help in seed growth, root development, and water movement in a beneficial way without harming the plants. When CNTs mix with soil, they slowly release specific nutrient to plants, thereby controlling the overuse of fertilizers. They also convert soil as a better place to store nutrients and water by reducing erosions, and minimize the amount of traditional fertilizer needed. CNTs are very strong and show easy movements to different parts of the plant including fruits, leaves, and brunches, thereby connecting with plant cells (Yadav et al. 2023; Aslani et al. 2014). Table 9.1 summarizes some studies based on nanofertilizers.

9.3.1.3 Nanopesticides

Nanotechnology is widely applied in plant protection products to enhance crop productivity. Farmers often use large amounts of pesticides, including fungicides, herbicides, and insecticides. But, over 90% of these pesticides do not get into targeted pests and get lost in the environment. This can badly affect the ecosystem and

Table 9.1 Application of nanofertilizers in plant growth

Nanofertilizers	Positive effects	Reference
Ammonium-based zeolites	Improve phosphorous uptake by plants	Dwivedi et al. (2016)
C-based nanomaterials	Minimize the leach of potassium nitrate	Shalaby et al. (2016)
Nanocalcite with nano SiO_2 (4%), MgO (1%), and ferrous (1%)	Increase the uptake of Ca, Mg, Fe, and P with micronutrients	Lateef et al. (2016)
Nanogypsum	Reduce the pH level and Na adsorption ratio	Patle and Sharma (2022)
Nanofertilizers extract from banana peel	Used to growth of peppers, tomatoes, or flowers	Sivarethinamohan and Sujatha (2021)
ZnO NPs	Growth of chickpea	Sivarethinamohan and Sujatha (2021)
TiO_2 NPs	Growth of spinach	Sivarethinamohan and Sujatha (2021)
Au- and S-based nanofertilizer	Growth of grape plants	Sivarethinamohan and Sujatha (2021)
TiO_2 NPs and SiO_2 NPs	Improve the action of nitrate and increase the plant absorption capacity of water and fertilizer	Otles and Sahyar (2018), Singh Sekhon (2014)
Urea-modified hydroxyapatite NPs-encapsulated *Gliricidia sepium* nanocomposites	Slow and sustained release of nitrogen over time at three different pH values	Kottegoda et al. (2011)
CNTs-based superfertilizer	Penetrate potato seeds and affect their germination and enhance plant growth	Khodakovskaya et al. (2009)
CNTs-based fertilizer	Increase the grain yields of rice, maize, soybean, wheat, and some vegetables	Liu et al. (2009)
Hydoxyapatite nanoparticles	Enhanced plant growth and produce essential oils	Chakravarty et al. (2023)
Cu-based nanofertilizers	Can penetrate through the plant cells and deliver sufficient nutrients to the root systems directly	Yadav et al. (2023)

increase the overall cost (Nuruzzaman et al. 2016). To prevent this and protect the plants, it is important to have the right concentration at the right place. In nanofarming, tiny pesticide particles are being used. These engineered nanostructures also have pest-fighting properties (Prasad et al. 2019). The application of a coating containing active pesticide ingredients at the nanoscale is known as nanoencapsulation. The pesticide is known as the core material, while the external coating is referred to as the external phase. This innovative approach gives a solution to pests attacks (Nuruzzaman et al. 2016). In advance, nanopesticides can increase water solubility and bioavailability, protect agrochemicals against environment degradation, and control pathogens, weeds, and insects in plants which prevent plants' growth (Yadav et al. 2020). Pests cause damage to plants and lead to financial losses by reducing the quality and quantity of food (Syafrudin et al. 2021).

Nanopesticides help maintain the availability of active ingredients for a long time period in the soil and plants without losing their potency and efficiency. Moreover, nanopesticides can overcome the issue that traditional pesticides have, including low water solubility and non-target deliveries (Dwivedi et al. 2016). For example, Petosa et al. (2017) revealed that the combination of polymeric nanocapsules with pesticides significantly improved crop yields. The nanoformulations released the pesticide more effectively over time, even when fertilizers were added to the soil. Therefore, this approach offers an effective alternative for delivering pesticides directly to the target site, minimizing runoff (Petosa et al. 2017). They also offer strength increment, better permeability, high stability, and biodegradability. Biodegradability is another important factor in sustainable farming as it allows for controlled release of the active particles and minimizes the required amounts of the ingredients. This is another important approach to minimize harmful chemicals and protect the environment and other plants as well (Khan and Rizvi 2014; Rai and Ingle 2012). Researchers have investigated the antimicrobial properties of different metal NPs, including Cu and Ag, against the plants pathogens (Cioffi et al. 2004). For instance, polymer-based Cu NPs show a significant effect on plants pathogenic fungi. Another study shows the effect of Si-Ag NPs in controlling fungi. Another study demonstrated the effectiveness of silica–silver nanoparticles (Si-Ag NPs) in controlling fungal infections. The nanoparticle-based pesticide was tested on pumpkin leaves infected with pathogens, and the diseases were eliminated within three days of application (Hae-Jun et al. 2006).

Global agricultural waste is significantly high due to pest and disease attacks; in this context, NPs play a crucial role in promoting plant growth. Halloysite nanoclay is an eco-friendly and cost-effective material that can be used as a carrier for pesticides, enabling controlled, prolonged release while reducing environmental impact (Panpatte et al. 2016). Another approach is the application of Si NPs that can penetrate through the outer layer of insects and kill them (Nair et al. 2010). NPs are promising materials for controlling pests and plant pathogens. Therefore, nanomaterials show excellent behaviors toward pesticides, fungicides, and herbicides and encourage sustainable agriculture. Figure 9.1 shows the basic types of nanopesticides.

9.3.1.4 Seed Treatment Technologies

Ensuring food security in developing countries is challenging due to low agricultural productivity, high harvest losses, degradation, and rapid population growth. Nanotechnology plays a crucial role in input management, plant improvement, and seed treatments. It enhances crop and livestock productivity to improve food quality, prevent disease growth and pests, and monitor plant growth. Nanoparticles can speed up seed germination, enhance seeding strength, and improve seed quality through pre-sowing and pre-storage treatments (Shukla et al. 2019).

Crops that can adapt to any climate have to be grown so as to ensure a steady food supply for the growing population (Singh et al. 2021). While genetic engineering and genome editing have been used to create climate-resilient crops, every

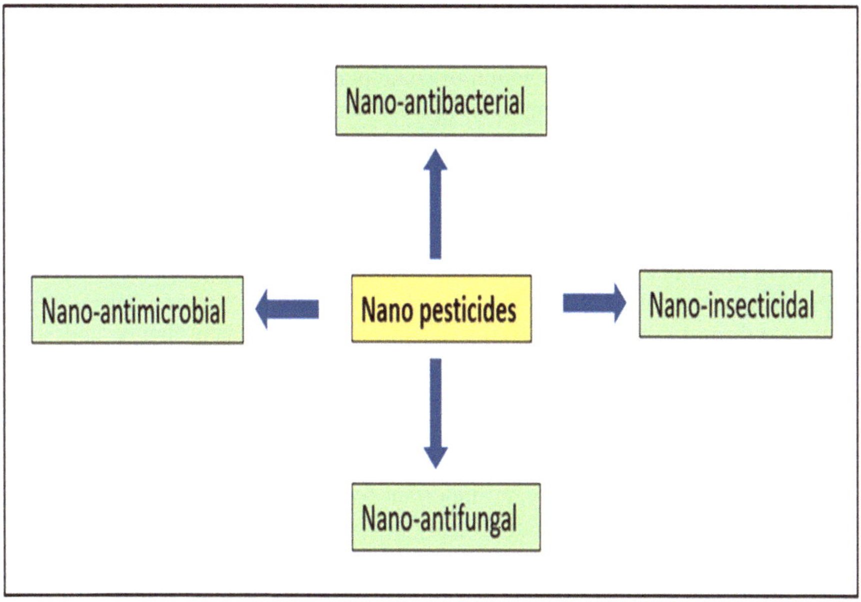

Fig. 9.1 Basic types of nanopesticides

country cannot depend on these techniques due to their low economy and low-scale farming patterns (Arora and Narula 2017). So, research have focused on new methods such as nano-priming of seeds. This method enhances seed germination, strengthens seeding, and encourages plant growth under abnormal conditions, including high temperatures, flooding, and freezing (Zhao et al. 2024). Moreover, nano-priming can improve the immunity system of plants, preventing diseases growth and reducing the need for pesticides. Some researchers state the ability of high-quality seeds and stress-resilient crops to withstand in climate changes (Acevedo et al. 2020). Seed priming-based treatments enhance seed germination, increase resistance toward abiotic and biotic stresses, and increase plants yield and quality (Zhao et al. 2024; Sharma et al. 2021a). Nano-based seed treatments are easy to demonstrate and are cost-effective which make them affordable. NPs due to their small size exhibit unique physicochemical and catalytic properties that promote their easy entrance to seeds (Khan et al. 2020; Zhao et al. 2024).

Seed priming is a hydrated technique under controlled temperature; this technique reduces the germination period and enhances both the germination rate and uniformity. Nano-seed priming mostly use Au NPs, Ag NPs, C NPs, Zn NPs, and TiO_2 as priming agents. Chickpea seeds priming with Fe NPs and Zn NPs showed an increment in seed weights and yield (Shukla et al. 2019). Treatment of watermelon seeds with Ag NPs, onion with gold NPs and Ag NPs, and carrot and beetroot with iron pyrite NPs for 12 hours showed increased germination rates and quality (Shweta et al. 2021). Seed treatment of *Coffea arabica* and *Triticum aestivum* with ZnO NPs showed improved yield, plant growth, and biomass accumulate (Rossi et al. 2019; Du et al. 2019). Similarly, SiO_2, TiO_2, and zeolite application showed enhanced seed germination (Chang et al. 2018; Lu et al. n.d.). Multi-walled CNTs have influenced the seed germination of some plants, including tomato, corn, soybean, barley, wheat, maize, peanut, and garlic (Shang et al. 2019).

9.3.2 Soil Health

Soil is a living system and plays a significant role in human life on earth (Rinot et al. 2019). Not only the air and water quality, but soil health also plays a crucial role in crop growth and sustainable farming. Factors such as climate change and some human activities including, industrial processes, mining, farming practice have harmful impact on the quality of soil. This reduces the crop production rates, food quality, and also the overall health and sustainability of our ecosystems (Hasan et al. 2020). Using nanomaterials in agriculture improves crop production and maintains soil health. Nanofertilizers play an important role in nutrient efficiency, chemical usage, and water pollution, which directly affect the soil health (Bünemann et al. 2018). Nanoscale fertilizers can reduce soil toxicity and harmful effects of conventional chemicals. This also reduces the amount of fertilizer needed. By introducing nanofertilizers into soil management systems, farmers can use less fertilizer (Neme et al. 2021).

9.3.2.1 Nanomaterials for Soil Remediation

Nowadays, nanotechnology has become an innovative step for addressing soil contamination in agricultural areas. The uniqueness of NPs such as high surface area and reactivity makes them an excellent tool for cleaning up polluted soil. Nanotechnology could change traditional soil remediation methods by offering more effective, precise, and eco-friendly solution to agricultural contamination (Prasad et al. 2017). Nanotechnology helps improve plants' ability to clean contaminated soil and water. Soil pollution and degradation are global issues that directly affect agricultural productivity, food security, and human health. Contamination from heavy metals, pesticides, and organic pollutants makes the situation more worse (Cheng et al. 2016; Qingming et al. 2009).

NPs show special advantages over traditional soil cleanup methods because of their tiny size and large surface area. These NPs can help plants absorb and break down pollutants and harmful chemicals in the soil (Dhanapal et al. 2024). Iron NPs show 100% efficiency in removing hexavalent chromium % from soil (Prema et al. 2022). CNTs are another example used to remove heavy metals and other organic-inorganic materials due to their high adsorption capacity (Liang et al. 2022). The in situ immobilization mechanism is a cost-effective and eco-friendly method for cleaning contaminated soil (Ma et al. 2022). The choice of nanomaterials depends on the type of contamination and soil conditions. Carbon- and metal-based NPs are commonly used in this process. Among them, carbon-based NPs such as fullerenes, carbon nanotubes (CNTs), and graphene are especially popular because they can attract and retain organic contaminants, effectively removing harmful substances from the soil (Zhang et al. 2022). Their water-repellent ability leads to easy adsorption of harmful organic substances. For example, CNTs in the soil can slow down polylactic aromatic hydrocarbons movements, making them less available to crop and microorganism to adsorb. On the other hand, presence of oxygen affects the adsorption of contaminants. In CNTs, presence of -OH group increases this property (Qiu et al. 2023). Engineered nanomaterials like iron oxide NPs can adsorb and immobilize heavy metals, such as Cd, Cr, Pb in the soil, reducing their mobility and bioavailability while decreasing their toxicity (Lin et al. 2019).

Nanoscale zerovalent iron (nZVI) particles are another types of NPs that can break down different types of pollutants. In addition, these particles can be customized to target specific contaminants, making them prominent materials to improve soil quality. TiO_2 NPs promote trapping and transforming pollutants, which helps reduce the negative effects of farming on soil health (Ahmad et al. 2020; Dhanapal et al. 2024).Another promising application is nanobiosorbents. Due to the rapid growth of industries, global pollution is increasing at a high rate. Nanobiosorbents are materials made out with renewable and natural sources that can help remove contaminants. Nanosorbents are considered the most efficient materials for cleaning wastewater. Many types of biosorbents and nanoadsorbents are designed to eliminate impurities from water. These materials work together to improve the removal of pollutants from water (Tofan and Suteu 2023; Khan et al. 2022).

9.3.2.2 Enhancing Soil Nutrients

Nanotechnology is widely used to enhance soil nutrients and boost agricultural production through green technology. Nanotechnology has improved the nutrient delivery to plants and enhanced the soil fertility by stimulating soil enzymes. NMs interact with bacteria in the root zones, which can boost plant growth and soil health. Nanofertilizers synthesized with coated NMs positively affect microbial activities in soil and the overall plant growth. Some studies report that engaging NMs with soil can change the soil structure and nutrient level in roots (Tian et al. 2019; Pérez-de-Luque 2017). Different types of NPs, including $TiO_{2,}$ ZnO, MgO, Au, and Ag are used to enhance soil fertility (Kale et al. 2024). Nanotechnology is becoming an important key to enhance crop production. Scientists use CNTs, nanofibers, and quantum dots as biosensors to check soil quality and how fertilizers spread. Vermiculite, nanoclay, and zeolite-like materials are used to make fertilizers work better and boost up crop production in agriculture, especially in sandy soils. Adding these materials to sandy loam soils can diminish the loss of nitrogen and improve the activity of nitrogen fertilizers (Mazloomi and Jalali 2019). Nnaoclays, including montmorillonite, bentonite, kaolinite, and halloysite based structures, can enhance soil quality and agricultural practices (sivarethinamohan and sujatha 2021). Table 9.2 represents the nanomaterials used to enhance soil nutrients.

Nanomaterials can enhance the physical structure of the soil, promoting better aeration, water retention, and root penetration. This can provide healthier soil ecosystems while improving the microbial activities and plant growth. Some NPs can facilitate the interactions between beneficial soil microbes that contribute to nutrient

Table 9.2 Application of nanomaterials for soil nutrient enhancements

Nanofertilizers	Positive effects	Reference
Nano HA	Increase soil content	Jia et al. (2022)
Biochar-based nanofertilizers	Protect water and nutrient level of soil	Saxena et al. (2014)
Zn-based nanofertilizers	Recovers the lost nutrient from soil	Elemike et al. (2019)
Zn-based nanofertilizers	Recovers the lost nutrient from soil	Elemike et al. (2019)
TiO_2-based nanofertilizers	Enhance nitrogen level of soil	Elemike et al. (2019)

availability. Nanomaterials can promote the growth of nitrogen-fixing bacteria, thereby increasing the nitrogen level of soil (Neme et al. 2021).

9.3.3 Pest and Disease Management

Herbicides containing atrazine encapsulated in poly(ε-caprolactone) were tested on *Brassica* species and proved to be successful. Ag-NPs combined with paraquat in a chitosan matrix tested on water hyacinth plant—considered a worst aquatic weed in the world— showed more effectiveness in killing this weed without harming the other plants or the soil (current reading). Moreover, using nanocapsules based on carboxymethyl cellulose to deliver the herbicide Clodinafop-propargyl resulted in improved crop yields compared to traditional herbicides. NPs have antimicrobial properties that help prevent diseases caused by pathogens like fungi, viruses, and bacteria. They can slow down or stop by fighting off harmful microorganisms, which reduces the spread of plant diseases and promotes healthier crops. Most of the NPs can enhance plant disease resistance. They can damage the cell membranes of these pathogens and prevent their entrance to plant tissues. In addition, these NPs enhance the nutrient uptake in plants to make healthier plants that are more resistant to disease due to their stronger cell walls and immune system.

9.3.3.1 Nano-based Biosensors for Disease Detection

Nanobiosensors are sensors made of NPs. Due to their tiny size, they are very effective traditional biosensors. They have better optical and mechanical properties and can be customized. In the field of agriculture, nanobiosensors are important for detecting and monitoring plant health and environmental conditions (Pal 2020; Pudake et al. 2019). They can identify pathogens, pests, and diseases before they widely spread, which helps farmers take actions quickly and reduce crop losses (Li et al. 2020). One of the biggest challenges in growing crops is keeping track of their health. They can respond quickly to disease growing and have many advantages—they are nontoxic, portable, and show high accuracy (Rai et al. 2012). To detect Fenitrothion organophosphate (OP) pesticides, nanobiosensors based on nano TiO_2 and Nafion are used because of their high surface area and excellent biocompatibility (Kumaravel and Chandrasekaran 2011). CNT-based nanobiosensors are used for detecting organophosphate pesticides (Joshi et al. 2005). Au NP-based biosensors are used to detect five types of pesticides: azinphos-methyl, chlorpyrifos, fenamiphos, pirimiphos-methyl, and phosalone. Also, AU, AG, CNTs, silica-coated NPs, and aptasensors are widely used to fabricate nanobiosensors (Miguel-Rojas and Pérez-De-Luque 2023).

Catechol-based sensors are widely used in agriculture for measuring plant health by pH, conductivity, or specific ion concentration in the plant. Moreover, they are used to ensure food safety, detect harmful chemicals, and prevent overfertilization, waterway pollution, and crop damages. This can help detect disease outbreaks and protect crops from damage (Li et al. 2023). One of the most important applications of this biosensor is it can be used to monitor pesticides like organophosphates, carbamates, and pyrethroids (Tsong and Khor 2023). H_2O_2 is another type of biosensor used to control pests, such as aphids, apider mites, whiteflies, and fungus gnats. Not only that but it can also be used for plant disease prevention, including powdery mildew, downy mildew, and black spot. H_2O_2 also is used to improve soil health by killing unnecessary bacteria and fungi while increasing beneficial microorganisms (Erkmen et al. 2022).

Carbon dots (CDs) are a new type of material that can be used as an alternative to quantum dots (QDs). They have low toxicity and high compatibility with living organisms, making them as nanobiosensors to detect microorganisms (Miguel-Rojas and Pérez-De-Luque 2023). QD-based nanobiosensors have been developed to detect *Citrus tristeza* virus, which affects citrus plants (Safarnejad et al. 2017). Also, specially developed QDs are used to deliver natural antibiotic kasugamycin for the effective management of plant diseases (Liang et al. 2019). Ag, Au, and metal oxide NPs like inorganic materials are used to modify nanobiosensors. (Khaledian et al. 2017) In this study, Au NPs combined with single-stranded DNA are used to detect the DNA of *Ralsonia solanacearum* that affects potato plants. Au NPs are effective in finding plant pathogen *Pantoea stewartii* using a special detection method involving enzymes (Zhao et al. 2014). Ag nanorods are also used to detect plant pathogen *Phytophthora ramorum* from real samples using a surface-enhanced Raman spectroscopy (Yüksel et al. 2015). Siddiquee et al. (2014) in their study developed ZnO NPs, chitosan, and DNA-based biosensor to detect *Trichoderma harazianum*, which is a soil-borne fungus that can be used as a biological control agent against various plant pathogens.

9.3.3.2 Targeted Delivery Systems for Agrochemicals

Bio-based NPs can be used to deliver agrochemicals, including pesticides and herbicides. These NPs can hold active ingredients and release them over time, which helps target their application and reduce environmental pollution. This controlled release makes agrochemicals allow for precise dosing, waste reduction, and minimize the risk of other creatures (Fincheira et al. 2023; Shen et al. 2023). They also help lower the chances of resistance caused too much or too little by active ingredients. Fe_3O_4-based magnetic nanocarrier CAN hold both insecticide and herbicides. These chemicals were coated with chitosan to control their release in acidic conditions (Shen et al. 2023).

The main goal of encapsulating fertilizers is to improve how well plants use nutrients and to reduce waste. N, P, K are the widely used polymer-coated controlled-release fertilizers in agriculture (Kumaraswamy et al. 2018). Chitosan is a natural polymer with many benefits for agriculture. It can promote plant growth, can prevent microbe activities, and is useful in nutrients deliveries. Some studies have shown that chitosan NPs loaded with N, P, and K can speed up the growth of wheat (Abdel-Aziz et al. 2016). Chitosan NPs improved gain yield even under drought conditions. Hydroxyapatite is another useful material found in bones and teeth. Also, it can be used as a beneficial way to deliver nutrients to soil. Then, HA have modified with urea to enhance the N efficiency of regular fertilizers. Ability to store nanofertilizers in wood pores, and have checked their nitrogen release. This method enabled a consistent release of nitrogen for over two months (Kottegoda et al. 2011). It illustrated the improvement of the yield is higher than conventional urea fertilizer. This method showed promising results when applied to rice fields, leading to increased crop yields.

The main advantage of nanoencapsulation is that it significantly reduces the amount of pesticide needed. This method involves dissolving the pesticide and releasing it slowly at the target site, helping to prevent pest resistance and reduce soil and water contamination. Another method called nanocomplexation uses cyclodextrins, which are derived from starch and are hydrophilic outside and hydrophobic inside (Sampathkumar et al. 2020). This allows them to form strong and more complex molecules. Some researchers have used cyclodextrins to create complexes with plant compounds like carvacrol and linalool, which act as insecticide (de Oliveira et al. 2018; de Oliveira et al. 2019). Cyclodextrins were mixed with chitosan to make NPs that are easier to apply to crops. (Sandhya et al. 2017). Natural polymer-based nanocarrier systems were developed for the efficient delivery of agrochemicals. For instance, application of chitosan and pectin provided a sustained and controlled carbendazim (fungicide) release, which was effective against fungi like *Fusarium oxysporus* and *Aspergillus parasiticus*. Application of carbendazim-loaded nanocarriers resulted in better germination and root growth in cucumber and corn seeds. (Maruyama et al. 2016) This study states that the toxicity level of imazapyr and imazapic herbicides could be reduced by 50% while improving their effectiveness against weeds in corn and peanut. Moreover, nanocapsules made by combining chitosan and alginate helped slow down the release of water-soluble insecticide acetamiprid. This carrier allows controlled release over 36 hours of chemicals (Kumar et al. 2015). These studies demonstrate that nanoformulations offer significant advantages by using smaller amounts of herbicides and pesticides while maintaining their effectiveness. They facilitate target delivery, precisely deliver these chemicals, and help prevent the harmful impacts on the environment.

9.3.4 Food Safety and Quality

There is a growing trend globally about food safety and security, making it crucial to ensure reliable supply of safe and nutritious food. Nanotechnology offers promising solutions to improve food safety and increase agricultural productivity in sustainable manner. Nanotechnology can be used to easily detect and reduce contamination in the food chain. It can also extend the shelf life of perishable items, which helps reduce food wastage and make better use of resources (Kiran et al. 2024). With the population growth, ensuring food safety becomes more challenging. Also, changes in individual diet patterns affect the availability of foods (yin 2024). To overcome the challenges in food security and quality, we need to improve food production efficiency, enhance distribution systems, decrease food waste, and increase plant yields. Meanwhile, it is important to use environmental friendly practices to protect the environment. Use of nanotechnology to boost crop productivity while enhancing the precision farming is a significant improvement in modern world (Singh et al. 2017).

Nowadays, nanotechnology offers faster, more accurate, and cost-effective ways to detect harmful microorganisms compared to the alternative methods. Especially, they can identify pathogens in different stages of food processing. Nanotechnology also reduces the time needed for incubation and measurements while providing high sensitivity. For example, using magnetic iron NPs with sugar molecules allowed for the isolation of 88% of *E.coli* in just 45 minutes with certified food safety and quality (Duncan 2011; Grumezescu and Holban 2018). Some studies state that combination of two or more nanoparticles can exhibit stronger antimicrobial effect (Nile et al. 2020). For instance, combining Ag NPs, TiO_2, and CNTs showed high effective against *E. coli* and *Bacillus cereus* spores (Gomes Silva et al. 2011). Additionally, mixing Ag NPs with sodium dodecyl sulfate enhanced their effectiveness against *E. coli* and *S. aureus* (Cho et al. 2005). Combining various NPs like Ag, chitosan, iron, Au, Zn, and CNTs with natural essential oils showed stronger antimicrobial effect than using NPs alone (Nile et al. 2020). For example, combining eugenol and cinnamaldehyde with certain NPs gives very effective bond against *Salmonella* and *Listeria* (Gomes Silva et al. 2011). Combining essential oils and NPs boosts their ability to fight food pathogens. This also helps work against drug-resistant microorganisms (Nile et al. 2020; Bajpai et al. 2018). Nanotechnology can be used to create antimicrobial coatings on surfaces to prevent the growth of biofilms. The environment of food industry always favor the growth of microorganisms. These bacteria attach to the surface and form biofilm and therefore are hard to clean (Myszka and Czaczyk 2011). Such biofilms contaminate foods, and in this regard, nanocoated films are very useful in maintaining food quality.

9.3.4.1 Nanopackaging for Food Preservation

Fresh foods are very sensitive to oxygen, water, and ethylene, which affects food quality. Due to this, good packaging is essential. The main purpose of packaging is to prevent foods from spoilage and contamination while providing physical protection from temperature changes, bad environment conditions, and microbial infections. Packaging materials work as a barrier to oxygen and other contaminants which lead to spoilage (Pradhan et al. 2015; Auffan et al. 2009). Nanotechnology plays a crucial role in food packaging by providing NPs and polymer-based composites, which helps overcome these issues. Biodegradable packaging reduces environmental pollution. Application of nanocoated biopolymer on the surface of food packages has shown good results in maintaining freshness. Nanocoatings such as

Ag NPs, TiO_2, and ZnO are used as surface coating agents in food industry. A good food package should have high barrier for plastic and microbes. Food preservation is the process of treating and handling foods with minimum spoilage and high quality (Mohammad et al. 2022).

For instance, Ag NPs, TiO_2, and silicate NPs are used in films to control oxygen exchange through package (Chaudhary et al. 2020). The NPs help prevent moisture leakages and keep foods fresh for longer time 58. SiO_2 and TiO_2 play a crucial role in food packaging, especially SiO_2 works as a food colorant and anticaking agent and it keeps foods dry by absorbing water molecules (Zhao et al. 2008). TiO_2 NPs can be used as a colorant for dairy products and cheese (Acosta 2009) and provide UV barrier in food packaging. Most of the studies state that Ag NPs are outstanding materials for food industry because they show high antimicrobial activities due to their surface area (Zhao et al. 2008). Ag NPs have shown activity against both gram-negative and gram-positive bacteria, but they are more active against gram-negative bacteria (Zhao et al. 2008; Dakal et al. 2016). Ag NPs also improve the shelf life of foods while absorbing and decomposing ethylene (Jiang et al. 2018). CNTs are used as a food packaging ingredient, but due to their toxicity level, some researchers don't recommend them for many applications (Ameta et al. 2020; Nile et al. 2020). Another drawback of TiO_2 is it has antimicrobial properties but it is active only when exposed to UV light. Also it works as a bactericide for different harmful pathogens. It can directly affect phospholipids by damaging the bacterial cell membrane. During the process, its color changes into blue if it had oxygen (Sánchez-López et al. 2020; Arora and Padua 2010).

Another application is chitosan; chitosan NPs have a variety of applications in food industry. Chitosan and PVA-based antimicrobial films showed activity against *E. coli, S. aureus*, and *B. subtilis*; they also increased the shelf life of tomatoes (Tripathi et al. 2009). Incorporation of chitosan-based nanocomposites in hydroxypropyl methylcellulose improves the mechanical and barrier properties of packaging (de Moura et al. 2009). Chitosan-based films also show antifungal, antimicrobial, and antioxidant activities, which is very helpful for food packaging (Yuan et al. 2016). They provide better shelf life without affecting any of the properties of food (No et al. 2007).

9.3.4.2 Nanotechnology in Food Traceability

The term "traceability" means being able to identify or track something either one step back or one step forward in a process. In food systems, this means monitoring everything from food systems, i.e., from food production to distribution to customers, ensuring that the quality of raw materials is preserved throughout the food chain. They give alerts under any hazard conditions that may occur in the process (Espiñeira and Santaclara 2016). Food and agriculture traceability system consists of six basic steps: product traceability, process traceability, genetic traceability, input traceability, disease and pest traceability, and measurement traceability (Aadil et al. 2019).

Nanotechnology is a modern technology that has transformed the food industry in various ways, especially in food processing, packaging, and safety. In food packaging, adding nanomaterials improves the strength of the packaging items. Nanosensors are very tiny tools that can detect and measure things at the nanoscale; nanosensors play a crucial role in monitoring the entire food chain and confirming the safety and quality. These portable tools allow food producers, manufacturers, storage facility suppliers, and consumers to quickly check food quality during transportation, storage, and use (Andreescu and Mustafaa 2020). In addition, food traceability system offers more benefits, such as to ensure the biosecurity protection, improve the efficiency of food safety investigation, minimize the recall of food production, and accelerate the improvements in strategical ways in food supply chain (King et al. 2018).

Advances in nanobiotechnology have led to the development of biosensors, which offer a better way to detect various contaminants, such as drug residues, heavy metals, pesticides, pathogens, and toxic substances. Biosensor is a combination of bioreceptor and a transducer that monitors biological events and converts them into useful data. Most of the biosensors are developed by using C NPs, Au NPs, Ag NPs, and semi-conductor materials. They have unique properties due to their small size and high surface area, which improves their ability to detect and bind to substance, while giving better sensitivity. Nanosensors can be used as nanotracers in food and agricultural fields. They can detect food poisoning and other threats; nanotracers can be added to packaging materials that maintain the quality of the foods over time. During their transportation and storage, nanosensors monitor the conditions of food products, especially foods with short life cycle, and ensure safety and quality (Aggarwal and Idrishi 2022).

Nanosensors show rapid response throughout the food supply chain, have high sensitivity, and identify different pathogens in low costs. For example, nanosensors play a crucial role in smart packaging, which provides data on pH, humidity, temperature, light exposure, and gas levels. They can also identify pathogens and toxic residues and track the conditions (Fraceto et al. 2016; Anon n.d.). Mostly, these traceability systems consist of numerous automatic identification tools that have the ability to show the data flow through the supply chain. These devices act as data carrier devices; they can be barcodes and RFID tags (Mirza Alizadeh et al. 2021). RFID tags have been used in kiwifruit supply chain to show more accurate information (Gautam et al. 2017). Another study showed usage of electronic traceability system in virgin olive oil production chain (Violino et al. 2020).

9.3.5 Precision Agriculture

Precision agriculture is changing farming around the world by offering many benefits, including improved crop quality, better environmental protection, sustainability, food safety, economic growth, higher profits, and increased productivity. It uses information technology to manage commercial farming with the goal of increasing crop yields while minimizing inputs like herbicides, pesticides, and fertilizers. Figure 9.2 shows the basic steps of precision farming. To identify specific geographical areas, certain technologies are used in precision agriculture, including

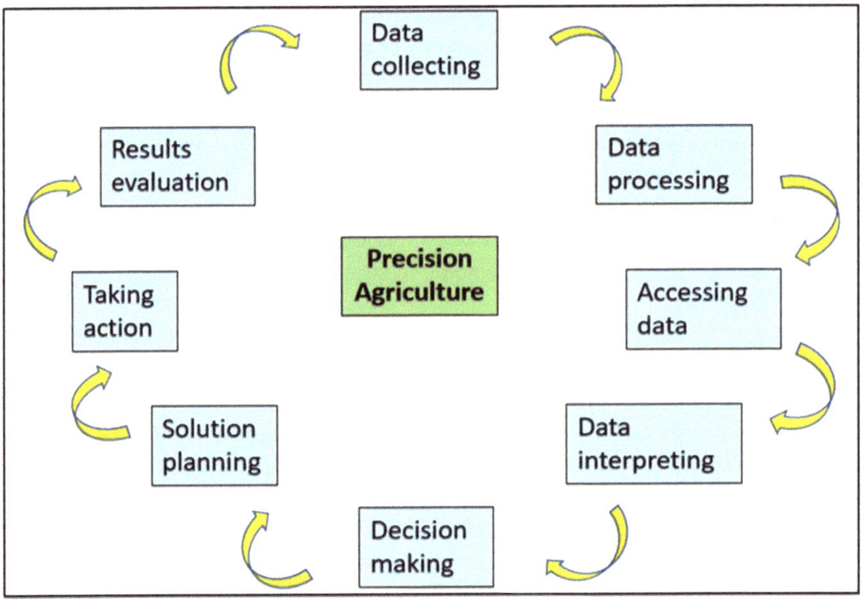

Fig. 9.2 Basic steps of precision farming

Geographical Information Systems (GIS), Global Positioning Systems (GPS), and remote sensing (Bharti et al. 2021). Nanotechnology plays a crucial role in agriculture by providing smart delivery systems and nanosensors help farmers reduce resource use such as water, agrochemicals, and nutrients while increasing the productivity under better monitoring (Duhan et al. 2017). For instance, nanosensors connected to GPS allow farmers to efficiently check soil conditions, nutrient availability, root growths, and crop yields. Some of these nanosensors help identify environmental factors that directly affect the overall growth of the plant. The use of nanotechnology in precision agriculture is growing as a way to boost productivity (Zain et al. 2024).

In addition, the development of nanobiosensors has brought significant progress in areas like analysis of protein, measurement of metabolites, and detection of stress and pathogen in plants. Nanosensors are highly sensitive and selective and capable of handling multiple tasks. However, creating smart nanobiotech sensors to monitor crop health is a challenge in terms of their effectiveness, precision, and reliability in real farming conditions (Muthumalai et al. 2021; Giraldo et al. 2019).

9.3.5.1 Use of Nanosensors for Environmental Monitoring

Nanobiosensors are advanced tools used for monitoring the environmental conditions. They combine nanotechnology with biological elements to detect pollutants quickly and accurately. These sensors use NMs like Au NPs, CNTs, and graphene to improve how they transmit signals, while biological components such as enzymes, antibodies, and nuclei acids help them to identify target pollutant particles. Recent advancements are allowed to monitor air, soil, and water quality in real time and on site. Nanosensors monitor air quality by detecting small amounts of pollutants like ammonia, hydrogen sulfide, nitrogen oxides, carbon monoxide, and volatile organic compounds (Sekhon 2014; Johnson et al. 2021). This helps reduce urban smog and industrial emissions, leading to healthier environments. In water quality monitoring, nanosensors are very useful as they can identify contaminants like heavy metals, nitrates, and pathogens, helping protect drinking water and aquatic ecosystems. They allow for quick analysis onsite, enabling a fast response to potential dangers (Suthari 2024).

Nnaosensors are also effective in monitoring soil quality. They can detect toxic substances and check for essential nutrients used for farming. This information is important for promoting sustainable farming practices and preventing contaminations (Sharma et al. 2021b). For example, graphene- based sensors are useful for detecting heavy metals and organic pollutants. CNT-based nanosensors which are functionalized with DNA can detect mercury ions in water. Also, multi-walled CNTs can identify volatile organic compounds and other pollutants. Quantum dot-based sensors are other highly sensitive nanosensors that can easily detect heavy metal particles. For instance, cadmium sulfide quantum dots have been used to recognize cadmium and lead ions in water samples. Au NPs exhibit useful sensing applications, which are developed to identify pesticides and heavy metals in the soil and water samples (Suthari 2024).

Heavy metals like lead, mercury, and cadmium cause serious environment and health dangers because they can remain in the environment for a long time period (Kundu et al. 2019). Some nanobiosensors use graphene oxide modified with DNA to help detect lead in water. This technology offers a quick and effective way to monitor lead levels. Mercury ions are very toxic and widespread pollutants, and they can be easily detected on-site and food samples by using nanosensors (Long et al. 2013). Kumar and Guleria (2024)showed the use of enzyme-based nanobiosensors developed for detecting pesticides in agricultural runoff. These sensors offer rapid and sensitive monitoring methods (Justino et al. 2017). DNA-based nanobiosensors help detect pathogens in environmental matrices, especially in water. DNA and Au NPs can detect *E. coli* in water and ensure water quality.

9.3.5.2 Drone and Nanotechnology in Crop Management

Precision agriculture, empowered with drone technology, is changing the perspectives of modern farming practices. Drones comprise advanced sensors, cameras, and GPS technologies that allow farmers to monitor their fields with great precision and efficiency. They update about real-time data based on crop health, nutrient levels, and pest problems (Ben-Gal et al. 2009). By capturing high-resolution images, drones help farmers identify issues early, enabling them to take required actions and use resources like water, fertilizers, and pesticides more effectively. This improves crop yields while reducing environmental impact and lowering production costs. Drones also allow farmers to create maps of their agricultural fields and analyze data closely, helping them make better decisions for specific areas of their lands. Overall, drone technology improves sustainable farming practices, resources efficiency, and profits for farmers. Drone technology continues to develop and has the potential to transform crop cultivation and management worldwide (Aggelopoulou et al. 2011).

Water management in agriculture has significantly improved with the use of drone technology. Drones typically consist of sensors and imaginary tools that help farmers assess and optimize water use in the field. Aerial surveys from drones provide important information about soil moisture levels, irrigation efficiency, and overall crop health. They can easily recognize the areas which are either underwater or overwatered, allowing farmers to make precise adjustments to their irrigation practices (Kulkarni and Natu 2016). By providing real-time data, drones help farmers decide when, where, and how much water is needed. These approaches help increase the crop yield while maintaining high quality. In addition, drones can recognize leaks or other abnormalities in irrigation systems, further improving water management. Also, they ensure the efficient use of water resources and boost agricultural productivity. As technology continues to improve, drones are expected to play an even bigger role in addressing water scarcity and improving farming practices (Zhang and Kovacs 2012).

The presence of weeds in the field can be easily identified using drones so that they will be quickly removed from the field. Drone technology can be used to spray agrochemicals instead of spraying unnecessary amounts, they can scan the amount of chemical need and can spray in real-time for even coverage. Studies show that the speed of this method is five times higher than the conventional methods (Ahirwar et al. 2019). Figure 9.3 shows the uses of drone technology in agriculture.

9.4 Challenges, Limitations, and Risks

While nanotechnology offers many benefits in food processing and agriculture, it also poses potential risks to the nature, society, and health. NPs can enter the ecosystem through pesticide use in agriculture or through packaging in processed foods, raising concerns about their toxicity. The risks come with NMs due to their reactivity and their increased ability to affect our bodies, leading to long-term health issues 99. NPs can enter our food chain in different ways; for example, NPS like nanoemulsions, nanocapsules, and nanoantimicrobial films can directly contaminate the foods, and on the other hand, NPs like nanolaminates, nanosensors, and CNTs can indirectly contaminate foods as they are used in food production, processing, preservation, and tracking. The risk of exposure is higher in areas where NMs are directly added to food as carriers for new ingredients 115.

9.4.1 Impact on Human Health and Environment

Their nanoscale gives NPs ability to permeate through barriers of living organisms, causing harmful biological-based reactions. These particles can enter the human body through the lungs, intestinal tract, or skin cells and may cause brain toxicity,

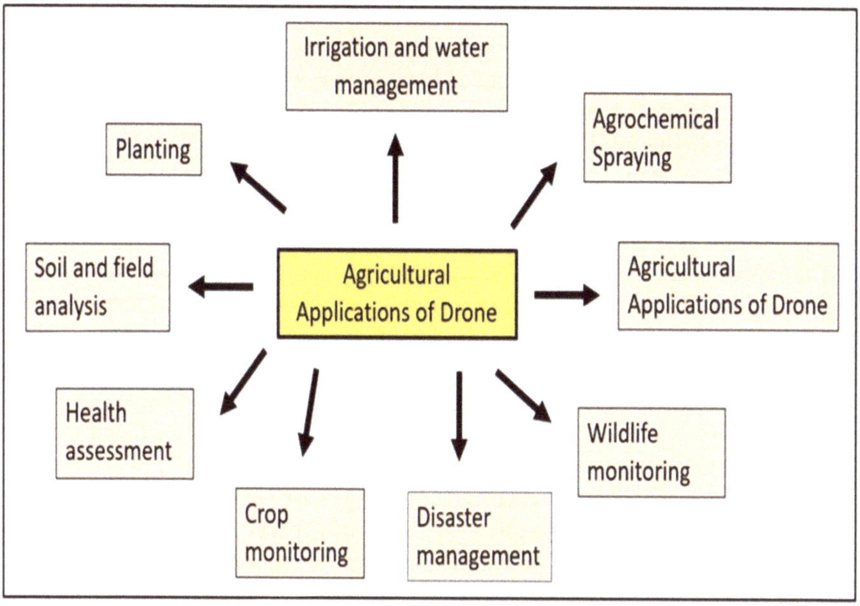

Fig. 9.3 Drone technology in agriculture

lung inflammation, and heart problems (Oberdorster et al. 2005). Some NPs can cause permanent call damages and organ injuries via oxidative stresses. The C-based NPs can be toxic to lung cancer cells, and the level of toxicity can depend on factors like materials, size, surface properties, structure, and the clumping method (Magrez et al. 2006). In addition, they can alter genetic materials and disrupt important signal processing systems. Inhaled Ps might even reach the brain and potentially cause neurodegenerative diseases (Win-Shwe and Fujimaki 2011). Some studies have shown that some NPs can be toxic to nerve cells. Titanium dioxide has been widely used in many products, such as food colorant, and other food processing methods. Exposure to TiO_2 NPs showed DNA and chromosome damage, heart disease, cancer, and neurological diseases (Donaldson et al. 2005).

Nanotechnology has many applications that could be harmful to plants and animals, so it needs careful evaluation. Due to their unique properties, NPs can accumulate in the organs and tissues of living organisms. Animals can inhale NPs resulting in dangerous lung problems (Savolainen et al. 2010). NPs can settle on plant leaves and flower petals while creating a toxic layer, which disrupts the absorption of water and essential minerals (Tarafdar 2015). Au NPs are nanoparticles that are widely used in food industry due to their antimicrobial properties; however, they cause cell proliferation and DNA damage (Rticles and Ern n.d.). Silver NPs are another promising NPs in food and agricultural industries due to their antibacterial and antifungal properties, but they can have harmful consequences when deposited on skin and in soft tissues, liver, kidney, and spleen (Fung and Bowen 1996;

Danscher 1981). Also, they can pollute water and harm aquatic life such as fish and algae, with their toxicity varying with the size, shape, concentration, and the pH level of the environment. Some studies have shown that CuO NPs and iron oxide NPs have varying levels of toxicity in soil samples (Samrot et al. 2017). At times, ZnO NPs can be harmful to plants, leading to stunt growth and changes in gene expression (Arora et al. 2024). Contamination of water with organic pollutants such as Co, Cu, Pb, Zn, As, Cd, and Hg can spread diseases like typhoid, cholera, and diarrhea (Prabhakar 2020).

On the other hand, CNTs are essential NMs in nanofertilizers and many applications; however, under some conditions, they can induce harmful effects such as inflammatory and fibrotic reactions (Rticles and Ern n.d.).Understanding the effects of nano-solutions used in agriculture on the environment and human health is challenging because the potential of risk associated with NPs is high. Biosolids, which contain tiny metal particles such as Ag, TiO_2, Au, and ZnO, are included in some products and are often used as fertilizers, disposed of in landfills, or incinerated. They can also leach into the environment through irrigation with wastewater or contaminated water. The toxic effect of these aged NMs can interact with other toxic particles and may increase the risks for human health and the environment. Moreover, these NPs can directly enter our food webs, which could affect the overall ecosystems (Chen et al. 2015; Muraisi et al. 2022). Figure 9.4 shows the health impacts of using nanoparticles in farming.

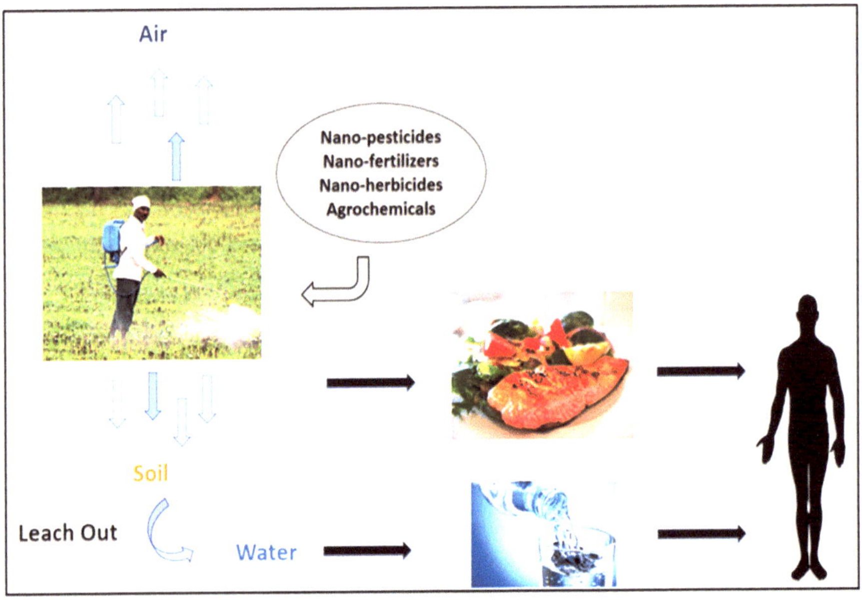

Fig. 9.4 Health impacts of using nanoparticles in farming

9.4.2 Regulatory Framework

A review of regulation around the world shows that the European Union and Switzerland have specific regulations toward nanomaterials. According to the European Commission's regulation state that products with nanomaterials should be labeled according to the policies. In 2004, the EU began to develop regulatory policies for these materials due to safety concerns. Because of these risks, regulations are needed for the use of NPs in agriculture and food industry. In 2011, the European Food Safety Authority (EFSA) issued guidelines for assessing the potential risks of using NPs in food processing and pesticide management. Meanwhile in USA, the Environmental Protection Agency (EPA) proposed to gather information about NMs in pesticides and their potential effects on human health and environment. Also, the National Institute for Occupational Safety (NIOSH) issued warning against exposure to CNTs and nanofibers during research ((CDC-NIOSH) 2019).

In the USA, they have not yet established specific regulations for the use of nanomaterials, but the US Food and Drug Administration has provided guidance for usage of NPs in food industry. Furthermore, the Canadian Food Inspection Agency (CFIA) has created rules and regulation system for using NMs inside Canada, and they have published another guidelines for oversees food regulation through Public Health Agency of Canada. In India, there are no specific regulation towards nanomaterials under Food Safety and Standard Act of 2006. The current studies show that regulations for nanomaterials are unclear and inconsistent. Different definitions exist based on how NMs are used in industry or research. According to the International Organization for Standardization (ISO), nanotechnology is controlling materials in the range between 1 to 100 nano-meter. This scale is the reason materials exhibit unique properties that differ from those of larger-sized materials.

The framework refers to a way of conducting risk assessments for NMs. Before stablishing a framework, life cycle of NMs, pre-assessment phase, exposure situations, adaptable structures, applications, and transparency should be considered. Among these criteria, the specific requirements for NMs are a top priority for research and development. This is due to the unique properties of nanoscale practices that require new testing methods and protocols to assess their hazards. Due to their high toxicity, it is important to have specific methods for characterizing and assessing the toxicity of materials at the nanoscale. Lack of clear guidelines for the production, use, and disposal of NMs in agriculture makes it difficult to assess their safety and environmental impact. Overall, current studies indicate the need for standardized testing methods for NMs that are consistent across countries, which are established by professional regulatory bodies.

9.4.3 Economic Consideration

Farmers need to invest a lot to introduce new technological devices to their fields, which may be more expensive when compared to conventional options. This includes purchasing specialized gadgets and systems for their fields. Implementation of nanotechnology practices such as imparting knowledge to labors and staff on new technologies and safety protocols may attract additional costs. To align with regulations and to maintain new standards and certifications are also costly. Moreover, adoption of nanotechnology depends on the awareness of farmers and acceptance of NMs (Ingale and Pathak 2023).

Food producers have invested enormous money to use nanotechnology-based products in food industry. Most of them are in the field of food contacting materials, such as smart packages, sensors, and detectors. No nanoproducts have yet been approved as food additives for direct use by humans, except for TiO_2 and iron oxide, which are currently used as food pigments and colorants (Neme et al. 2021).

9.5 Future Directions

Current nanotechnology studies show impressive progress. Nanotechnology has unique properties that allows them to be used in various areas, improving the quality of life. In agriculture and food industries, nanotechnology plays a significant role offering numerous benefits. Farmers can use nanotechnology for nanopriming of seeds and precision agricultural practices, which promotes sustainability and eco-friendliness. Nanotechnology also plays an important role in genetic engineering and delivers genetic materials. This may help to prevent poverty, malnutrition, and food shortages. Sustainable developments are essential to overcome growth challenges with opportunities. To achieve this, nanomaterials should be synthesized using eco-friendly methods rather than chemical methods. Application of nanopesticides, nanoherbicides, and nanofertilizers has improved agricultural output. However, more studies are required to understand these nanoproducts.

The interactions of NMs with living things depend on their characteristics, like size, structure, shape, and crystallinity. The environmental and toxic effects of engineered nanomaterials are influenced by their properties and the conditions they are exposed to. Therefore, it is important to analyze NMs toxic level before use. We need to assess how these materials affect soil, water, food, and plants over time and study their degradation in different environmental conditions.

Also, nanotechnology has made prominent improvement in food science to improve food quality and safety. Innovative ideas like smart packaging and active packaging have made nanocomposite films to become a reality. Carbon-neutral biodegradable nanocomposites are used in food packaging, and nanosilica has potential as a coating material that improves barrier properties. Artificial intelligence (AI) and machine learning (ML) methods could help control the quality and safety with self decision-making skills, cost reduction, and improved business strategies. For examples AI technology can be used to identify pH levels of foods, especially dairy foods, and waste, and nanosensors in packaging can detect gases from spoiled foods.

Every technology has its limitations. We still need more knowledge on long-term effects of NMs on the environment and should consider the future risks. By minimizing the negative impacts while maximizing the positive impacts, nanotechnology can be of more benefits to the world. Therefore, advancing nanotechnology in sustainable ways is important.

9.6 Conclusion

Nanotechnology offers transformative potential for agribusiness, by giving innovative solutions that can improve agriculture productivity and food safety. By utilizing NMs and nanosystems, farmers can benefit from pest managements, target deliveries, water management, and soil health, which overall will increase crop yields. Moreover, advancements in nanotechnology facilitate precision farming that allows the use of AI technologies in monitoring environment and crop growth and helps ensure a harvest that is free from pathogens. Despite these advantages, it is crucial to overcome disadvantages. Future research are necessary to fully investigate the long-term environmental and health impacts of NMs and consider about the regulatory frameworks to address safety concerns. As the population grows, the demand for food increases. Nanotechnology presents a promising avenue to achieve sustainable agricultural practices. By handling nanotechnology responsibly and ethically, agribusiness can harness the power of nanotechnology to create a more efficient, productive, and sustainable food system for the future.

References

(CDC-NIOSH), United States Centers for Disease Control and Prevention National Institute for Occupational Health and Safety (2019) Continuing to protect the Nanotechnology Workforce: NIOSH Nanotechnology Research Plan for 2018–2025. (February):Publication 2019-116

Aadil RM, Madni GM, Roobab U, ur Rahman U, Zeng XA (2019) Chap 1 - Quality control in beverage production: An overview. In: Grumezescu AM, Holban BI (eds) Quality control in the beverage industry. Academic Press, pp 1–38

Abdel-Aziz HMM, Hasaneen MNA, Omer AM (2016) Nano chitosan-NPK fertilizer enhances the growth and productivity of wheat plants grown in Sandy soil. Span J Agric Res 14(1):e0902–e0902

Acevedo M, Pixley K, Zinyengere N, Meng S, Tufan H, Cichy K, Bizikova L, Isaacs K, Ghezzi-Kopel K, Porciello J (2020) A scoping review of adoption of climate-resilient crops by small-scale producers in low- and middle-income countries. Nat Plants 6(10):1231–1241. https://doi.org/10.1038/s41477-020-00783-z

Acosta E (2009) Bioavailability of nanoparticles in nutrient and nutraceutical delivery. Curr Opin Colloid Interface Sci 14(1):3–15

Aggarwal D, Idrishi R (2022) Nanotechnology applications for food traceability. In: Nanotechnology applications for food safety and quality monitoring, pp 457–472. https://doi.org/10.1016/B978-0-323-85791-8.00011-2

Aggelopoulou KD, Pateras D, Fountas S, Gemtos TA, Nanos GD (2011) Soil spatial variability and site-specific fertilization maps in an apple orchard. Precis Agric 12(1):118–129. https://doi.org/10.1007/s11119-010-9161-x

Ahirwar S, Swarnkar R, Bhukya S, Namwade G (2019) Application of drone in agriculture. Int J Curr Microbiol App Sci 8:2500–2505. https://doi.org/10.20546/ijcmas.2019.801.264

Ahmad BM, Mahmood A, Ahmad H (2020) Role of nanotechnology in agriculture: a review

Alizadeh M, Adel MM, Shakooie M, Khajavi MZ, Farhoodi M (2021) Trends and applications of intelligent packaging in dairy products: a review. Crit Rev Food Sci Nutr 62(2):383–397. https://doi.org/10.1080/10408398.2020.1817847

Ameta SK, Rai AK, Hiran D, Ameta R, Ameta SC (2020) Use of nanomaterials in food science. In: Biogenic nano-particles and their use in agro-ecosystems, pp 457–488

Andreescu S, Mustafaa F (2020) Nanotechnology-based approaches for food sensing and packaging applications, vol 33. R Soc Chem
Anon (n.d.) Nanosensors applications in agriculture and food industry
Arora L, Narula A (2017) Gene editing and crop improvement using CRISPR-Cas9 system. Front Plant Sci 8(November). https://doi.org/10.3389/fpls.2017.01932
Arora A, Padua GW (2010) Nanocomposites in food packaging. J Food Sci 75(1):R43–R49
Arora R, Roy T, Adak P (2024) A review of the impact of nanoparticles on environmental processes. BIO Web Conf 86. https://doi.org/10.1051/bioconf/20248601001
Ashraf MA, Peng W, Zare Y, Rhee KY (2018) Effects of size and aggregation/agglomeration of nanoparticles on the interfacial/interphase properties and tensile strength of polymer nanocomposites. Nanoscale Res Lett 13(1):214. https://doi.org/10.1186/s11671-018-2624-0
Aslani F, Bagheri S, Julkapli NM, Juraimi AS, Hashemi FSG, Baghdadi A (2014) Effects of engineered nanomaterials on plants growth: An overview. Sci World J 2014(1):641759
Auffan M, Rose J, Bottero J-Y, Lowry GV, Jolivet J-P, Wiesner MR (2009) Towards a definition of inorganic nanoparticles from an environmental, health and safety perspective. Nat Nanotechnol 4(10):634–641
Ayyub P, Chandra R, Taneja P, Sharma AK, Pinto R (2001) Synthesis of Nanocrystalline material by sputtering and laser ablation at low temperatures. Appl Phys A 73(1):67–73. https://doi.org/10.1007/s003390100833
Baig N, Kammakakam I, Falath W (2021) Nanomaterials: a review of synthesis methods, properties, recent progress, and challenges. R Soc Chem 2:1821–1871. https://doi.org/10.1039/D0MA00807A
Bajpai VK, Kamle M, Shukla S, Mahato DK, Chandra P, Hwang SK, Kumar P, Huh YS, Han Y-K (2018) Prospects of using nanotechnology for food preservation, safety, and security. J Food Drug Anal 26(4):1201–1214
Ben-Gal A, Victor Alchanatis Y, Cohen UY, Zipori I, Presnov E, Sprintsin M, Dag A (2009) Evaluating water stress in irrigated olives: correlation of soil water status, tree water status, and thermal imagery. Irrig Sci 27:367–376. https://doi.org/10.1007/s00271-009-0150-7
Bhagat Y, Gangadhara K, Rabinal C, Chaudhari G, Ugale P (2015) Nanotechnology in agriculture: a review. J Pure Appl Microbiol 9(1):737–747. https://doi.org/10.20546/ijcmas.2018.712.205
Bharti A, Kumar V, Singh M, Bhatt GD (2021) Role of nanosensors in agriculture sciences. Short communication role of nanosensors in agriculture sciences. Rec Adv Biol Med 7(1):1–3
Biswas R, Alam M, Animesh Sarkar M, Ismail Haque M, Hasan M, Hoque M (2022) Application of nanotechnology in food: processing, preservation, packaging and safety assessment. Heliyon 8(11):e11795. https://doi.org/10.1016/j.heliyon.2022.e11795
Bratovcic A (2019) Different applications of nanomaterials and their impact on the environment. Int J Mater Sci Eng 5(1):1–7. https://doi.org/10.14445/23948884/ijmse-v5i1p101
Bünemann EK, Bongiorno G, Bai Z, Creamer RE, De Deyn G, de Goede R, Fleskens L, Geissen V, Kuyper TW, Mäder P, Pulleman M, Sukkel W, Willem J, van Groenigen, and Lijbert Brussaard. (2018) Soil quality—a critical review. Soil Biol Biochem 120(January):105–125. https://doi.org/10.1016/j.soilbio.2018.01.030
Buzea C, Pacheco II, Robbie K (2007) Nanomaterials and nanoparticles: sources and toxicity. Biointerphases 2(4):MR17–MR71. https://doi.org/10.1116/1.2815690
Chakravarty P, Deka H, Chowdhury D (2023) Anthracene removal potential of green synthesized titanium dioxide nanoparticles (TiO2-NPs) and Alcaligenes Faecalis HP8 from contaminated soil. Chemosphere 321:138102
Chang PH, Li Z, Jiang WT, Sarkar B (2018) Clay minerals for pharmaceutical wastewater treatment
Chaudhary P, Fatima F, Kumar A (2020) Relevance of nanomaterials in food packaging and its advanced future prospects. J Inorg Organomet Polym Mater 30(12):5180–5192
Chen C, Unrine JM, Judy JD, Lewis RW, Guo J, McNear Jr DH, Tsyusko OV (2015) Toxicogenomic responses of the model legume Medicago Truncatula to aged biosolids containing a mixture of nanomaterials (TiO2, Ag, and ZnO) from a pilot wastewater treatment plant. Environ Sci Technol 49(14):8759–8768

Cheng HN, Doemeny LJ, Geraci CL, Schmidt DG (2016) Nanotechnology: delivering on the promise: volume 2. American Chemical Society

Chhipa H (2019) Applications of nanotechnology in agriculture, vol 46, 1st edn. Elsevier Ltd.

Cho K-H, Park J-E, Osaka T, Park S-G (2005) The study of antimicrobial activity and preservative effects of nanosilver ingredient. Electrochim Acta 51(5):956–960

Cioffi N, Torsi L, Ditaranto N, Sabbatini L, Zambonin PG, Tantillo G, Ghibelli L, D'Alessio M, Bleve-Zacheo T, Traversa E (2004) Antifungal activity of polymer-based copper nanocomposite coatings. Appl Phys Lett 85(12):2417–2419. https://doi.org/10.1063/1.1794381

Dakal TC, Kumar A, Majumdar RS, Yadav V (2016) Mechanistic basis of antimicrobial actions of silver nanoparticles. Front Microbiol 7:1831

Danks AE, Hallb ASR, Schnepp Z (2016) The evolution of 'sol–gel' chemistry as a technique for materials synthesis. Mater Horiz 2

Danscher G (1981) Light and electron microscopic localization of silver in biological tissue. Histochemistry 71(2):177–186

de Moura MR, Aouada FA, Avena-Bustillos RJ, McHugh TH, Krochta JM, Mattoso LHC (2009) Improved barrier and mechanical properties of novel hydroxypropyl methylcellulose edible films with chitosan/tripolyphosphate nanoparticles. J Food Eng 92(4):448–453

de Oliveira JL, Campos EVR, Pereira AES, Pasquoto T, Lima R, Grillo R, de Andrade DJ, Santos FAD, Fraceto LF (2018) Zein nanoparticles as eco-friendly carrier systems for botanical repellents aiming sustainable agriculture. J Agric Food Chem 66(6):1330–1340. https://doi.org/10.1021/acs.jafc.7b05552

de Oliveira JL, Campos EVR, Germano-Costa T, Lima R, Vechia JFD, Soares ST, Junior D, de Andrade K, Gonçalves C, do Nascimento J, Polanczyk RA, Fraceto LF (2019) Association of Zein nanoparticles with botanical compounds for effective pest control systems. Pest Manag Sci 75(7):1855–1865. https://doi.org/10.1002/ps.5338

Dhanapal AR, Thiruvengadam M, Vairavanathan J, Venkidasamy B, Easwaran M, Ghorbanpour M (2024) Nanotechnology approaches for the remediation of agricultural polluted soils. ACS Omega 9(12):13522–13533. https://doi.org/10.1021/acsomega.3c09776

Donaldson K, Tran L, Jimenez LA, Duffin R, Newby DE, Mills N, MacNee W, Stone V (2005) Combustion-derived nanoparticles: a review of their toxicology following inhalation exposure. Part Fibre Toxicol 2:1–14

Du W, Yang J, Peng Q, Liang X, Mao H (2019) Comparison study of zinc nanoparticles and zinc sulphate on wheat growth: from toxicity and zinc biofortification. Chemosphere 227:109–116

Duhan JS, Kumar R, Kumar N, Kaur P, Nehra K, Duhan S (2017) Nanotechnology: the new perspective in precision agriculture. Biotechnol Reports 15:11–23. https://doi.org/10.1016/j.btre.2017.03.002

Duncan TV (2011) Applications of nanotechnology in food packaging and food safety: barrier materials, antimicrobials and sensors. J Colloid Interface Sci 363(1):1–24

Dwivedi S, Saquib Q, Al-Khedhairy AA, Musarrat J (2016) Microbial inoculants in sustainable agricultural productivity

Ealias AM, Saravanakumar MP (2017) A review on the classification, characterisation, synthesis of nanoparticles and their application. IOP Conf Ser 263(3):032019. https://doi.org/10.1088/1757-899X/263/3/032019

Elemike EE, Uzoh IM, Onwudiwe DC, Babalola OO (2019) The role of nanotechnology in the fortification of plant nutrients and improvement of crop production. Appl Sci (Switzerland) 9(3):1–32. https://doi.org/10.3390/app9030499

El-Ramady H, Abdalla N, Sári D, Ferroudj A, Muthu A, Prokisch J, Fawzy ZF, Brevik EC, Solberg S (2023) Nanofarming: promising solutions for the future of the global agricultural industry. Agronomy 13(6). https://doi.org/10.3390/agronomy13061600

Erkmen C, Unal DN, Kurbanoglu S, Uslu B (2022) Basics of electrochemical sensors. In: Organic electrodes: fundamental to advanced emerging applications. Springer, pp 81–99

Espiñeira M, Santaclara FJ (2016) 1 - What is food traceability? In: Espiñeira M, Santaclara T (eds) Woodhead publishing series in food science, technology and nutrition. Woodhead Publishing, pp 3–8

Fincheira P, Hoffmann N, Tortella G, Ruiz A, Cornejo P, Diez MC, Seabra AB, Benavides-Mendoza A, Rubilar O (2023) Eco-efficient systems based on nanocarriers for the controlled release of fertilizers and pesticides: toward smart agriculture. Nano 13(13):1–22. https://doi.org/10.3390/nano13131978

Fraceto LF, Grillo R, de Medeiros GA, Scognamiglio V, Rea G, Bartolucci C (2016) Nanotechnology in agriculture: which innovation potential does it have? Front Environ Sci 4(MAR):1–5. https://doi.org/10.3389/fenvs.2016.00020

Fung MC, Bowen DL (1996) Silver products for medical indications: risk-benefit assessment. J Toxicol Clin Toxicol 34(1):119–126

Gade A, Ingle A, Whiteley C, Rai M (2010) Mycogenic metal nanoparticles: progress and applications. Biotechnol Lett 32(5):593–600. https://doi.org/10.1007/s10529-009-0197-9

Gajanan K, Tijare SN (2018) Applications of nanomaterials. Mater Today 5(1, Part 1):1093–1096. https://doi.org/10.1016/j.matpr.2017.11.187

Garg V, Mote RG, Jing F (2020) Facile fabrication of functional 3D micro-nano architectures with focused ion beam implantation and selective chemical etching. Appl Surf Sci 526:146644. https://doi.org/10.1016/j.apsusc.2020.146644

Gautam R, Agnisha Singh K, Karthik S, Pandey FS, Tiwari MK (2017) Traceability using RFID and its formulation for a kiwifruit supply chain. Comput Ind Eng 103:46–58. https://doi.org/10.1016/j.cie.2016.09.007

Giraldo JP, Honghong W, Newkirk GM, Kruss S (2019) Nanobiotechnology approaches for engineering smart plant sensors. Nat Nanotechnol 14(6):541–553. https://doi.org/10.1038/s41565-019-0470-6

Grumezescu AM, Holban AM (2018) Impact of nanoscience in the food industry, vol 12. Academic Press

Hae-Jun P, Ho KS, Hwa-Jung K, Choi S (2006) A new composition of nanosized silica-silver for control of various plant diseases. Plant Pathol J 22:295. https://doi.org/10.5423/PPJ.2006.22.3.295

Hasan MK, Shopan J, Ahammed GJ (2020) Chapter 12 - Nanomaterials and soil health for agricultural crop production: current status and future prospects. In: Husen A, Jawaid M (eds) Micro and nano technologies. Elsevier, pp 289–312

Ikhmayies SJ (2014) Characterization of nanomaterials. JOM 66(1):28–29. https://doi.org/10.1007/s11837-013-0826-6

Ingale V, Pathak S (2023) Agri articles (September)

Jain P, Patidar B, Bhawsar J (2020) Potential of nanoparticles as a corrosion inhibitor: a review. J Bio- and Tribo-Corros 6. https://doi.org/10.1007/s40735-020-00335-0

Jia X, Shi N, Tang W, Ziyao S, Chen H, Tang Y, Sun B, Zhao L (2022) Nano-hydroxyapatite increased soil quality and boosted beneficial soil microbes. Plant Nano Biol 1:100002. https://doi.org/10.1016/j.plana.2022.100002

Jiang J, Pi J, Cai J (2018) The advancing of zinc oxide nanoparticles for biomedical applications. Bioinorg Chem Appl 2018(1):1062562

Johnson MS, Sajeev S, Nair RS (2021) Role of nanosensors in agriculture. In: 2021 International Conference on Computational Intelligence and Knowledge Economy (ICCIKE), pp 58–63

Joshi KA, Tang J, Haddon R, Wang J, Chen W, Mulchandani A (2005) A disposable biosensor for organophosphorus nerve agents based on carbon nanotubes modified thick film strip electrode. Electroanalysis 17(1):54–58

Joudeh N, Linke D (2022) Nanoparticle classification, physicochemical properties, characterization, and applications: a comprehensive review for biologists. J Nanobiotechnol 20(1):262. https://doi.org/10.1186/s12951-022-01477-8

Justino CIL, Duarte AC, Rocha-Santos TAP (2017) Recent progress in biosensors for environmental monitoring: a review. Sensors (Switzerland) 17(12). https://doi.org/10.3390/s17122918

Kale SS, Chauhan R, Nigam B, Gosavi S, Chaudhary IJ (2024) Effectiveness of nanoparticles in improving soil fertility and eco-friendly crop resistance: a comprehensive review. Biocatal Agric Biotechnol 56:103066. https://doi.org/10.1016/j.bcab.2024.103066

Kaushal M, Wani SP (2017) Nanosensors: Frontiers in precision agriculture. In: Nanotechnology: an agricultural paradigm, pp 279–291
Khaledian S, Nikkhah M, Shams-bakhsh M, Hoseinzadeh S (2017) A sensitive biosensor based on gold nanoparticles to detect Ralstonia Solanacearum in soil. J Gen Plant Pathol 83:231–239
Khan MR, Rizvi TF (2014) Nanotechnology: scope and application in plant disease management. Plant Pathol J 13(3):214–231. https://doi.org/10.3923/ppj.2014.214.231
Khan I, Saeed K, Khan I (2019) Nanoparticles: properties, applications and toxicities. Arab J Chem 12(7):908–931. https://doi.org/10.1016/j.arabjc.2017.05.011
Khan I, Raza MA, Awan SA, Shah GA, Rizwan M, Ali B, Tariq R, Hassan MJ, Alyemeni MN, Brestic M, Zhang X, Ali S, Huang L (2020) Amelioration of salt induced toxicity in pearl millet by seed priming with silver nanoparticles (AgNPs): the oxidative damage, antioxidant enzymes and ions uptake are major determinants of salt tolerant capacity. Plant Physiol Biochem 156:221–232. https://doi.org/10.1016/j.plaphy.2020.09.018
Khan AH, Khan NA, Zubair M, Shaida MA, Manzar MS, Abutaleb A, Naushad M, Iqbal J (2022) Sustainable green nanoadsorbents for remediation of pharmaceuticals from water and wastewater: a critical review. Environ Res 204:112243
Khodakovskaya M, Dervishi E, Mahmood M, Yang X, Li Z, Watanabe F, Biris AS (2009) Carbon nanotubes are able to penetrate plant seed coat and dramatically affect seed germination and plant growth. ACS Nano 3(10):3221–3227
Khot LR, Sankaran S, Maja JM, Ehsani R, Schuster EW (2012) Applications of nanomaterials in agricultural production and crop protection: a review. Crop Prot 35:64–70. https://doi.org/10.1016/j.cropro.2012.01.007
King T, Osmond-McLeod MJ, Duffy LL (2018) Nanotechnology in the food sector and potential applications for the poultry industry. Trends Food Sci Technol 72:62–73. https://doi.org/10.1016/j.tifs.2017.11.015
Kiran V, Harini K, Thirumalai A, Girigoswami K, Girigoswami A (2024) Nanotechnology's role in ensuring food safety and security. Biocatal Agric Biotechnol 58:103220. https://doi.org/10.1016/j.bcab.2024.103220
Kolahalam LA, Kasi Viswanath IV, Diwakar BS, Govindh B, Reddy V, Murthy YLN (2019) Review on nanomaterials: synthesis and applications. Mater Today 18:2182–2190. https://doi.org/10.1016/j.matpr.2019.07.371
Kottegoda N, Munaweera I, Madusanka N, Karunaratne V (2011) A green slow-release fertilizer composition based on urea-modified hydroxyapatite nanoparticles encapsulated wood. Curr Sci:73–78
Kulkarni S, Natu A (2016) Adoption and utilization of drones for advanced precision farming: a review. https://doi.org/10.17762/ijritcc.v4i5.2237
Kumar V, Guleria P (2024) Application of DNA-nanosensor for environmental monitoring: recent advances and perspectives. Curr Pollut Rep 10(4):765–785. https://doi.org/10.1007/s40726-020-00165-1
Kumar S, Chauhan N, Gopal M, Kumar R, Dilbaghi N (2015) Development and evaluation of alginate-chitosan nanocapsules for controlled release of Acetamiprid. Int J Biol Macromol 81:631–637. https://doi.org/10.1016/j.ijbiomac.2015.08.062
Kumaraswamy RV, Kumari S, Choudhary RC, Pal A, Raliya R, Biswas P, Saharan V (2018) Engineered chitosan based nanomaterials: bioactivities, mechanisms and perspectives in plant protection and growth. Int J Biol Macromol 113:494–506. https://doi.org/10.1016/j.ijbiomac.2018.02.130
Kumaravel A, Chandrasekaran M (2011) A biocompatible nano TiO2/Nafion composite modified glassy carbon electrode for the detection of Fenitrothion. J Electroanal Chem 650(2):163–170
Kundu M, Krishnan P, Kotnala RK, Sumana G (2019) Recent developments in biosensors to combat agricultural challenges and their future prospects. Trends Food Sci Technol 88:157–178. https://doi.org/10.1016/j.tifs.2019.03.024
Lateef A, Nazir R, Jamil N, Alam S, Shah R, Khan MN, Saleem M (2016) Synthesis and characterization of zeolite based nano–composite: an environment friendly slow release fertilizer. Microporous Mesoporous Mater 232:174–183

Li C, Li Y, Li Y, Guozhan F (2018) Cultivation techniques and nutrient management strategies to improve productivity of rain-fed maize in semi-arid regions. Agric Water Manag 210:149–157
Li Z, Tao Y, Paul R, Fan J, Yang Y, Wei Q (2020) Agricultural nanodiagnostics for plant diseases: recent advances and challenges. Nanoscale Adv 2(8):3083–3094
Li L, Chen D, Chen J, Yang C, Zeng Y, Jin T, Zhang Y, Sun X, Mao H, Zhixiang M (2023) Gelatin and catechol-modified quaternary chitosan cotton dressings with rapid hemostasis and high-efficiency antimicrobial capacity to manage severe bleeding wounds. Mater Des 229:111927
Liang Y, Duan Y, Fan C, Dong H, Yang J, Tang J, Tang G, Wang W, Jiang N, Cao Y (2019) Preparation of Kasugamycin conjugation based on ZnO quantum dots for improving its effective utilization. Chem Eng J 361:671–679
Liang W, Wang G, Peng C, Tan J, Wan J, Sun P, Li Q, Ji X, Zhang Q, Yonghong W (2022) Recent advances of carbon-based nano zero valent iron for heavy metals remediation in soil and water: a critical review. J Hazard Mater 426:127993
Lin J, Sun M, Binglin S, Owens G, Chen Z (2019) Immobilization of cadmium in polluted soils by phytogenic iron oxide nanoparticles. Sci Total Environ 659:491–498. https://doi.org/10.1016/j.scitotenv.2018.12.391
Liu J, Zhang YD, Zhang ZM (2009) The application research of nano-biotechnology to promote increasing of vegetable production. Hubei Agric Sci 1:20–25
Long F, Zhu A, Shi H (2013) Recent advances in optical biosensors for environmental monitoring and early warning. Sensors (Switzerland) 13(10):13928–13948. https://doi.org/10.3390/s131013928
Lu CM, Zhang CY, Wen JQ, Wu GR, Tao MX (n.d.) Research of the effect of nanometer materials on germination and growth enhancement of. Glycine Max:168–172
Ma W, Sun T, Yingming X, Zheng S, Sun Y (2022) In–situ immobilization remediation, soil aggregate distribution, and microbial community composition in weakly alkaline cd–contaminated soils: a field study. Environ Pollut 292:118327
Machado S, Pacheco JG, Nouws HPA, Albergaria JT, Delerue-Matos C (2015) Characterization of green zero-valent iron nanoparticles produced with tree leaf extracts. Sci Total Environ 533:76–81. https://doi.org/10.1016/j.scitotenv.2015.06.091
Magrez A, Kasas S, Salicio V, Pasquier N, Seo JW, Celio M, Catsicas S, Schwaller B, Forró L (2006) Cellular toxicity of carbon-based nanomaterials. Nano Lett 6(6):1121–1125
Malandrino G (2009) Chemical vapour deposition. Precursors, processes and applications. Edited by Anthony C. Jones and Michael L. Hitchman. Angew Chem Int Ed 48:7478–7479. https://doi.org/10.1002/anie.200903570
Maruyama CR, Guilger M, Pascoli M, Natalia B-J, Abhilash PC, Fraceto LF, De Lima R (2016) Nanoparticles based on chitosan as carriers for the combined herbicides Imazapic and Imazapyr. Sci Rep 6(2015). https://doi.org/10.1038/srep19768
Mazloomi F, Jalali M (2019) Effects of vermiculite, nanoclay and zeolite on ammonium transport through saturated Sandy loam soil: column experiments and modeling approaches. Catena 176:170–180
McKee MS, Filser J (2016) Impacts of metal-based engineered nanomaterials on soil communities. Environ Sci Nano 3(3):506–533. https://doi.org/10.1039/c6en00007j
Miao YF, Wang ZH, Li SX (2015) Relation of nitrate N accumulation in dryland soil with wheat response to N fertilizer, vol 170, p 119
Miguel-Rojas C, Pérez-De-Luque A (2023) Nanobiosensors and Nanoformulations in agriculture: new advances and challenges for sustainable agriculture. Emerg Top Life Sci 7(2):229–238. https://doi.org/10.1042/ETLS20230070
Millán G, Agosto F, Vázquez M, Botto L, Lombardi L, Juan L (2008) Use of clinoptilolite as a carrier for nitrogen fertilizers in soils of the Pampean regions of Argentina. Ciencia e Investigacion Agraria 35(3):245–254
Mishra S, Singh A, Keswani C, Singh HB (2014) Nanotechnology: exploring potential application in agriculture and its opportunities and constraints. Biotech Today 4(1):9. https://doi.org/10.5958/2322-0996.2014.00011.8

Mohammad ZH, Ahmad F, Ibrahim SA, Zaidi S (2022) Application of nanotechnology in different aspects of the food industry. Discov Food 2(1). https://doi.org/10.1007/s44187-022-00013-9
Muraisi L, Hariyadi DM, Athiyah U, Pathak Y (2022) Eco-friendly nanotechnology in agriculture: opportunities, toxicological implications, and occupational risks. In: Sustainable nanotechnology: strategies, products, and applications, pp 287–296. https://doi.org/10.1002/9781119650294.ch18
Muthumalai K, Gokila N, Haldorai Y, Kumar RTR (2021) Nanosensors for crop protection: design and fabrication. In: Nanosensors for smart agriculture. Elsevier
Myszka K, Czaczyk K (2011) Bacterial biofilms on food contact surfaces - a review. Pol J Food Nutr Sci 61(3):173–180. https://doi.org/10.2478/v10222-011-0018-4
Nair R, Varghese SH, Nair BG, Maekawa T, Yoshida Y, Sakthi Kumar D (2010) Nanoparticulate material delivery to plants. Plant Sci 179(3):154–163. https://doi.org/10.1016/j.plantsci.2010.04.012
Naz MY, Shukrullah S, Ghaffar A (2021) Sensors detecting controlled fertilizer release. In: Controlled release fertilizers for sustainable agriculture. Elsevier, pp 131–153
Neme K, Nafady A, Uddin S, Tola YB (2021) Application of nanotechnology in agriculture, postharvest loss reduction and food processing: food security implication and challenges. Heliyon 7(12):e08539. https://doi.org/10.1016/j.heliyon.2021.e08539
Nile SH, Baskar V, Selvaraj D, Nile A, Xiao J, Kai G (2020) Nanotechnologies in food science: applications, recent trends, and future perspectives. Nano-Micro Lett 12:1–34
No HK, Meyers SP, Prinyawiwatkul W, Zheheng X (2007) Applications of chitosan for improvement of quality and shelf life of foods: a review. J Food Sci 72(5):R87–R100
Nuruzzaman M, Rahman MM, Liu Y, Naidu R (2016) Nanoencapsulation, nano-guard for pesticides: a new window for safe application. J Agric Food Chem 64(7):1447–1483. https://doi.org/10.1021/acs.jafc.5b05214
Oberdorster G, Oberdorster E, Oberdorster J (2005) Nanotoxicology: an emerging discipline evolving from studies of ultrafine particles. Environ Health Perspect 113(7):823–839
Otles, S., and B. Y. Sahyar. 2018. "Nanotechnology application and emergence in agriculture." Emerging trends in agri-nanotechnology: fundamental and applied aspects 2 (6):204–213. doi: https://doi.org/10.1079/9781786391445.0204
Pal K (2020) Nanofabrication for smart nanosensor applications. Elsevier
Panpatte DG, Jhala YK, Shelat HN, Vyas RV (2016) Microbial inoculants in sustainable agricultural productivity. In: Nanoparticles: the next generation technology for sustainable agriculture. Springer, New Delhi, pp 289–300
Patle T, Sharma SK (2022) Synthesis of nano-gypsum: a computational approach to encounter soil salinity and land degradation. Comput Theor Chem 1217:113909. https://doi.org/10.1016/j.comptc.2022.113909
Pérez-de-Luque A (2017) Interaction of nanomaterials with plants: what do we need for real applications in agriculture? Front Environ Sci 5:12
Petosa AR, Rajput F, Selvam O, Öhl C, Tufenkji N (2017) Assessing the transport potential of polymeric nanocapsules developed for crop protection. Water Res 111:10–17
Prabhakar PK (2020) Bacterial siderophores and their potential applications: a review. Curr Mol Pharmacol 13(4):295–305. https://doi.org/10.2174/1874467213666200518094445
Pradhan N, Singh S, Ojha N, Shrivastava A, Barla A, Rai V, Bose S (2015) Facets of nanotechnology as seen in food processing, packaging, and preservation industry. Biomed Res Int 2015(1):365672
Prasad R, Bhattacharyya A, Nguyen QD (2017) Nanotechnology in sustainable agriculture: recent developments, challenges, and perspectives. Front Microbiol 8(JUN):1–13. https://doi.org/10.3389/fmicb.2017.01014
Prasad R, Kumar V, Kumar M, Choudhary DK (2019) Nanobiotechnology in bioformulations. Springer
Prema P, Nguyen V-H, Karthikeyan Venkatachalam JM, Murugan HM, Ali MZM, Salem BR, Balaji P (2022) Hexavalent chromium removal from aqueous solutions using biogenic iron nanoparticles: kinetics and equilibrium study. Environ Res 205:112477

Pudake RN, Chauhan N, Kole C (2019) Nanoscience for sustainable agriculture, vol 711. Springer
Qingming S, Qianhao M, Jianjun S, Liping J, Jian-Rong Z, Wenhua H, Jun-Jie Z (2009) Morphology-controlled synthesis of palladium nanostructures by sonoelectrochemical method and their application in direct alcohol oxidation. J Phys Chem C 113(4):1267–1273. https://doi.org/10.1021/jp807881s
Qiu C, Jiang L, Gao Y, Sheng L (2023) Effects of oxygen-containing functional groups on carbon materials in supercapacitors: a review. Mater Des 230:111952. https://doi.org/10.1016/j.matdes.2023.111952
Rai M, Ingle A (2012) Role of nanotechnology in agriculture with special reference to management of insect pests. Appl Microbiol Biotechnol 94(2):287–293. https://doi.org/10.1007/s00253-012-3969-4
Rai V, Acharya S, Dey N (2012) Implications of nanobiosensors in agriculture, vol 03, p 315
Rameshaiah GN, Pallavi J, Shabnam S (2015) Nano fertilizers and nano sensors–an attempt for developing smart agriculture. Int J Eng Res Gen Sci 3(1):314–320
Rinot O, Levy GJ, Steinberger Y, Svoray T, Eshel G (2019) Soil health assessment: a critical review of current methodologies and a proposed new approach. Sci Total Environ 648:1484–1491. https://doi.org/10.1016/j.scitotenv.2018.08.259
Roduner E (2006) Size matters: why nanomaterials are different. Chem Soc Rev 35(7):583–592. https://doi.org/10.1039/B502142C
Rossi L, Fedenia LN, Sharifan H, Ma X, Lombardini L (2019) Effects of foliar application of zinc sulfate and zinc nanoparticles in coffee (Coffea Arabica L.) plants. Plant Physiol Biochem 135:160–166
Rticles A, Ern C (n.d.) Nanoparticles: is toxicity a concern. 22:92–101
Safarnejad MR, Samiee F, Tabatabie M, Mohsenifar A (2017) Development of quantum dot-based nanobiosensors against Citrus Tristeza Virus (CTV). Sens Transducers 213(6):54
Sampathkumar K, Tan KX, Loo SCJ (2020) Developing nano-delivery systems for agriculture and food applications with nature-derived polymers. IScience 23(5):101055. https://doi.org/10.1016/j.isci.2020.101055
Samrot AV, Justin C, Padmanaban S, Burman U (2017) A study on the effect of chemically synthesized magnetite nanoparticles on earthworm: Eudrilus Eugeniae. Appl Nanosci 7(1):17–23. https://doi.org/10.1007/s13204-016-0542-y
Sánchez-López E, Gomes D, Esteruelas G, Bonilla L, Lopez-Machado AL, Galindo R, Cano A, Espina M, Ettcheto M, Camins A (2020) Metal-based nanoparticles as antimicrobial agents: an overview. Nano 10(2):292
Sandhya SK, Kumar D, Dilbaghi N (2017) Preparation, characterization, and bio-efficacy evaluation of controlled release carbendazim-loaded polymeric nanoparticles. Environ Sci Pollut Res Int 24(1):926–937. https://doi.org/10.1007/s11356-016-7774-y
Savolainen K, Pylkkänen L, Norppa H, Falck G, Lindberg H, Tuomi T, Vippola M, Alenius H, Hämeri K, Koivisto J, Brouwer D, Mark D, Bard D, Berges M, Jankowska E, Posniak M, Farmer P, Singh R, Krombach F, Bihari P, Kasper G, Seipenbusch M (2010) Nanotechnologies, engineered nanomaterials and occupational health and safety—a review. Saf Sci 48(8):957–963. https://doi.org/10.1016/j.ssci.2010.03.006
Saxena M, Maity S, Sarkar S (2014) Carbon nanoparticles in 'biochar' boost wheat (Triticum Aestivum) plant growth. RSC Adv 4(75):39948–39954. https://doi.org/10.1039/c4ra06535b
Sekhon BS (2014) Nanotechnology in agri-food production: an overview. Nanotechnol Sci Appl:31–53
Setyowati Retno Djiwanti SK (2019) Nanopesticide: future application of nanomaterials in plant protection. In: Plant nanobionics nanotechnology in the life sciences, pp 255–298. https://doi.org/10.1007/978-3-030-16379-2_10
Shalaby TA, Bayoumi Y, Abdalla N, Taha H, Alshaal T, Shehata S, Amer M, Domokos-Szabolcsy É, El-Ramady H (2016) Nanoparticles, soils, plants and sustainable agriculture. Nanosci Food Agric 1:283–312
Shang Y, Hasan K, Ahammed GJ, Li M, Yin H (2019) Shang-2019-Applications-of-Nanotechnology-in-p (1).Pdf. Molecules 24(2558):1–24

Sharma D, Afzal S, Singh NK (2021a) Nanopriming with phytosynthesized zinc oxide nanoparticles for promoting germination and starch metabolism in rice seeds. J Biotechnol 336:64–75. https://doi.org/10.1016/j.jbiotec.2021.06.014

Sharma P, Pandey V, Sharma MMM, Patra A, Singh B, Mehta S, Husen A (2021b) A review on biosensors and nanosensors application in agroecosystems. Nanoscale Res Lett 16(1):136. https://doi.org/10.1186/s11671-021-03593-0

Shen M, Liu S, Jiang C, Zhang T, Chen W (2023) Recent advances in stimuli-response mechanisms of nano-enabled controlled-release fertilizers and pesticides, vol 2. Eco-Environ Health, p 161

Shukla P, Chaurasia P, Younis K, Qadri OS, Faridi SA, Srivastava G (2019) Nanotechnology in sustainable agriculture: studies from seed priming to post-harvest management. Nanotechnol Environ Eng 4(1):1–15. https://doi.org/10.1007/s41204-019-0058-2

Shweta SS, Sharma A, Chadha S, Guleria V (2021) Nanotechnology: a cutting-edge technology in vegetable production. J Hortic Sci Biotechnol 96(6):682–695

Siddiquee S, Rovina K, Yusof NA, Rodrigues KF, Suryani S (2014) Nanoparticle-enhanced electrochemical biosensor with DNA immobilization and hybridization of Trichoderma Harzianum gene. Sens Bio-Sens Res 2:16–22

Silva G, Cláudia RJ, Marino T, Molinari R, García H (2011) Influence of excitation wavelength (UV or visible light) on the photocatalytic activity of Titania containing gold nanoparticles for the generation of hydrogen or oxygen from water. J Am Chem Soc 133(3):595–602

Singh Sekhon B (2014) Nanotechnology in agri-food production: an overview. Nanotechnol Sci Appl 7(2):31–53. https://doi.org/10.2147/NSA.S39406

Singh T, Shukla S, Kumar P, Wahla V, Bajpai VK (2017) Application of nanotechnology in food science: perception and overview. Front Microbiol 8(AUG):1–7. https://doi.org/10.3389/fmicb.2017.01501

Singh AK, Shikha K, Shahi JP (2021) Chapter 3 - Hybrids and abiotic stress tolerance in horticultural crops. In: Rai AC, Rai A, Rai KK, Rai VP, Kumar A (eds) Stress tolerance in horticultural crops. Woodhead Publishing, pp 33–50

Singh R, Dutt S, Sharma P, Sundramoorthy AK, Dubey A, Singh A, Arya S (2023) Future of nanotechnology in food industry: challenges in processing, packaging, and food safety. Global Chall 7(4). https://doi.org/10.1002/gch2.202200209

Sivarethinamohan R, Sujatha S (2021) Unlocking the potentials of using nanotechnology to stabilize agriculture and food production, vol 2327. AIP Publishing

Suthari VK (2024) Nano biosensors for prediction of environmental pollutants. (October):0–4

Suttiponparnit K, Jiang J, Sahu M, Suvachittanont S, Charinpanitkul T, Biswas P (2010) Role of surface area, primary particle size, and crystal phase on titanium dioxide nanoparticle dispersion properties. Nanoscale Res Lett 6(1):27. https://doi.org/10.1007/s11671-010-9772-1

Syafrudin M, Kristanti RA, Yuniarto A, Hadibarata T, Rhee J (2021) Pesticides in drinking water — a review. Int J Environ Res Public Health 18:468

Tarafdar J (2015) Nanoparticle production, characterization and its application to horticultural crops, pp 222–229

Tian H, Kah M, Kariman K (2019) Are nanoparticles a threat to Mycorrhizal and Rhizobial symbioses? A critical review. Front Microbiol 10:1660

Tofan L, Suteu D (2023) Renewable resource biosorbents for pollutant removal from aqueous effluents in column mode. Separations 10(2). https://doi.org/10.3390/separations10020143

Tripathi S, Mehrotra GK, Dutta PK (2009) Physicochemical and bioactivity of cross-linked chitosan–PVA film for food packaging applications. Int J Biol Macromol 45(4):372–376

Tsong JL, Khor SM (2023) Modern analytical and bioanalytical technologies and concepts for smart and precision farming. Anal Methods 15(26):3125–3148

Violino S, Pallottino F, Sperandio G, Figorilli S, Ortenzi L, Tocci F, Vasta S, Imperi G, Costa C (2020) A full technological traceability system for extra virgin olive oil. Foods 9(5). https://doi.org/10.3390/foods9050624

Win-Shwe T-T, Fujimaki H (2011) Nanoparticles and neurotoxicity. Int J Mol Sci 12(9):6267–6280. https://doi.org/10.3390/ijms12096267

Yadav TP, Yadav RM, Singh DP (2012) Mechanical milling: a top down approach for the synthesis of nanomaterials and nanocomposites. Nanosci Nanotechnol 2(3):22–48. https://doi.org/10.5923/j.nn.20120203.01

Yadav RK, Singh NB, Singh A, Yadav V, Bano C, Khare S, Niharika. (2020) Expanding the horizons of nanotechnology in agriculture: recent advances, challenges and future perspectives. Vegetos 33:203–221

Yadav A, Yadav K, Abd-Elsalam K (2023) Nanofertilizers: types, delivery and advantages in agricultural sustainability. Agrochemicals 2(2):296–336. https://doi.org/10.3390/agrochemicals2020019

Yang H, Mingmin X, Koide RT, Liu Q, Dai Y, Liu L, Bian X (2016) Effects of ditch-buried straw return on water percolation, nitrogen leaching and crop yields in a rice-wheat rotation system. J Sci Food Agric 96(4):1141–1149. https://doi.org/10.1002/jsfa.7196

Yuan G, Chen X, Li D (2016) Chitosan films and coatings containing essential oils: the antioxidant and antimicrobial activity, and application in food systems. Food Res Int 89:117–128

Yüksel S, Schwenkbier L, Pollok S, Weber K, Cialla-May D, Popp J (2015) Label-free detection of Phytophthora ramorum using surface-enhanced Raman spectroscopy. Analyst 140(21):7254–7262

Zain M, Ma H, Rahman SU, Nuruzzaman M, Chaudhary S, Azeem I, Mehmood F, Duan A, Sun C (2024) Nanotechnology in precision agriculture: advancing towards sustainable crop production. Plant Physiol Biochem 206:108244. https://doi.org/10.1016/j.plaphy.2023.108244

Zhang C, Kovacs JM (2012) The application of small unmanned aerial systems for precision agriculture: a review. Precis Agric 13(6):693–712. https://doi.org/10.1007/s11119-012-9274-5

Zhang J, Chaker M, Ma D (2017) Pulsed laser ablation based synthesis of colloidal metal nanoparticles for catalytic applications. J Colloid Interface Sci 489:138–149. https://doi.org/10.1016/j.jcis.2016.07.050

Zhang D, Ye K, Yao Y, Liang F, Tao Q, Ma W, Yang B, Dai Y, Watanabe T (2019) Controllable synthesis of carbon nanomaterials by direct current arc discharge from the inner wall of the chamber. Carbon 142:278–284. https://doi.org/10.1016/j.carbon.2018.10.062

Zhang Y, Luo J, Zhang H, Li T, Hongxia X, Sun Y, Xueyuan G, Xin H, Gao B (2022) Synthesis and adsorption performance of three-dimensional gels assembled by carbon nanomaterials for heavy metal removal from water: a review. Sci Total Environ 852:158201

Zhao R, Torley P, Halley PJ (2008) Emerging biodegradable materials: starch-and protein-based bio-nanocomposites. J Mater Sci 43(9):3058–3071

Zhao Y, Liu L, Kong D, Kuang H, Wang L, Chuanlai X (2014) Dual amplified electrochemical Immunosensor for highly sensitive detection of Pantoea Stewartii Sbusp. Stewartii. ACS Appl Mater Interfaces 6(23):21178–21183

Zhao L, Zhou X, Kang Z, Peralta-Videa JR, Zhu Y-G (2024) Nano-enabled seed treatment: a new and sustainable approach to engineering climate-resilient crops. Sci Total Environ 910:168640. https://doi.org/10.1016/j.scitotenv.2023.168640

Zhuang S, Lee ES, Lei L, Nunna BB, Kuang L, Zhang W (2016) Synthesis of nitrogen-doped graphene catalyst by high-energy wet ball milling for electrochemical systems. Int J Energy Res 40(15):2136–2149. https://doi.org/10.1002/er.3595

MIX
Papier aus verantwortungsvollen Quellen
Paper from responsible sources
FSC® C105338

If you have any concerns about our products,
you can contact us on
ProductSafety@springernature.com

In case Publisher is established outside the EU,
the EU authorized representative is:
Springer Nature Customer Service Center GmbH
Europaplatz 3, 69115 Heidelberg, Germany

Printed by Libri Plureos GmbH
in Hamburg, Germany